Classical and Molecular Plant Cytogenetics

Timir Baran Jha • Mihir Halder

Classical and Molecular Plant Cytogenetics

A Laboratory Manual

Springer

Timir Baran Jha
Department of Botany
Maulana Azad College
Kolkata, West Bengal, India

Mihir Halder
Department of Botany
Barasat Government College
Kolkata, West Bengal, India

ISBN 978-981-95-8264-8 ISBN 978-981-95-8265-5 (eBook)
https://doi.org/10.1007/978-981-95-8265-5

This Springer imprint is published by the registered company Springer Nature Singapore Pte Ltd.
The registered company address is: 152 Beach Road, #21-01/04 Gateway East, Singapore 189721, Singapore

The authors dedicate this book to their beloved mentor teacher
Late Professor Sumitra Sen
Dept. of Botany, University of Calcutta

Preface

Studying and practicing with plant chromosomes, along with visualizing these distinct genomic structures under a microscope, is truly rewarding. The field of plant cytogenetics has a rich history spanning over a century, captivating both students and researchers alike. Chromosomes—remarkable species-specific condensed structures of genomic material—encapsulate the identity of every organism, serving as the fundamental blueprint for life's DNA. Within the vast diversity of the plant kingdom, chromosome counts can vary dramatically, ranging from $2n = 4$ to over 1400. These structures are particularly notable for their visibility under a microscope, offering essential insights into genetic information.

Keeping the above simple yet very significant words in mind, we decided to prepare a hands-on training manual on "Classical & Molecular Plant Cytogenetics." This book aims to equip readers with the essential knowledge and hands-on training in plant chromosome preparation, observation, and analysis, utilizing both classical and molecular cytogenetic techniques. It outlines core concepts of cell division and chromosome structure, with karyotype analysis and illustrated examples to enhance understanding. One of the book's standout features is its comprehensive methodology for chromosome preparation, alongside practical protocols for cytogenetic practices, supported by numerous diagrams and an extensive collection of chromosome plates derived from our instructional work. A dedicated chapter on fluorescent in situ hybridization (FISH) provides conceptual foundations, protocols, and applications to ignite interest in research on plant chromosomes.

Additionally, the inclusion of methodologies for preparing meiotic and mitotic (including pollen mitosis) chromosomes, complemented by our illustrations, is intended to refine practical skills. A conceptual discussion on genome size and the C-value paradox, paired with flow cytometric analysis and genomic sorting techniques, presents new avenues of inquiry for readers. To bolster the learning and analytical experience, two specialized chapters on instruments and biostatistics will further enhance documentary and analytical proficiencies. Each chapter is accompanied by references and suggestions for additional reading, ensuring a well-rounded scholarly approach. The anticipated use of this book is as a comprehensive kit designed for both learning and practicing the intricacies of plant chromosomes. Its structure and content aim to facilitate a deeper understanding of chromosomal structure and function and to provide a valuable resource for students and researchers alike in plant genetics.

While we have endeavored to maintain an impartial approach, it is possible that certain significant aspects may not have received adequate consideration. We welcome suggestions from our readers to deepen and broaden our future edition.

The authors sincerely thank Springer Nature Publishers for their full cooperation in completing the process on time.

We would like to express our sincere gratitude to the former and current Principals of Maulana Azad College, Dr. Subhasis Dutta and Dr. Krishnendu Dutta, respectively, for their invaluable support. We also extend our thanks to Dr. Santanu Chakroborti, Principal of Barasat Government College, for his encouragement. Lastly, we wish to acknowledge the steadfast support of our spouses, whose understanding and encouragement have been instrumental in our pursuits. Their unwavering belief in us has played a crucial role in our achievements.

Kolkata, West Bengal, India
February 2026

Timir Baran Jha
Mihir Halder

Competing Interests The authors have no competing interests to declare that are relevant to the content of this manuscript.

Contents

About the Authors

Timir Baran Jha, Ph.D., is a distinguished former Professor of Botany with over 40 years of experience in research and education, specializing in plant Cytogenetics and Biotechnology. After earning his B.Sc., M.Sc., and Ph.D. in Botany from the University of Calcutta, he joined the West Bengal Education Service (WBES) and later the Senior Education Service (WBSES) under the Government of West Bengal, India. He held the position of Professor and Head in the Postgraduate Department of Botany at both Darjeeling Government College and Presidency College in Kolkata. In addition to supervising Ph.D. students, he has authored nearly 100 research papers and contributed to numerous scientific publications in national and international journals, demonstrating his extensive expertise. Currently, from an Indian perspective, he is focused on developing a chromosomal database for valuable Indian crop plants, including Rice, Lentil, and Capsicum, utilizing molecular cytogenetics. Furthermore, he has authored a book on Plant Biotechnology.

Mihir Halder, Ph.D., is currently serving as an Assistant Professor in the Post-Graduate Department of Botany at Barasat Government College in West Bengal, India. He obtained his B.Sc. with honors in Botany in 2007 from Presidency College, Kolkata, followed by an M.Sc. in Botany in 2009 from the University of Calcutta. In 2017, he earned his Ph.D. in Biochemistry from the University of Calcutta. With over a decade of experience, he has been actively involved in teaching courses on Genetics, Cell and Molecular Biology, and Plant Biotechnology. He has made notable research contributions in plant cytogenetics and plant biotechnology, published 25 research papers and book chapters. Furthermore, he has edited a book published by Springer and is presently working on another edited volume for Springer.

Abbreviations

μL	Microliter
A	Adenine
ACL	Average chromosome length
ADC	Analogue-to-digital converter
AMD	Actinomycin D
BAC	Bacterial artificial chromosome
BOD	Biochemical oxygen demand
BSA	Bovine serum albumin
C	Cytosine
CAR	Chimeric antigen receptors
CCDs	Charge-coupled devices
CD	Cluster of Differentiation
cDNA	Complementary DNA
CFC	Chlorofluorocarbon
CI	Centromeric index
CMA	Chromomycin A3
DA	Distamycin A
DAPI	4′, 6-Diamidino-2-phenylindole
DI	Disparity index
EDTA	Ethylenediaminetetraacetic acid
EMA method	Enzymatic maceration and air drying method
Exm	Expansion microscopy
FACS	Fluorescence-activated cell sorting
FCM	Flow cytometry
FDA	Fluorescein diacetate
FISH	Fluorescence in situ hybridization
FSC	Forward scatter
IGS	Intergenic spacer
ITSs	Internal transcribed spacers
G	Guanine
gm	gram
HCL	Haploid chromosome length
ETSs	External transcribed spacers
HLA	Human leukocyte antigen

L	Liter
ISH	In situ hybridization
Kb	Kilo base pair
M	Molar
mM	Millimolar
Mol. Wt.	Molecular weight
MMC	Megaspore mother cell
NA	Numerical aperture
NOR	Nucleolar organizing region
NTS	Non-transcribed spacer
PCR	Polymerase chain reaction
PD	Photodiode
PI	Propidium iodide
PMC	Pollen mother cell
PMT	Photomultiplier tube
rDNA	Ribosomal DNA
rpm	Revolutions per minute
SD	Standard deviation
SDS	Sodium dodecyl sulfate
SE	Standard error
SEM	Scanning electron microscope
SFC	Spectral flow cytometry
sRNA	Small regulatory RNA
SSC	Saline sodium citrate
SSC	Side scatter
T	Thymine
TCL	Total chromosome length
TEM	Transmission electron microscope
UPGMA	Unweighted Pair Group Method with Arithmetic Mean
VC	Variation coefficient
YAC	Yeast artificial chromosome

Introduction

1

Darlington, a pioneering cytologist in the early twentieth century, defined "cytogenetics" as the study of chromosomes, genetics, and evolution. Since that time, the field has concentrated on understanding the structures, properties, and behaviors of nuclear genomes at the chromosome level during cell division, including mitosis and meiosis. Furthermore, researchers have investigated how structural and numerical variations in chromosomes affect phenotype and the factors that contribute to chromosomal changes.

Contemporary scientists are dedicated to uncovering the specific functions of cytogenetics. Long before Watson and Crick unveiled the structure of DNA in 1953, Bateson had introduced the concept of genes in 1916. In 1924, Feulgen and Rossenbock demonstrated the presence of DNA within chromosomes through histochemical staining. From 1921 to 1926, Belling developed an iron-aceto-carmine staining method for chromosomes, while Barbara McClintock further refined it by adding a heating step. She became the first to identify all 10 haploid maize chromosomes from maize microspores and to create a semi-diagrammatic representation known as the ideogram of haploid chromosome complements. The discovery of the structure of DNA solidified the unique relationship between chromosomes and genes. It confirmed that chromosomes are condensed, discrete genomic units that carry thousands of genes, which represent various levels of genetic information. Across the globe, many plant scientists committed themselves to visualizing chromosomes in a wide range of plant materials, recognizing that the successful visualization of these chromosomes under a microscope significantly depends on how well the samples are prepared.

Research on chromosomes in the early twentieth century began with the use of microtome sectioning, later evolving into the squash method following various technical enhancements. Science has identified roughly 340,000 to 390,000 vascular plant species. Since the early 1900s, classical cytogenetics has successfully established the somatic chromosome number in thousands of plant species, although many more remain to be examined. Plants exhibit remarkable chromosomal

T. B. Jha, M. Halder, *Classical and Molecular Plant Cytogenetics*,
https://doi.org/10.1007/978-981-95-8265-5_1

diversity, with counts ranging from $2n = 4$ chromosomes in *Xanthisma gracile* to as much as $2n = 1440$ chromosomes in *Ophioglossum reticulatum*. The quest for plant species with $2n = 2$, similar to *Ascaris megalocephala*, is ongoing. Chromosome research heavily depends on human expertise and requires a global workforce to advance the program. In contemporary fields of molecular biology and genomics, there is a prevailing view that classical cytogenetics has lost its relevance. Nevertheless, the study of chromosomes remains a cost-effective approach for understanding genomic structure within cells. Despite significant advancements in DNA research, chromosome studies remain vital in all areas of plant sciences, particularly in systematics and evolutionary biology. Unfortunately, undergraduate and postgraduate students often receive inadequate training in classical plant cytogenetics, which is essential for visualizing fascinating genomic structures, such as chromosomes under a microscope.

For researchers focused on plant chromosomes, the primary goal is to prepare high-quality plant chromosomes. This task is challenging due to the diverse composition of plant cell walls across different species. In contrast, animal cells do not have cell walls, which alleviate many of the obstacles associated with chromosome preparation. Plant cytologists have faced difficulties in preparing chromosomes from numerous plant species because of variations in chromosome numbers and sizes, as well as differences in cytoplasmic contents.

The field of plant science has made considerable advancements over time. Prior to the discovery of DNA, Emsweller and Stuart (1944) proposed the use of enzymes to break down plant tissues for chromosome analysis. While this method was initially employed in human cytology, it was not adapted for plants until the early 1970s due to the inherent differences between plant and animal cells. Progress in chromosome biology led to the development of enzymatic maceration and the air-drying technique on glass slides, which eliminated the need for cover glasses and enabled multiple experiments to be conducted on a single slide. This innovation laid the groundwork for molecular cytogenetics, a vital field for genomics and breeding programs. The first detailed karyotype was created by Kurata and Omura (1978) using this methodology, which has since become a widely accepted modern approach to plant chromosome analysis. Studying plant chromosomes is essential for understanding taxonomy, phylogeny, evolution, genetics, breeding, and biotechnology. Accurately interpreting chromosome numbers and morphology provides crucial information for students and researchers, enhancing their understanding of genome structure and organization.

The objective of this practical manual is to ignite students' interest in plant chromosome research by integrating classical and molecular cytogenetics. It aims to highlight the benefits of combining traditional chromosome processing techniques with modern molecular biology methods. This integration has the potential to reveal molecular characteristics that may be difficult to identify using current molecular biology approaches. It establishes a foundation for exploring a new dimension of molecular cytogenetics and offers exciting research opportunities in this field.

This book is written in a clear and accessible style, using a hands-on visual training approach to bridge the gap between theory and practical application of plant chromosome analysis. It covers essential topics, including the cell cycle, cell division, and the morphological and molecular characteristics of plant chromosomes to build a holistic knowledge base in this field. The most engaging chapters provide detailed, step-by-step methodologies for processing, preparing, and analyzing somatic and gametic plant chromosomes using hands-on kits in both classical and molecular cytogenetics. These sections include numerous mitotic and meiotic chromosome plates of common teaching materials prepared from our research and designed explicitly for educational purposes.

Moreover, a concise, student-friendly chapter covers flow cytometric estimation of genomic DNA, furnish with protocols, and an introduction to fluorescence-activated cell sorting (FACS) and karyotyping. A chapter on biostatistics offers comprehensive parameters for karyotype analysis. Additionally, the book contains dedicated chapters on various commonly used instruments in cytogenetics and advanced molecular cytogenetics techniques like fluorescent in situ hybridization (FISH), all aimed at enhancing the learning experience for undergraduate students. The approach and content presented in this book are unique compared to other resources available in the market. The authors are confident that it will foster students' interest, develop their skills, and deepen their understanding of chromosome science in alignment with their needs. It is affordably priced and will serve as a valuable resource for students, researchers, and educators interested in hands-on learning about plant chromosome analysis.

The introduction will conclude with a quote from Barbara McClintock, one of the most esteemed cytogeneticists in history, who highlighted the significance of observing organisms in their entirety and grasping their unique variations. She advocated for a deeply observational approach to science. McClintock stated, "No two plants are exactly alike. They're all different, and as a consequence, you have to know the difference. I start with the seedling, and I don't want to leave it. I don't feel I really know the story if I don't watch the plant all the way through. So, I know every plant in the field. I know them intimately. And I find it a great pleasure to know them."

An Overview of Cell Division and Plant Chromosomes

2

Abstract

This chapter provides a comprehensive overview of cell division and plant chromosomes, tracing the historical development of cell theory and its implications for understanding life. The concept of the cell originated with Robert Hooke in the seventeenth century when he observed box-shaped structures in cork, which he named "cells." This discovery laid the groundwork for classical cell theory, formulated by Theodor Schwann and Matthias Schleiden in 1839, asserting that all organisms are composed of cells and that cells arise from preexisting cells, as emphasized by Rudolf Virchow's contribution in 1858. Subsequent advancements in technology confirmed the cell as the fundamental unit of life, whether unicellular or multicellular, prokaryotic or eukaryotic. The chapter highlights crucial milestones, including Friedrich Miescher's identification of DNA as the hereditary material in the nineteenth century and the groundbreaking work of Avery, MacLeod, and McCarty in 1944 that provided conclusive evidence of DNA's genetic role. The introduction of terms such as "chromosome" by Heinrich Wilhelm Gottfried Waldeyer in 1888 and "genome" by Hans Winkler in 1920 established important connections in the study of genetics. Pioneers like Thomas Hunt Morgan contributed to our understanding of the relationship between genes and chromosomes, while Watson and Crick's elucidation of DNA structure in 1953 bridged the gap between genetics and cellular biology. Along with the conceptual ideas, the chapter has given special emphasis on the classical and molecular structures of the chromosome, highlighting the integral roles of DNA and chromosomes in heredity, growth, and development. Plant cytogeneticists are acknowledged as key figures in genome research, underscoring the importance of cellular studies in advancing biological understanding.

T. B. Jha, M. Halder, *Classical and Molecular Plant Cytogenetics*,
https://doi.org/10.1007/978-981-95-8265-5_2

2.1 Introduction

Life is defined by metabolically active cells capable of growth and reproduction. The term "cell" was first introduced by Robert Hooke (1635–1703), who sought to explore the microscopic world. When he examined a piece of cork under his improved compound microscope, he observed astonishing, box-shaped tiny pores that comprised the cork. He referred to these structures as "cells" because they reminded him of the chambers in monasteries. The discovery of cells marked a significant milestone in the life sciences and paved the way for the formulation of the classical cell theory. Proposed by Theodor Schwann and Matthias Schleiden in 1839, this theory asserts that all organisms are composed of cells and that cells are the fundamental units of life. Rudolf Virchow later contributed to this theory in 1858 by stating that all cells arise from preexisting cells, encapsulated in the phrase "omnis cellula e cellula." The concept of cell division emerged with the establishment of cell theory, emphasizing that new cells originate from existing ones (National Geographic Society 2023, https://education.nationalgeographic.org).

Since the formulation of classical cell theory, significant advancements in technology and understanding have demonstrated that life is impossible without cells, whether they are structurally simple (unicellular) or complex (multicellular), or whether they are prokaryotic or eukaryotic in nature. Every cell, whether surrounded by a cell wall (as in plants and microbes) or lacking one (as in animals), contains a life-supporting substance known as cytoplasm. This cytoplasm houses various subcellular organelles, including the nucleus, mitochondria, and chloroplasts (the latter found only in plants). The nucleus is regarded as the control center of any eukaryotic cell, housing a substantial amount of genetic material necessary for regulating life processes. The milestones in this field are further illustrated in Fig. 2.1.

In 1869, the Swiss physiological chemist Friedrich Miescher first identified what he called "nuclein" in the nuclei of human white blood cells. Later, the term "nuclein" was changed to "nucleic acid" and eventually to "deoxyribonucleic acid," or "DNA" (Sharma and Sharma 1980). This recognition established the cell as the fundamental unit of all life forms, structurally and functionally.

After seven decades, the landmark discovery of Avery, MacLeod, and McCarty, based on their clinical research with the bacterium *Streptococcus pneumoniae* in 1944, provided convincing proof that, regardless of its nature (prokaryotic or eukaryotic), DNA is the hereditary material present in the cell. The scientific community took some time to accept the new "dogma" of DNA as genetic material. It was further established that DNA is present inside the eukaryotic cells' vital cellular component, the nucleus (Sharma and Sharma 1980).

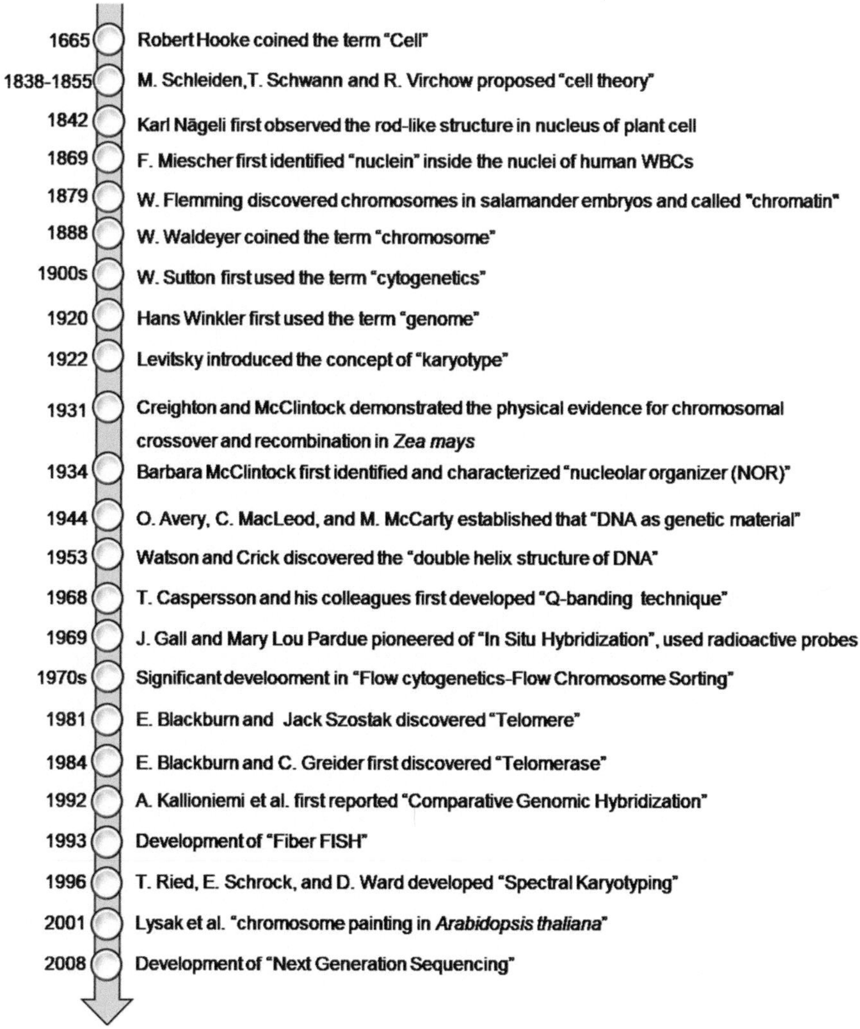

Fig. 2.1 A timeline showing milestones in the history of cytogenetics

Edouard Balbiani, Walther Flemming, Edouard van Beneden, and Theodor Boveri (1870–1890) were the first to describe cell division and fertilization, although they did not properly identify the elements involved in these processes. The term "chromosome"was coined by Heinrich Wilhelm Gottfried Waldeyer in 1888, derived from Greek words meaning "colored body," thus translating loosely to "staining body." Over a century ago, Walter Sutton referred to the study of chromosomes as cytogenetics, a term that merges cytology and genetics (Cremer and Cremer 1988). Additionally, in 1920, Hans Winkler introduced the term "genome" to synthesize the study of genes and chromosomes (Deakin et al. 2019). The naming and visualization of chromosomes, or distinct genomic structures, began long before the structure of DNA was discovered. In this regard, plant cytogeneticists should be recognized as the pioneers of genome research.

The connections between DNA and chromosomes were further elucidated by the observations of Thomas Hunt Morgan and his colleagues, who demonstrated that genes reside on chromosomes. In 1931, Creighton and McClintock provided the first chromosome morphology and experimental evidence that genes are physically positioned on chromosomes in maize plants. Later, in 1944, Avery and his team presented evidence that DNA contains the biological factors known as genes. Nevertheless, scientists remained uncertain about how DNA facilitated the transfer of genetic information between cells and organisms. It was not until 1953 that Watson and Crick unveiled the structure of DNA, highlighting its novel features with significant biological implications. Their model established a vital link between genes and chromosomes, as well as processes such as heredity, growth, and development, thus paving the way for new avenues of research and understanding.

The aforementioned discoveries have led to the enhancement of modern cell theory, which retains the foundational principles of classical cell theory while integrating new insights. Notably, these include the revelations that (i) DNA is transmitted between cells during cell division, (ii) the cellular composition of organisms within similar species exhibits considerable structural and chemical homogeneity, and (iii) energy flow is a fundamental process occurring within cells.

In 1953, the groundbreaking work of James Watson and Francis Crick, which synthesized X-ray diffraction data collected by Rosalind Franklin and Maurice Wilkins, elucidated the structure and composition of DNA molecules. This monumental discovery is often regarded as one of the most significant breakthroughs in the field of biology. Professor Arne Tiselius, the then President of the Nobel Foundation, characterized it as "a revolution far greater in its potential significance than the atomic or hydrogen bomb." This achievement not only provided a physical framework for understanding the storage and transmission of genetic information but also reinforced the intricate relationship between DNA and chromosomes.

Subsequent investigations into the structural complexity of chromosomes have revealed that they are composed of DNA molecules that are meticulously coiled around histone proteins, forming a dynamic, complex structure known as chromatin. During the process of cell division, this chromatin undergoes significant condensation, ultimately resulting in the formation of the distinct and observable structures recognized as chromosomes. These groundbreaking discoveries have fundamentally transformed our understanding of heredity and have catalyzed advancements in various disciplines, including molecular biology, cytogenetics, and genomics. Today, these insights continue to shape our comprehension of life and the mechanisms of inheritance.

The domain of plant cytogenetics represents a particularly compelling area of study involving the analysis of chromosomes in plants during cell division. A foundational understanding of cell division and chromosomal architecture is paramount for success in this field. Such knowledge is especially crucial for novices seeking to enhance their proficiency in chromosome preparation, karyomorphological analysis, and data interpretation. This chapter presents a conceptual framework that addresses the essentials of cell division and chromosome structure, thereby equipping readers with a robust foundation to thrive in the field of plant cytogenetics.

2.2 Cell Division

Cell theory stipulates that cells originate from preexisting cells. Every cell has an inherent desire to transmit its genetic information to the next generation, and sustaining the gene pool is a fundamental principle of biology. To achieve this goal, a parent cell divides into two (mitosis) or four (meiosis) daughter cells through a process called cell division. While amitosis is the direct cell division of prokaryotic cells (Fig. 2.2a), mitosis and meiosis are the two primary processes of cell division in all eukaryotes. Somatic cells undergo mitotic cell division, in which homologous chromosomes with duplicated chromatids are separated, ultimately producing two daughter cells that are genetically identical to the parent cell (Fig. 2.2b). This type of cell division is called equational cell division. In unicellular organisms, this process is equivalent to reproduction, whereas in multicellular organisms, it is responsible for growth, development, tissue maintenance, repair, regeneration, and asexual reproduction. Unlike animal cells, for chromosome analysis, primary sites in plants are specifically restricted to meristematic tissues at shoot tips, root tips, and leaf tips, not throughout the body.

On the other hand, meiosis occurs in specialized cells called germ cells in sexually reproducing organisms. It is a two-step process in which a parent cell undergoes two successive cell divisions (meiosis I and meiosis II) without an intervening round of DNA replication that leads to the formation of four daughter cells or gametes with half the number of chromosomes compared to the parent cell (Fig. 2.2c). The meiosis I is called reductional cell division, whereas meiosis II is similar to mitotic cell division, where number of chromosome remains same (Lima-de-Faria 2010). This process of meiosis ensures the development of genetic diversity in offspring through genetic recombination (crossing over) during prophase I, followed by independent chromosome segregation during anaphase I and II. This genetic variation is crucial for the adaptation and evolution of species over time. Additionally, meiosis helps prevent the accumulation of harmful mutations by allowing the repair of damaged DNA during recombination.

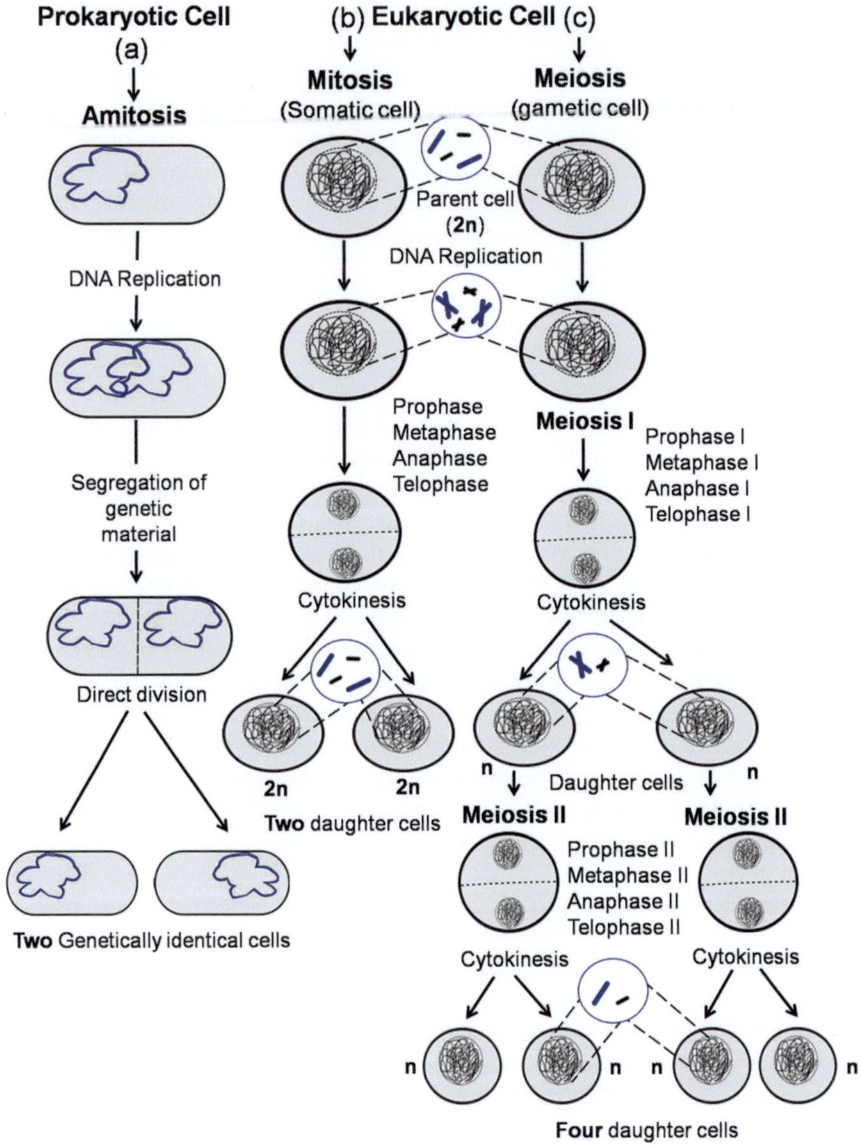

Fig. 2.2 Overview of three types of cell division. (**a**) amitosis, (**b**) mitosis, and (**c**) meiosis

2.2.1 Cell Cycle

Scientists were aware of cell division for more than a century. However, it was Alma Howard and Stephen Pelc in the 1950s who, for the first time, reported the stages of the mitotic cell cycle in root tip cells of the broad bean *(Vicia faba)*, and these stages were later found to be universal in all eukaryotes (Dubrovsky and Ivanov 2003). Plant growth results from the coordinated interaction of the mitotic cell cycle and cell expansion. The cell cycle is one of the most fundamental processes in all living organisms, consisting of a series of well-coordinated stages by which actively dividing eukaryotic cells transmit their genetic and epigenetic information to their daughter cells. However, the cell cycle is defined as the period between successive cell divisions. In eukaryotic cells (both in animals and in plants), the cell cycle is broadly divided into interphase and the M (division) phase (Fig. 2.3).

Interphase is the phase between two consecutive division phases during which the cell prepares itself for the subsequent division. It generally lasts for a relatively long period. It is further subdivided into three distinct phases: G1 phase or Gap 1 phase or postmitotic interphase, S phase or DNA synthesis phase, and G2 phase or Gap 2 phase or premitotic interphase (Fig. 2.3). The G1 phase is the gap period that occurs between the completion of the previous mitosis and the start of the S phase. During this phase, the cell grows and synthesizes the proteins required for DNA replication and for the formation of organelles. During the S phase, DNA replication occurs, resulting in two identical sister chromatids per chromosome. Following DNA replication, the cell enters the G2 phase, during which it continues to grow and synthesizes additional proteins and organelles in preparation for the M phase.

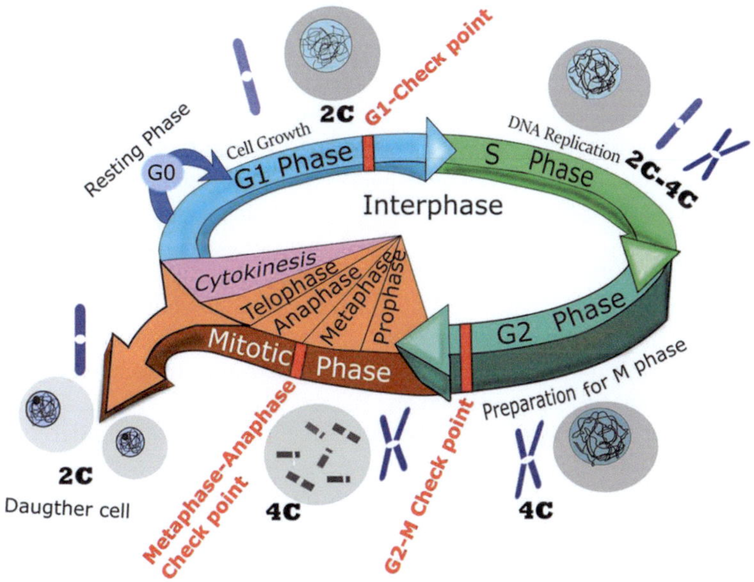

Fig. 2.3 An overview of the eukaryotic cell cycle that consists of interphase (G1-, S-, G2 phase) and M phase (prophase, metaphase, anaphase, telophase, cytokinesis) and three check points of cell cycle

The division phase is also known as the M phase. Irrespective of mitotic or meiotic cell division, it involves two main processes: karyokinesis (nuclear division) and cytokinesis (cytoplasmic division). During karyokinesis, replicated chromosomes condense, align, and segregate into two daughter nuclei, while during cytokinesis, the cytoplasm divides. It starts during telophase. Mitotic karyokinesis has four distinct phases: prophase, metaphase, anaphase, and telophase, which ultimately result in the formation of two daughter cells, each containing the same amount of genetic material as the parent cell (Fig. 2.3). Their characteristic features are described in the next section.

The remarkable diversity in the numbers and types of cells found in eukaryotic organisms raises intriguing questions about whether the rates of cell division and the durations of the different stages of the cell cycle are consistent across cell types within a single organism. The answer is affirmative; however, the length of the cell cycle can vary widely among organisms and cell types. Notable variations in cell cycle duration have been observed in the embryos of *Xenopus* and sea urchins, as well as in both fusion and budding yeasts. For instance, *Drosophila melanogaster* (fruit fly) embryonic cells exhibit an exceptionally rapid cell cycle, lasting approximately 8 minutes. During this time, cell division occurs without cytokinesis, and it takes about 13 rounds of nuclear division before a plasma membrane forms around each nucleus. This process ultimately leads to the separation of the syncytial cytoplasm into individual cells (Morgan 2007).

In humans and plants, cell division occurs asynchronously. Most somatic cells in the human body spend the majority of their time in a metabolically active but nondividing state, known as the G_0 phase. However, when these cells receive appropriate signals, they enter the active stages of the cell cycle. The cell cycle in rapidly dividing mammalian cells grown in culture—under optimal conditions outside the body—lasts approximately 24 hours. This cycle is divided roughly into 9 hours for the S phase, 7 hours for the G1 phase, 6 hours for the G2 phase, and 1 to 2 hours for the M phase. In contrast, the duration of the cell cycle in vivo varies significantly among different cell types. For instance, cells involved in early embryonic development can divide within a few hours, often without a gap phase and with a very short S phase. Epithelial cells may take 2 to 5 days to divide, while liver cells remain quiescent for extended periods, typically dividing only once every 1 to 2 years. Additionally, cortical neurons and cardiac muscle cells may remain in the G_0 phase for the entire human lifespan (Blanco and Blanco 2017).

Cell division in plant systems is generally asynchronous, with minimal coordination of division timing between cells (Dewitte and Murray 2003). The initial investigation into cell cycle durations in plants was reported by Brown in 1951 (Ivanov and Dubrovsky 1997), who studied pea (*Pisum sativum* L.) roots. He found that the duration of interphase was 23 hours, while prophase lasted 2 hours, metaphase 25 minutes, anaphase 5 minutes, and telophase 22 minutes. Since then, numerous researchers have explored cell cycle durations using various methods. Minimum mitotic cycle times for several plant species, namely *Trillium erectum* (29 hours),

Tulipa kaufmanniana (23 hours), *Vicia faba* (13 hours), *Helianthus annuus* (9 hours), *Tradescantia paludosa* (18 hours), and *Pisum sativum* (10 hours) were reported (Ivanov and Dubrovsky 1997). In *Arabidopsis thaliana*, the durations of mitotic and endoreduplication cell cycles in the root meristem and elongation zones (the site of endoreduplication) were measured using 5-ethynyl-2′-deoxyuridine (EdU). The results indicated that cell cycle duration in root meristematic cells was 17.1 hours, while in elongation zones, it was 30.0 hours; the duration of the S-phase was measured at 2.9 hours and 8.7 hours, respectively (Hayashi et al. 2013).

2.2.2 Cell Cycle Checkpoints

The cell cycle has checkpoints that act as critical control points where internal and external signals are integrated. These checkpoints ensure the presence of essential factors or the error-free completion of previous events before proceeding to the next step. If any checkpoint detects a problem, the cell cycle is arrested to allow for repair or signaling for cell death if repair is impossible. The cell cycle has three checkpoints that precisely regulate its progression to prevent the transmission of genetic errors to the next generation (Fig. 2.3).

The G1 checkpoint, also known as the start checkpoint or the restriction point (Fig. 2.3), assesses cell size and nutrient availability. If all parameters are within normal limits, the cell commits to division and enters the S phase. The G2-M checkpoint occurs during the transition between the G2 and M phases (Fig. 2.3) and monitors DNA replication status and integrity. This checkpoint plays a crucial role in maintaining genetic fidelity and tissue homeostasis by faithfully transmitting genetic information that ultimately promotes a healthy and functional organism.

The metaphase-anaphase transition checkpoint, or spindle checkpoint (Fig. 2.3), monitors the proper attachment of the spindle fibers with each chromosome from two opposite poles and the proper alignment of chromosomes before starting anaphase. It ensures proper segregation of sister chromatids or homologous chromosomes to the two opposite poles so that each daughter cell receives an equal and complete set of chromosomes. It is very crucial for the fidelity of chromosome segregation and prevents the development of chromosomal abnormalities.

2.2.3 Cell Cycle Regulators

The cell cycle is a dynamic process that regulates cell growth, DNA replication, and cell division. The timing of events within the cell cycle is precisely controlled by a complex interplay of internal and external signals that ensure accurate duplication of genetic material and proper cell division. This regulation is essential for preventing errors that could lead to mutations and uncontrolled growth. Internal signals are cyclin-dependent kinases (CDKs), cell size, and growth conditions. In contrast, external signals comprise growth factors like epidermal growth factor (EGF) and platelet-derived growth factor (PDGF), as well as nutrient and oxygen levels. In a cultured medium, cells stop dividing due to contact inhibition, influenced by these factors.

The maturation-promoting factor (MPF) serves as a central regulator of cell cycle progression. MPF is composed of a heterodimer formed by specific cyclin-dependent kinases (CDKs) as the catalytic subunit and specific cyclins as the regulatory subunit. The activation and inactivation of MPF govern the transition into and out of M phase, which encompasses both mitosis and meiosis. Activation of MPF relies on several critical factors: an optimal concentration of cyclin protein, dephosphorylation of threonine 14 (Thr14) and tyrosine 15 (Tyr15) residues of CDK1 by phosphatases such as Cdc25, phosphorylation of threonine 161 (Thr161) of CDK1 by CDK-activating kinase (CAK), and the absence of CDK inhibitors. The degradation of cyclin B is facilitated by the anaphase-promoting complex/cyclosome (APC/C), an E3 ubiquitin ligase that targets cyclin B for proteolysis by the 26S proteasome, thereby enabling exit from M phase (Fig. 2.4).

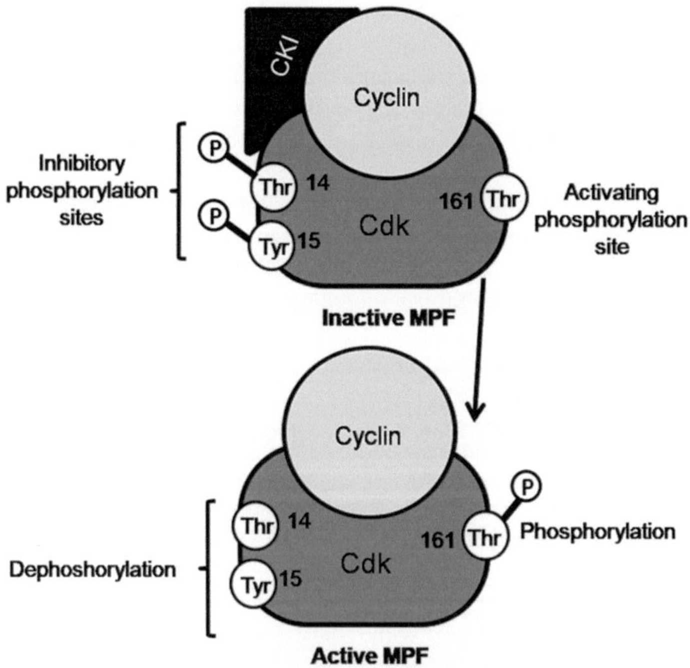

Fig. 2.4 Active and inactive forms of maturation-promoting factor (MPF)

2.2.4 Mitosis

2.2.4.1 Prophase

Prophase marks the beginning of the M phase, which includes the start of supercoiling and condensation of interphase chromatin into visible chromosomes, mitotic spindle formation, the disintegration of the nuclear envelope, and the disappearance of the nucleolus (Fig. 2.5). During this phase, highly dispersed chromatin begins to condense into more compact, thread-like visible chromosomes, aided by topoisomerase II and condensins. This process helps to ensure the even /accurate distribution of chromosomes between the daughter cells. Topoisomerase II uses ATP to untangle and condense two sister chromatids. At the same time, condensins bind to the DNA of a chromatid at multiple sites and induce the formation of coils and loops by twisting DNA, contributing to chromosome condensation. The disappearance of nucleolus accompanies this due to chromosome condensation and the breakdown of the nuclear envelope due to the depolymerization of atomic lamins and other nuclear membrane proteins following their phosphorylation by cyclin B-Cdk. Centrioles, if present, divide and migrate to opposite poles.

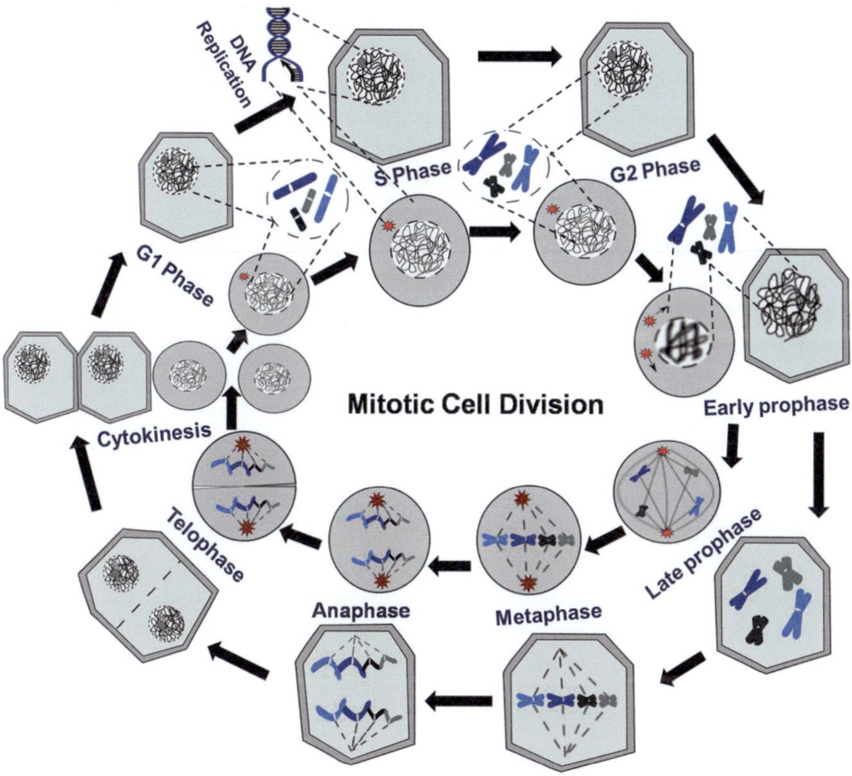

Fig. 2.5 Different stages of mitotic cell division in animal and plant

2.2.4.2 Metaphase

Metaphase represents the second stage of the M phase, which is characterized by the highest level of chromosome condensation, complete breakdown of the nuclear membrane, and the alignment of chromosomes in an orderly fashion along the metaphase plate (a hypothetical plane located equidistant between the two poles of the cell), or equatorial plane of the cell (Fig. 2.5). At this stage, the complete disappearance of the nuclear membrane allows direct interaction between the MTOCs or centrosomes-generated mitotic spindle and the chromosomes at the kinetochore regions. This interaction facilitates proper chromosome alignment and movement during division. Each metaphase chromosome contains two chromatids, tightly attached with the help of cohesin proteins. The attachment of chromosomal microtubules to the kinetochores of each sister chromatid during metaphase is crucial for precise chromosome alignment along the metaphase plate and for proper chromosome distribution during anaphase.

2.2.4.3 Anaphase

Anaphase is characterized by the splitting of centromeres and the starting of anaphasic movement of sister chromatids toward opposite poles of the cell (Fig. 2.5). The anaphase-promoting complex (APC/C) or cyclosome plays a crucial role in triggering the separation of sister chromatids and mitotic exit. The APC/C acts as an E3 ubiquitin ligase, an enzyme complex that targets specific proteins for degradation by the 26S proteasome. This "tagging" for degradation is achieved by attaching a small protein called ubiquitin to the target protein. During anaphase, the APC/C targets three key proteins: securin, S, and M cyclins.

As securin inhibits the protease separase, cohesins are able to hold sister chromatids together until anaphase. By degrading securin, activated APC/C allows separase to degrade cohesins. This cleavage and constriction of spindle fibers allow each sister chromatid to move toward opposite poles during anaphase. Degradation of S and M cyclins by the APC/C inactivates CDKs, which is a crucial step in exiting mitosis. Dysregulation of APC/C can lead to premature anaphase or metaphase arrest. Interestingly, the APC/C itself is tightly regulated by specific activator subunits (Cdc20 or Cdh1), which bind the complex and determine its substrate specificity.

Anaphase is further divided into two sub-phases known as anaphase A and anaphase B (Fig. 2.6). Anaphase A refers to the gradual contraction and shortening of microtubules attached to the kinetochores, which results in the movement of two sister chromatids toward the opposite poles of the cell (Fig. 2.6). This is achieved by removing tubulin dimers from the plus ends attached to the kinetochores. The chromosomes move with their centromeres leading toward the poles and their arms trailing.

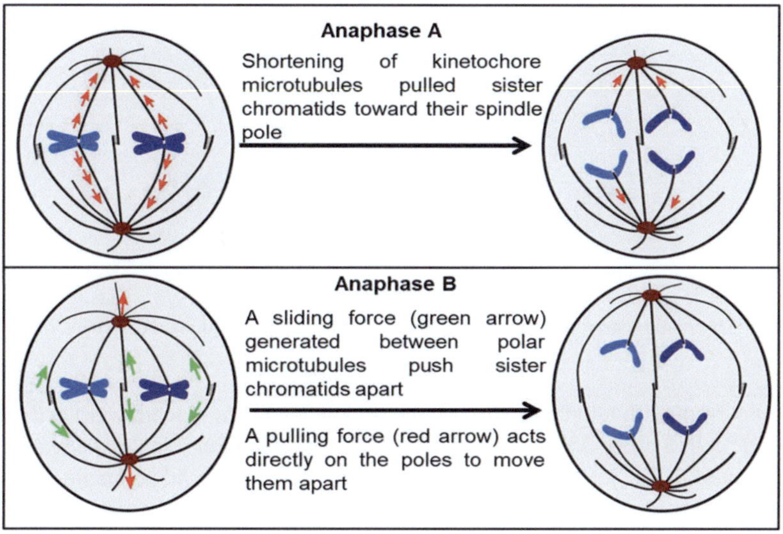

Fig. 2.6 Mechanism of anaphase A and anaphase B

Anaphase is essential for maintaining genomic stability. It ensures accurate distribution (each daughter cell receives a complete set of chromosomes) of the duplicated genetic material into two genetically identical daughter cells. Any errors or abnormalities in chromosome segregation during anaphase can lead to numerical (aneuploidy) and structural instability.

2.2.4.4 Telophase

Telophase represents the final stage of the M phase, characterized by the reappearance of the nuclear envelope and nucleoli, chromatin decondensation, and initiation of cytokinesis (Fig. 2.7). During this phase, the separated sister chromatids reach opposite poles of the cell, and the highly condensed separated sister chromatids begin to decondense, relax, and unravel back into their extended chromatin form. This decondensation allows for the resumption of gene transcription and normal cellular activities in the daughter cells. As the sister chromatids reach opposite poles of the cell, the nuclear membrane starts to form around each set of chromosomes, marking the reconstitution of two separate nuclei within each daughter cell (Fig. 2.7). Nucleoli reappear (Fig. 2.7) as the nuclear membranes reform around the chromosomes, indicating the reinitiation of protein synthesis and cellular metabolism in the newly formed daughter cells.

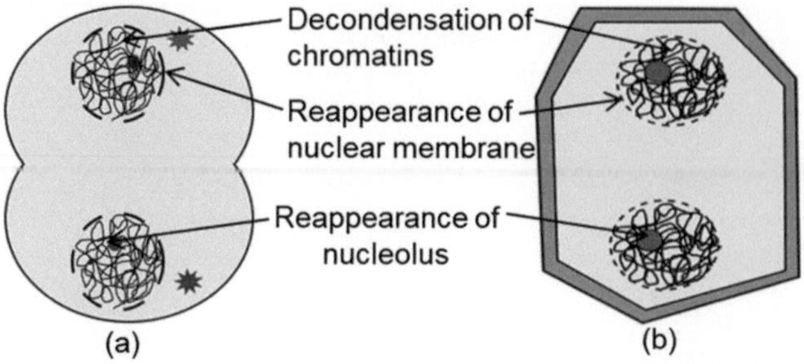

Fig. 2.7 Schematic representation of telophase in an animal (**a**) and a plant cell (**b**)

Concurrent with telophase, the cell membrane begins to pinch inward at the center of the cell, a process known as cleavage furrow formation. This marks the beginning of cytokinesis, the physical separation of the cytoplasm and organelles between the two daughter cells. In plant cell, cell plate formation between the two daughter nuclei initiate during this phase.

2.2.5 Meiosis

The term meiosis derives from the Greek word meioun, meaning "to make small," referring to a specialized cell division process that occurs in the reproductive cells or gametic cells and produces gametes in animals and microspores in plants. The term meiosis was first used by J.B. Farmer and Moore in 1905. But in the late nineteenth century, it was first studied in sea urchins by Hertwig in 1876 and in the nematode *Ascaris megalocephala* by Van Beneden in 1883 (Farmer and Moore 1905). Thus, plant and animal meiosis is often mentioned as sporic and gametic. Meiosis, unlike mitosis, is a unique cell division that involves a single round of interphase (DNA replication), quite similar to mitotic interphase, with G1, S, and G2 phases. The interphase is followed by two consecutive M phases (meiosis I and II). It results in four genetically different cells with half the DNA content of the mother cell. Before entering meiosis, reproductive cells are generally diploid with two homologous copies of each chromosome. Just before the initiation of meiosis, the cell duplicates its DNA, so it contains four copies of each chromosome, distributed into two pairs of homologous chromosomes. Meiosis also differs from mitosis in that it involves recombination, during which chromosomes exchange segments. As a result, the gametes produced during meiosis are genetically unique. The karyokinesis of meiosis I consists of extended prophase I, metaphase I, anaphase I, and telophase I, whereas meiosis II has prophase II, metaphase II, anaphase II, and telophase II.

Events of meiosis with the characteristic features of these phases are described below.

2.2.5.1 Meiosis I (Fig. 2.2)

Prophase I: Prophase I is the longest phase of meiosis I and is comparatively complex, characterized by the formation of bivalents (pairing of homologous chromosomes), followed by crossing over (exchange of genetic material) and terminalization of chiasmata. It has five substages: leptotene, zygotene, pachytene, diplotene, and diakinesis.

Leptotene: The nuclear membrane and the nucleolus cannot be observed conspicuously. Chromosomes appear as long, thread-like structures interwoven together and uniformly distributed. Chromosomes display a beaded appearance, called chromomeres. The ends of chromosomes are drawn toward the nuclear membrane near the centriole. In some plants, chromosomes may form synthetic knots. Chromosomes, each consisting of two sister chromatids, begin to condense.

Zygotene: During this substage, homologous chromosomes with duplicated chromatids are paired up over the entire length of the chromosomes to form bivalents by the process called synapsis. Each pair of homologous chromosomes is known as a bivalent.

Pachytene: During this substage, the closely attached chromatids of the bivalents exchange DNA via crossing over. It produces chromosomes with new combinations of maternal and paternal alleles (genetic recombination). The x-shaped physical appearance of crossing over is called chiasmata, they are the site of exchange of parts between two homologous chromosomes.

Diplotene: In this stage, terminalization of chiasmata, i.e., movement of chiasmata toward the ends of the synapsed chromosomes, can be seen.

Diakinesis: The chromosomes become more condensed and appear shorter and thicker in comparison to earlier stages. Terminalization of chiasmata is completed, and bivalents appear as eye-shaped. The complete disappearance of the nucleolus and nuclear membrane facilitates the attachment of spindle fibers to the highly condensed bivalents.

Metaphase I: The bivalents (pair of homologous chromosomes) align themselves at the metaphase plate. The centromere of each chromosome of a terminalized tetrad is directed toward the opposite poles. The chromosomal microtubular spindle fibers remain attached to the centromeres and homologous chromosomes, ready to separate.

Anaphase I: In anaphase I, each chromosome consisting of two chromatids separates from its homologue and moves toward opposite poles of the cell, ensuring that each daughter cell will have a unique combination of genetic material with half chromosomes compared to the parent cell. This process is crucial for creating genetic variation within a population.

Telophase I: The nuclear membrane and nucleolus reformed around each set of chromosomes. The chromosomes remain condensed. Cytokinesis may also occur to produce two haploid cells. This type of cell division is termed reduction division.

2.2.5.2 Meiosis II (Fig. 2.2)

Prophase II: Prophase II is the beginning of meiosis II in haploid cells produced by meiosis I. During this phase, chromosome condensation and nuclear envelope breakdown occur without further DNA replication.

Metaphase II: In metaphase II, chromosomes with two chromatids (one normal and one recombined) rearrange in the equatorial plate. The centromeres lie in the equator, while the arms are directed toward the poles.

Anaphase II: The two sister chromatids of each chromosome (one normal and one recombined) are separated from one another during anaphase II, resulting in four genetically unique haploid cells.

Telophase II: The chromosomes uncoil after reaching the opposite poles and become less distinct. The nuclear membrane and nucleolus reappear, resulting in the formation of four haploid daughter nuclei.

Cytokinesis: Cytoplasm divide by the formation of cell plate which separates each nucleus to two cells. Cytokinesis occurs twice, after telophase I and telophase II as a result four haploid cells are produced (Fig. 2.2).

2.3 Plant Chromosomes

Chromosomes are a universal term associated with eukaryotic organisms, with their numbers varying by species. In diploid plants, the recorded chromosome numbers range from $2n = 4$ to over 1400. Chromosomes are distinct genomic structures that represent the highest level of genetic organization, housing genes or DNA within organized packages inside the nucleus of eukaryotic cells. Essentially, they can be thought of as compact spools of DNA that contain the intricate blueprints for the genetic information vital to constructing and maintaining the structure and function of each cell.

Under a student's microscope, condensed chromosomes can be observed in countable quantities after undergoing chromosome processing with minimal resources. This process requires skill and hands-on training. Consistent chromosome numbers and unique karyomorphological features of each species or population are essential data points in cytogenetics, serving as keys to unlocking the mysteries of a species' life.

During cell division, several significant changes occur, including RNA and protein synthesis, DNA replication, centriole division, chromatin condensation, spindle fiber formation, and the cyclical disappearance and reappearance of the nucleus. However, only the alterations in the structural and morphological configurations of chromatin or chromosomes and the nucleus can be visualized under a compound microscope, which highlights the unique importance of chromosomes in cytogenetics. As cell division progresses, the long, thin, fibrous interphase chromatin gradually condenses into more organized, thread-like structures known as chromosomes during metaphase. These changes facilitate proper segregation of condensed chromosomes into each daughter cell during anaphase. Thus, a sound conceptual understanding of chromosomes is a prerequisite in the field of cytogenetics.

It is quite fascinating to observe that ~1 m long DNA double helix of the haploid human genome is packaged into 1–10 μm long chromosomes during the metaphase of meiosis or mitosis. DNA molecules of individual chromosomes assemble separately into distinct chromosomes without any untangling. The packaging process is rapid and incredibly accurate, rendering the lengthy DNA molecules fit inside the tiny nucleus of the cell. However, the compact DNA within chromosomes is not locked away and remains accessible to proteins and enzymes that need to interact with it. It ensures the instructions can still be read and used when required. It's like having a well-organized bookshelf where you can easily find the book you want to read.

DNA in decondensed chromatin is 2 nm wide, while chromatin in condensed chromosomes becomes 1400 nm wide. Kornberg (1974) described this excellent condensation method of chromatin through his nucleosome model. He considered the nucleosome to be the basic structural unit of chromatin, containing approximately 200 base pairs of DNA. Thus, when chromatin material is placed under an electron microscope, it looks like a beaded thread. Folding and refolding of the nucleosomes is the path of chromatin condensation toward forming individual chromosomes. The various compaction levels of the chromosomes have been shown in Fig. 2.8.

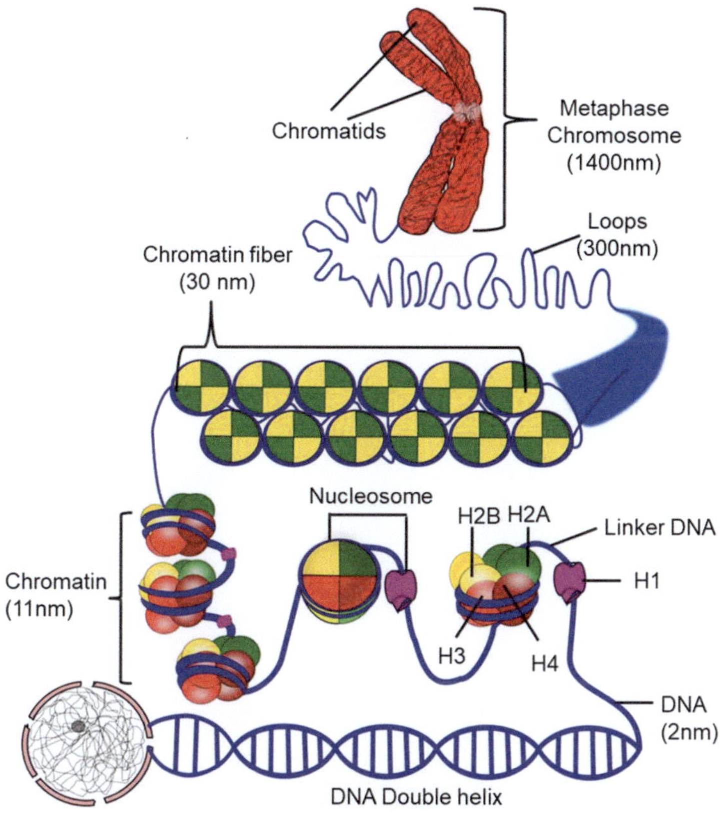

Fig. 2.8 Schematic illustration of different compaction levels of chromosomes

2.3.1 Structural Organization of Metaphase Chromosomes: Classical and Molecular

It is important to recognize that the morphological structure of chromosomes is dynamic and varies throughout the different phases of the cell cycle in terms of organization, condensation state, and visual appearance. For instance, during interphase, chromosomes are in a less condensed state, called chromatin, which comprises negatively charged DNA, basic histone proteins, and various chromosome-associated proteins. In the S phase of the cell cycle, each chromosome undergoes precise duplication, resulting in each chromosome consisting of two identical sister chromatids in the G2, prophase, and metaphase stages. Notably, these sister chromatids cannot be distinguished under a compound microscope, as they are tightly held together by cohesion proteins along their entire length and at the centromere. As chromatin continues to condense during G2 and prophase, its appearance changes from elongated, thin threads to rod-like, condensed chromosomes visible in metaphase.

The morphological structure of a chromosome refers to its physical appearance and organization during metaphase. This phase, which typically lasts only a few minutes, provides an optimal opportunity for visualizing and determining various karyomorphological features and the arrangement of chromosomes. Asynchronous cell division in plants poses significant challenges for studying the karyomorphological characteristics of any given species. In a metaphase chromosome, several morphological features can be clearly identified.

- *Chromosomal Arms*

The chromatids present on both sides of the centromere are called chromosomal arms (Fig. 2.9). The short arm is designated as *p (petit)* and the long arm is defined as *q (next letter in the alphabet)*. Variations of arm ratios with the chromosome complement resulted in metacentric, submetacentric, acrocentric, and telocentric chromosomes. When a karyotype is made, the q arm is always put on the bottom and the p arm on the top.

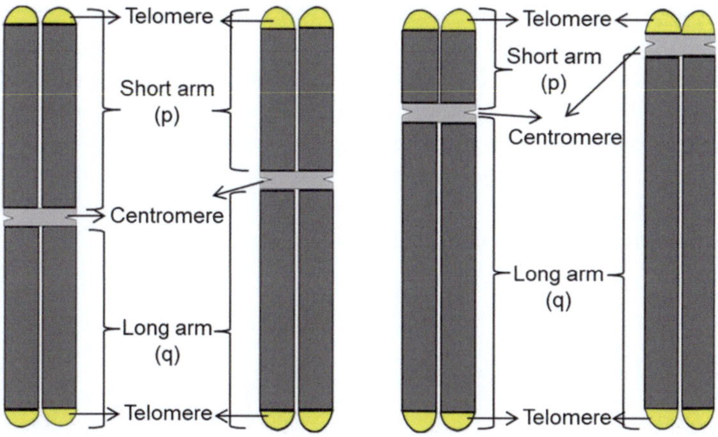

Fig. 2.9 Nomenclature of chromosomes based on arm ratios. (**a**) Metacentric chromosome, (**b**) submetacentric chromosome, (**c**) acrocentric chromosome, and (**d**) telocentric chromosome

- *Centromere*

Darlington (1936) and Schrader (1939) independently coined the terms "centromere" and "kinetochore" for the site where spindle fibers associate with the chromosomes during cell division (Fig. 2.10). The terms were long thought redundant, but it has recently proven useful to differentiate between them (Fukagawa and Earnshaw 2014). The centromeres are a crucial structural features of metaphase chromosomes, which specifies where the kinetochore will form and are primarily responsible for directing the faithful segregation of chromosomes during cell division. The chromosomes without a centromere (acentric chromosomes) cannot segregate properly during cell division. On the other hand, kinetochores are complex disk-like protein structures lacking DNA (Fukagawa and Earnshaw 2014), and spindle fibers are attached to them (Fig. 2.10).

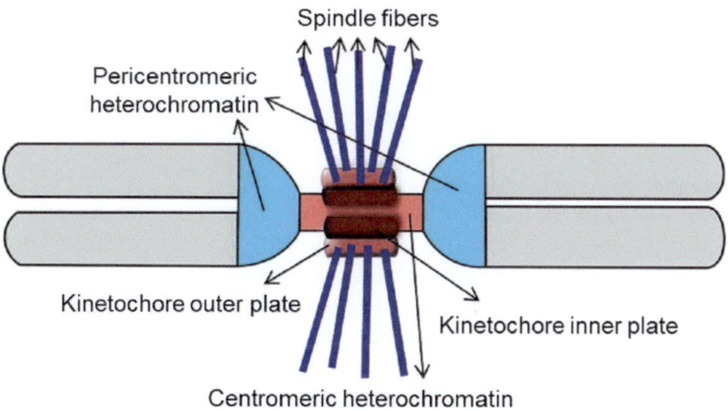

Fig. 2.10 Structure of centromere and kinetochore

Although monocentric chromosomes with a single centromere are very common, there is a wide variability in the number and structure of centromeres in different plant species (Fig. 2.11). Cytogenetic studies have revealed diverse centromere structures, including point centromeres (found in *Saccharomyces cerevisiae*), regional centromeres (found in *Schizosaccharomyces pombe, Hordeum vulgare, Secale cereale, Cuscuta japonica*), a meta-polycentric centromere (found in *Pisum sativum and Lathyrus*), and holocentromeres (found in *Rhynchospora pubera, Luzula elegans*), each with unique characteristics and evolutionary implications.

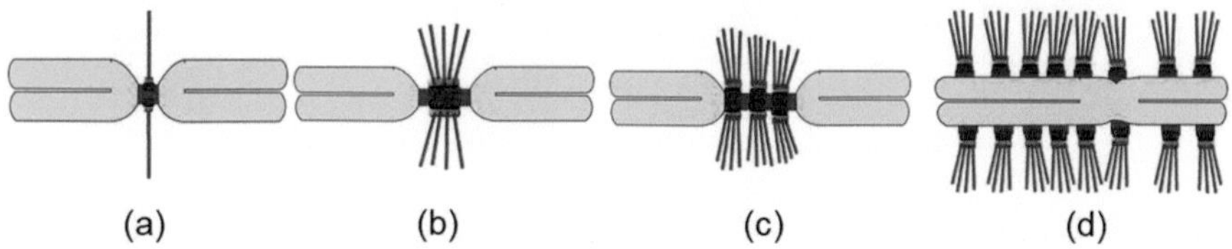

Fig. 2.11 Variation in centromere in plants. (**a**) Point centromere, (**b**) regional centromere, (**c**) meta-polycentric centromere, and (**d**) holocentromere

Centromeres are also recognized as the primary constriction of mitotic chromosomes. It represents a constitutive heterochromatic region within the mitotic chromosomes that does not take stain in the condensed metaphase stage, making it visible as a "gap" in the classical orcein-staining method. Molecular analysis has revealed that the centromeric region generally contains specialized centromeric DNA and satellite repeats. Its organization and function are highly conserved among eukaryotes yet the surprising diversity in their DNA sequence organization within the eukaryotes has been reported. With some exceptions (several megabase-long bulky satellites in potato or wheat), the centromeric region generally contains a tandem array of 150 to 180 bp satellite repeats in head-to-tail orientation, juxtaposed with centromeric retroelements in plants. In humans and many other organisms, a specific highly repetitive DNA sequence called the alpha satellite DNA forms the core of centromeres.

Molecular cytogenetic techniques, such as FISH, are invaluable tools for detecting centromere regions with a specific probe and for identifying abnormalities in centromere regions when the centromere-specific sequence is known. The application of AT/GC-specific fluorescence stains is also very useful for differentiating this region from the rest of the chromosome and providing a preliminary idea of its sequence specificity, whether AT-rich, GC-rich, or AT-GC neutral. Cytogenetic techniques, including immunofluorescence and chromatin immunoprecipitation, have allowed scientists to visualize and study these proteins within centromeres. Despite significant progress, many mysteries surrounding centromeres remain unresolved. Challenges in studying centromeres include their repetitive nature, dynamic organization, and variability across species. Future research in cytogenetics should be continued to unravel the complexities of centromeres, shedding light on their role in health and disease.

- *Telomeres and Telomerase*

Karyomorphologically, the ends of linear eukaryotic chromosomes are called telomeres (Fig. 2.12). Telomeric areas play a vital role in protecting chromosome ends from degradation or end-to-end fusion with other chromosomes and preventing double-stranded DNA breaks. Thus, they are essential for maintaining the genomic stability and integrity of chromosome structure.

They are very specialized end structures that comprise of nucleoprotein complexes. Like centromere, these regions are a part of constitutive heterochromatin containing repetitive DNA sequences in tandem clusters. It has been noted that telomere sequences are ancient and highly conserved across many species. The most frequently detected conserved repetitive telomeric DNA sequence is $T_xA_yG_z$. Research has shown that all tested vertebrates and many basal metazoans use TTAGGG. In contrast, arthropods, including insects, use TTAGG, and those of most plants contain the *Arabidopsis*-type telomere repeat $(TTTAGGG)_n$ (Fig. 2.12; Peska and Garcia 2020). Although the *Arabidopsis*-type telomeric sequence has been reported in large sections of plants, a novel sequence $(TTATGGGCTCGG)_n$ of neither *Arabidopsis* nor vertebrate type was reported in *Allium* of Allioideae, known as *Allium*-type telomere sequence. Later, the first vertebrate-type telomere sequence TTAGGG, unusual for land plants, was reported in *Aloe vera* and some other Asparagales. Despite all the telomere motif exceptions detected, the real diversity in telomeric sequences in land plants is probably greatly underestimated. In diverse plant kingdoms, only 10% of species have been analyzed for telomeric sequences so far. Unravelling the plant telomeric repeats in other crops may reveal interesting correlations between plant and animal kingdoms (Peska and Garcia 2020; Zavodnık et al. 2023).

Telomerase (telomere reverse transcriptase, TERT), an enzyme found in embryonic, germ, and cancer cells, plays a crucial role in ensuring proper telomere replication (Fig. 2.12). It consists of an internal template RNA, the telomerase RNA (TR) subunit, and proteins, such as TPP1 and POT1, telomeric repeat-binding factors (TRFs). During each round of replication, a short region of TR serves as a template that dictates the repeat motif in telomeres and prevents the shortening of somatic chromosomes (Fig. 2.12). However, telomerase inactivation in adult somatic cells results in telomere shortening during cell division. Some researchers have suggested a possible relationship between critical telomere length and organism longevity. Additionally, active telomerase has been identified in plant organs with high meristematic activity, including the meristem cells of seedlings, the tips of roots, the leaves of young and middle-aged plants, and floral meristems.

- *Nucleolus, NORs, and Secondary Constrictions*

Nucleus, the gene-bearing organelle within the cell, is demarcated by its own membrane from the cytoplasm. But, within the nucleus, there is no demarcated space for any specific functions. However, it exhibits structural domains in which particular macromolecular components are assembled to perform distinct roles (Sáez-Vásquez and Medina 2008). The presence of the nucleolus within the nucleus was first described by Fontana in 1781 as an ovoid body (Vincent 1955). In the nineteenth century, many cytologists using light microscopy began studying this tiny particle within the nucleus. It was noted that a smaller body, but not a smaller nucleus with a constant shape, is present inside the nucleus. However, its presence was not reported in bacteria. Within the nucleus, it was identified as an organelle with its own properties and shown to be related to the chromosomes (McClintock 1934; Sirri et al. 2008).

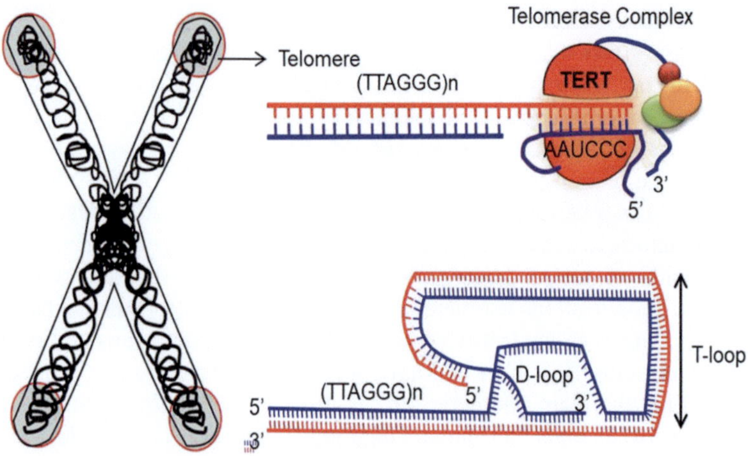

Fig. 2.12 Structure of telomeres and telomerase

It was also discovered that the nucleolus arises at a specific chromosomal locus. In continuation, Heitz (1931a) proposed that the nucleolus is particularly associated with the secondary constricted chromosomes, and the number of "secondary constrictions" within a karyotype indicates the number of nucleoli present within the interphase nuclei. It was also noted that the nucleoli is organized in the telophase and starts to disappear in late prophase along with the nuclear membrane. The importance of the nucleolus was further underscored by McClintock's (1934) observation that the nucleolus was organized by the activity of a distinct, deep-staining body with a definite position on one chromosome (the satellite chromosome). McClintock identified the deep-staining body as "the nucleolar-organizing body" (later renamed nucleolar organizer regions or NORs). Recent advancements in molecular and cellular biology have helped clarify the mysterious relationship between NORs and nucleoli and have identified them as the sites of ribosome manufacturing and ribosomal gene expression.

2.3.1.1 Staining and Visualization of the Nucleolus in Plant Cells

Nucleolus staining in plants is not a very simple method. On the other hand, many methods are available for animal systems. In plants, their appearance and disappearance in cell divisional stages need to be standardized. Nucleolus staining in plant cells was initiated long back (Rattenbury 1952) with fixative and carmine, but the disadvantage of the procedure was that it was heavily dependent on hydrolysis factors. With certain modifications of Rattenburys method, Morrison et al. (1959) provided another method for plant nucleolus staining. The method is presented in a student-friendly manner.

- *Requirements*:

- *Plant material*: Fresh root tips of *Allium cepa*.
- *Chemicals required*: Ethanol, Formalin (37–40% aqueous solution of formaldehyde), propionic acetic acid, and 1 N HCl.
- *Staining dye*: Carmine, Schiff's reagents (Feulgen), and Fast green.

- *Procedure*:

(i) *Collection of roots*: Fresh and healthy roots less than 1 cm in length were taken in water at appropriate time and then transferred to the fixative (a mixture of ethanol, formalin, and propionic acid in a ratio of 15:1:2) for one to two hours.
Preparation of Fixative
To prepare 90 mL of the fixative, mix 75 mL of ethanol +5 mL of formalin and 10 mL of propionic acid. Keep it in glass container at room temperature.
(ii) Fixed rot tips are hydrolyzed in 1 N HCl at 60 °C for 15 minutes.
(iii) After hydrolysis, root tips are incubated in Schiff's reagents (Feulgen stain) for 30–60 minutes.
(iv) Transfer tissues to aceto-carmine for 10–15 minutes.
(v) Put one to two drops of 0.05% Fast green FCF, dissolved in 45% acetic acid solution, on the clean glass slide.
(vi) Excise the root meristem part on it, cover it with a cover glass and squash the tissue.
(vii) Observe the slide under a microscope.

This method uses three stains. Feulgen stains chromosomes, nucleolus, and spindle. Fast green FCF better stains the nucleolus than the spindle and cytoplasm. At the same time, aceto-carmine further intensifies the nucleolar staining.

- *Molecular view of Nucleolar organiser regions (NORs)*

In a karyotype, NORs are considered chromosomal landmarks, and nucleoli at the molecular level are the sites of ribosome biogenesis. Eukaryotic ribosomes have four structural ribosomal RNAs: 5S, 5.8S, 18S, and 26S (Garcia et al. 2012). The 45S rDNA gene cluster is scattered throughout the genome. These loci have hundreds or thousands of tandem repeats of nearly identical rDNA sequences. These regions are highly conserved between species.

Figure 2.13 shows the structural organization of the 45S rDNA gene cluster and 5S rDNA unit in eukaryotes. Expression of the rRNA genes generates the nucleolus at interphase. The transcript of 18S-5.8S-26S (35S) genes is generated from a single operon using RNA polymerase I (Pol I), whereas the transcript of 5S RNA genes is separately generated by RNA polymerase III (Pol III). Pol I-induced transcription yields preribosomal RNA (pre-rRNA), and mature 18S, 5.8S, and 28S rRNAs are generated by the removal of external transcribed spacers (ETSs) and internal transcribed spacers (ITSs). In eukaryotes, each unit of NOR comprises 18S, 5.8S, and 25S rRNA genes, two internal transcribed spacers (ITS1 and ITS2), and an intergenic spacer (IGS) that contains a non-transcribed spacer (NTS) and an external transcribed spacer (ETS) (Davidian et al. 2021).

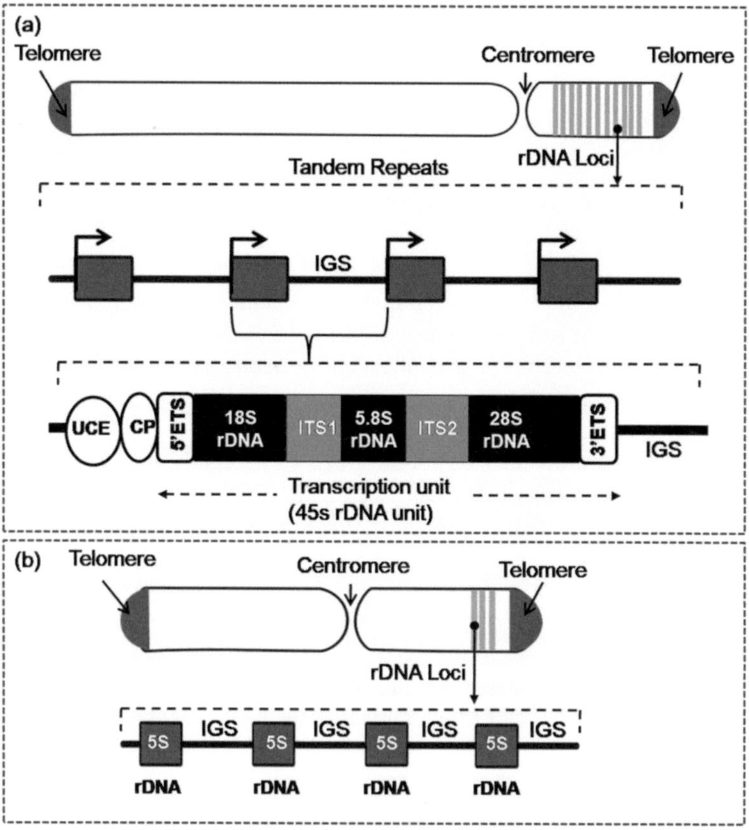

Fig. 2.13 Organization of eukaryotic rDNA genes: (**a**) 45S rDNA locus comprising 18S, 5.8S, and 28S rDNA, (**b**) tandem repeats of 5S rDNA (CP core promoter; ETS external transcribed spacer, IGS intergenic spacer, ITS internal transcribed spacer, UCE upstream control element)

However, not all rRNA genes are needed during the growth and development of organisms. In plants and mammals, specific rRNA genes that remain inactive are selectively silenced via changes in cytosine methylation and repressive histone modifications. The question of what dictates the fate of expression of an individual rRNA gene among nearly identical genes has long been a puzzle. Studies in *Arabidopsis thaliana* demonstrated that entire clusters of rRNA genes, known as nucleolus organizer regions, are silenced based on the chromosome on which they reside (Brian McStay 2016; Mohannath et al. 2016). This may explain why conventional methods have failed to visualize the required number of secondary constrictions in many diploid plant species. However, the use of molecular cytogenetic methods has facilitated the detection of NOR sites. The number, position, and structure of the 5S and 18S-5.8S-26S ribosomal DNA (rDNA) loci within the NOR region have become important species characteristics.

NORs are often recognized on the stained metaphase chromosome as a small achromatic gap in addition to the primary constriction (Heitz 1931b; McClintock 1934). Karyomorphologically, the nucleolar organizer region (NOR) and the constriction that it forms along the chromosome arm of metaphase chromosomes are known as a secondary constriction (Darlington 1926). Electron tomography of human NORs has revealed that this morphology arises from undercondensation of the arrays of rDNA repeats within active NORs (Heliot et al. 1997).

The number of secondary constriction-bearing chromosomes, their location/relative position, and the size of each chromosome within a karyotype are very important karyomorphological features for identification and comparison within and between species or populations, especially when chromosomal complements show very similar morphology (Murray et al. 1992). Lima-de-Faria (1976) opined those secondary constrictions were preferentially distributed on the short arms and in the subterminal region in most species of plants and animals. However, their presence in the proximal, interstitial, and terminal regions of the chromosome arm has been reported in many species.

NORs are evolutionarily conserved structures, but their organization can undergo significant changes during speciation and evolution. The way rDNA chromatin decondenses varies depending on the cell type and species being studied. Some studies have suggested that rDNA activity at different loci may change throughout the cell cycle, with their transcription being determined in a time- and region-specific manner (Li et al. 2006; Chandrasekhara et al. 2016). As a result, observing the exact numbers of secondary constrictions on mitotic chromosomes may be challenging unless other confirmatory, more powerful methods are utilized. Classical chromosome preparation methods often failed to identify the number and position of secondary constrictions properly. However, the enzymatic maceration and drying method (EMA) followed by Giemsa staining helped to produce cytoplasm-free metaphase plates and resulted in distinct chromosome morphology with well-marked secondary constrictions in terminal subterminal or interstitial regions. The application of fluorescent banding with AT/GC-specific fluorochromes on the same metaphase plates exclusively confirms the number and positions of secondary constrictions. Comparative cytogenetic studies of NORs provide valuable insights into evolutionary relationships among species.

Currently, by using the fluorescence in situ hybridization (FISH) technique, it is known that the secondary constrictions represent only the expression of rRNA genes and that other functional sites may not form secondary constrictions, especially if located too close to the terminal end of the chromosomes (Heslop-Harrison and Schwarzacher 2011).

- *Chromatin*
  A DNA molecule, together with histone and other associated non-histone proteins, forms the chromatin. It resides within the nucleus of eukaryotic cells. Chromatin is the base composition of condensed chromosomes, and histones are the significant chromatin proteins. Thus, at the molecular level, chromosomes primarily consist of DNA, histone proteins, and other proteins.
  - *DNA*: In chromosome or chromatin, deoxyribonucleic acid (DNA) is organized into long, double-stranded helical molecules with a diameter of 2 nm (Fig. 2.8). Each chromosome has a specific, unique sequence of DNA that carries all the information of living organisms.
  - *Histone proteins*: Histones are a group of small, positively charged basic proteins that associate with DNA to form nucleosomes (Fig. 2.8). The histone proteins are highly conserved across all eukaryotes and are composed of unusually high amounts of the basic amino acids arginine and lysine. The presence of very similar histone H_1, H_2A, H_2B, H_3, and H_4 proteins is detected across different species of eukaryotes. They play a crucial role in packaging and organizing DNA within the chromosome. It also regulates gene expression and other chromosomal functions.
  - *Nucleosomes*: Partial digestion of interphase chromatin with endonuclease yields the basic repeating structural units of chromatin called "nucleosomes." These nucleosomes help condense and package DNA into a more compact and organized form to fit within the limited space of the cell nucleus. Nucleosomes are composed of roughly 200 bp segments of DNA wrapped in a 1.65 turn around the "beads" of a histone octameric core nucleosome particle and threads like linker DNA, forming a "bead-like" structure (Fig. 2.8). Prolonged nuclease digestion of chromatin yields 146 bp DNA associated with a nuclease-resistant bead, called the core nucleosome structure. The packing ratio of DNA: nucleosome is 7:1. This core nucleosome protects DNA from endonuclease attack and is composed of two molecules each of four histone types, namely H_2A, H_2B, H_3, and H_4 (Fig. 2.8). The histone proteins are arranged into four heterodimers: two dimers of H_2A-H_2B and two of H_3–H_4. Dimerization is achieved by the C-terminal domains of histone proteins, whereas the C-terminal domain of H_2A projects out from the nucleosome as a long stretchy tail, further helping to wrap the DNA helix around the core particle. Apart from the core nucleosome structure, a group of other closely related histones, collectively called linker histones (H_1, $H_{1.5}$, H1oo, H_1t, and H_1x), attach to linker DNA to facilitate compaction and stabilization of neighboring nucleosomes. The core nucleosome and H_1 histone association are termed "chromatosome." However, some non-histone proteins also bind

to linker DNA to aid in compaction and affect gene expression in the surrounding region.

After this condensation, a 10 nm fiber coils to create a 30 nm interphase chromatin fiber, which is then coiled around non-histone scaffolds to form a 100–120 nm chromonema. This process is repeated until the maximally condensed metaphase chromosome (~1400 nm wide) thickness is achieved by assembling large supercoiled loops or domains around a central scaffold (Fig. 2.8).

- *Euchromatin and Heterochromatin*

Before the discovery of DNA, Heitz (1928–1935) studied chromosomes in the dividing cells of *Pellia epiphyla, Vicia equine,* and *Drosophila*. He elaborated that all chromosomes exhibit a longitudinal differentiation into two types of chromatin: euchromatin and heterochromatin. He observed that certain portions of chromosomes remained condensed throughout the cell cycle, which he named heterochromatin, to distinguish them from the rest of the chromosomes called euchromatin. He also noted that the distribution of heterochromatin may be species-specific as well as chromosome-specific, and it is located at corresponding positions on both homologous chromosomes.

Euchromatin and heterochromatin form the basic structure of eukaryotic chromosomes. Molecular analysis has firmly established that euchromatin is the more extended, transcriptionally active, and less densely packed form of chromatin distributed throughout the chromosomes. Ibraimov (2024) suggested that Heitz defined euchromatin as the parts of chromosomes that are no longer visible under an optical microscope at the end of the telophase. Euchromatin is loosely packed, making it easily accessible to the transcription machinery and is associated with actively transcribed genes. It constitutes the bulk of the genome and contains the majority of protein-coding genes and regulatory elements involved in gene expression, such as promoters, enhancers, and transcription factor binding sites (Ibraimov 2024). Euchromatin undergoes dynamic changes in structure and accessibility during different stages of the cell cycle and in response to cellular signals. It plays a vital role in gene regulation, DNA replication, recombination, and repair.

In contrast, heterochromatin is a densely packed, highly condensed, transcriptionally inactive form of chromatin that contains clusters of repetitive DNA sequences. Molecular analysis of nuclear DNA revealed that heterochromatin sites are compactly packed due to their repetitive DNA sequences and a higher protein-to-DNA ratio than euchromatin. Although centromeres and nucleolus-organizing regions represent heterochromatin, it is also distributed throughout the length of the chromosomes. Heterochromatin regions contain inactive genes. It appears darkly stained during interphase and lightly stained during M phase under a microscope. In contrast, euchromatin appears lightly stained during interphase and darkly stained during M phase.

Heterochromatin was further classified by Brown (1966) into "constitutive" and "facultative" heterochromatin, which remain debated among scientists (Ibraimov

2024). Constitutive heterochromatin comprises highly repetitive DNA sequences that remain condensed during the whole cell cycle. Preferably, this type of heterochromatin is AT-rich, GC-rich, or neutral.

The most prominent visible sites of constitutive heterochromatin within a karyotype are the centromeric regions or primary constrictions and secondary constriction sites of NOR. However, its presence has also been located in pericentromeric, telomeric, transposons, and some gene-poor regions of the genome. Constitutive heterochromatin consistently remains silenced in all cell types of an organism. On the other hand, facultative heterochromatin is cell-type-specific heterochromatic regions that retain their potential to switch into euchromatin under certain conditions and are frequently present in developmental genes.

Several studies on heterochromatin differentiation have revealed its distributions, variability, molecular composition, direct and indirect effects on the karyotype, putative function, etc. (Mizianty 1984a, 1984b, 1985; John 1988; Hennig 1999). The distribution patterns of heterochromatin blocks throughout the karyotype can suggest functional roles in genome organization and gene regulation (Guerra 2000). Maintaining the balance between heterochromatin and euchromatin is crucial for proper genome organization, regulation of gene expression, centromere function, genome stability, and cellular homeostasis (Lei and Berger 2019; Penagos-Puig and Furlan-Magaril 2020)

2.3.2 Banding of Plant Chromosomes

The identification of individual chromosomes is essential for detecting both normal chromosomal variations and chromosomal abnormalities. Moreover, the identification of individual chromosomes through chromosomal analysis can serve as an effective tool for distinguishing closely related species, especially when distinguishable morphological differences are either absent or difficult to discern. It also has a significant application in understanding evolutionary relationships.

The traditional karyotyping method involves classical chromosome preparation and staining, followed by pairing based on relative length and arm ratios, and arranging them according to length. Conventional cytogenetics has several limitations for identifying individual chromosomes compared to molecular techniques, such as fluorescence and non-fluorescence banding, FISH, and CGH. Conventional cytogenetics primarily relies on relative lengths, arm ratios, and secondary constriction for chromosome identification, which is not always effective for species with morphologically similar chromosomes. The development of plant chromosome banding techniques in the 1970s significantly addressed these limitations and greatly advanced the field of plant cytogenetics (Ansari et al. 2016).

Chromosome banding refers to the alternating light and dark regions along the length of a chromosome, which are visible after staining with specific dyes. A band is defined as a distinguishable chromosomal segment that appears darker or lighter than its adjacent segments through the application of specific banding techniques (Spinner and Ferguson-Smith 2019).

The development of modern chromosome banding techniques in animal and plant chromosomes began with Heitz's findings in 1928. He reported the longitudinal differentiation between euchromatin and heterochromatin in chromosomes, which can be visualized through specific chromosomal stains. However, limited knowledge of the chemical and physical properties of DNA, euchromatin, and heterochromatin delayed its potential application in chromosome research. The discovery of the DNA structure in 1953 and the subsequent understanding of the highly repetitive nature of AT or GC bases in euchromatin and heterochromatin regions, along with their distribution across the chromosome, facilitated this technique.

Chromosome banding techniques were initially applied to animal chromosomes in the 1970s using specific non-fluorescent stains such as Giemsa (Pardue and Gall 1970; Arrighi and Hsu 1971; Sumner et al. 1971), as animal cells lack cell walls. Schweizer (1974) modified the protocol and introduced the Giemsa C-banding technique for plant chromosome research. This method allows visualization of centromeric or constitutive heterochromatin regions within a chromosome, as in C-banding, by treating Pisum sativum and Secale cereale with barium hydroxide before Giemsa staining.

Later G-banding, which stains AT-rich regions and appears as dark bands, was achieved by pretreating animal chromosomes with trypsin before Giemsa staining. It became a standard method for the analysis of animal and human chromosomes. Chen et al. (1994) later adapted G-banding for plant chromosomes through pretreatment with trypsin and urea before Giemsa staining and successfully applied it to five plant species. Both C and G banding utilize non-fluorescent Giemsa dye.

The use of fluorescent dyes such as quinacrine (Q) or quinacrine mustard (QM) to label plant chromosomes was introduced by Caspersson et al. (1969) in *Vicia faba*. Schweizer (1976a, b) first utilized DNA base-specific (AT/GC) fluorescent dyes, such as 4–6-diamidino-2-phenylindole (DAPI) and chromomycin A3 (CMA), to produce fluorescent bands in the same mitotic cells of *V. faba*. He observed that DAPI created bright fluorescence ($DAPI^+$ signals) in AT-rich regions, while these same regions exhibited pale fluorescence with CMA (CMA^- signals). Additionally, nucleolus-organizing regions displayed bright fluorescence with CMA and pale fluorescence with DAPI (CMA^+ signals/ $DAPI^-$ signals).

2.4 Conclusion

From a cytological perspective, conventional chromosome analysis, such as squashing followed by aceto-orcein staining, can reveal karyomorphological features, including the length of chromosome arms, primary constrictions (centromeres), and secondary constrictions (Fig. 2.14). In contrast, molecular cytogenetics offers greater resolving power with more accurate and confident results. The application of chromosome banding is considered one of the most significant innovations in cytogenetics, countering the notion that chromosome science is obsolete in the era of molecular biology and genomics.

Molecular cytogenetics overcomes the limitations of traditional cytogenetics and expands the field of research. Plant chromosome banding analysis with both non-fluorescent and fluorescent dyes fulfills the aspirations of chromosome researchers. This technique can more precisely identify and confirm the locations of the centromere and secondary constriction. Fluorescent banding with DNA base-specific dyes, such as DAPI and CMA, has drastically changed how chromosomes appear under a microscope.

Fluorescence banding can also detect AT- or GC-rich heterochromatin regions found in terminal, subtelomeric, pericentromeric, and paracentric locations. Molecular methods categorize at least five distinct regions along each chromosome arm: the telomere, subtelomeric region, interstitial region, proximal region, and centromere. Depending on the composition of repetitive DNA, DAPI and CMA can produce CMA^+/DAP^-, CMA^-/DAP^+, or even CMA-DAPI neutral bands. These bands can be species-specific and even chromosome-specific, making it much easier to discriminate and identify morphologically similar chromosomes (Jha et al. 2024).

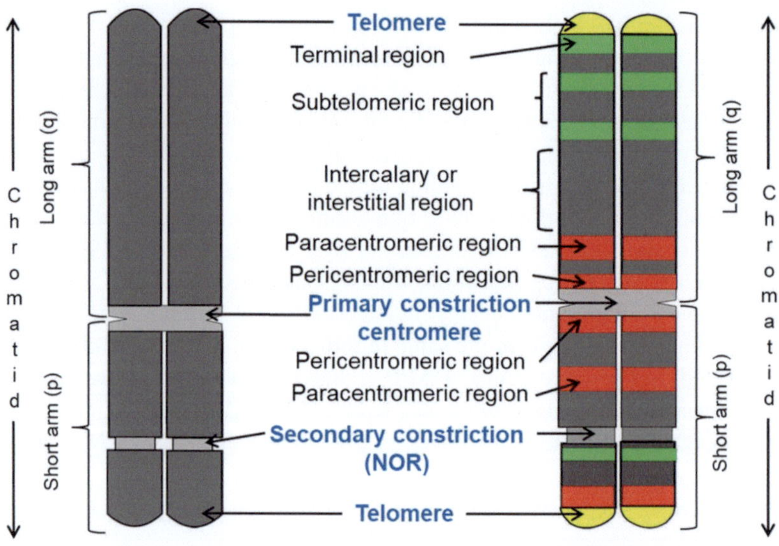

Fig. 2.14 Comparative view of chromosome structure showing the resolving power of conventional and molecular cytogenetics

Heterochromatic elements on the chromosomes can be visualized in centromeric, nucleolar, telomeric, subtelomeric, or interstitial regions. This technique reveals the nature of the constitutive heterochromatin profile within a karyotype, facilitating the construction of a "molecular karyotype." The method of fluorescent chromosome banding has become a powerful tool for analyzing plant chromosomes and has already been employed in a variety of plant species (Fukui 1996; Sheikh and Katsuhiko Kondo 1996; Moscone et al. 1996; Guerra 2000; Hizume et al. 2002;

Yamamoto et al. 2010, 2016; Heslop-Harrison and Schwarzacher 2011; Schneeweiss and Schneeweiss 2013; Bhowmick and Jha 2021; Jha et al. 2021; Jha and Bhowmick 2021; Jha and Halder 2023, 2025; Samanta et al. 2023; Santra et al. 2024; Jha et al. 2024). The distribution of genes, retrotransposons, satellite DNA, and any specific sequences within the karyotype can be located using FISH with a specific probe.

References

Ansari HA, Ellison NW, Bassett SA, Hussain SW, Bryan GT, Williams WM (2016) Fluorescence chromosome banding and FISH mapping in perennial ryegrass, *Lolium perenne* L. BMC Genom 17(1):977

Arrighi FE, Hsu TC (1971) Localization of heterochromatin in human chromosomes. Cytogenetics 10(2):81–86

Avery OT, MacLeod CM, McCarty M (1944) Induction of transformation by a desoxyribonucleic acid fraction isolated from pneumococcus type III Die Entdeckung der Doppelhelix 97

Bhowmick BK, Jha S (2021) A comparative account of fluorescent banding pattern in the karyotypes of two Indian *Luffa* species. Cytologia 86:35–39

Blanco A, Blanco G (2017) The genetic information. In: Blanco A, Blanco G (eds) Medical biochemistry, pp 465–492

Brown SW (1966) Heterochromatin. Science 131:417–425

Caspersson T, Zech L, Modest EJ, Foley GE, Wagh U, Simonsson E (1969) Chemical differentiation with fluorescent alkylating agents in *Vicia faba* metaphase chromosomes. Exp Cell Res 58(1):128–140

Chandrasekhara C, Mohannath G, Todd Blevins T, Pontvianne F, Pikaard CS (2016) Chromosome-specific NOR inactivation explains selective rRNA gene silencing and dosage control in *Arabidopsis*. Genes Dev 30(2):177–190

Chen R, Song W, Li X, An Z (1994) Chromosome G-banding in plants by inducing with trypsin and urea. Cell Res 4(1):79–87

Creighton HB, McClintock B (1931) A correlation of cytological and genetical crossing-over in Zea mays. Proc Natl Acad Sci USA 17(8):492–497

Cremer T, Cremer C (1988) Centennial of Wilhelm Waldeyer´ s introduction of the term "chromosome" in 1888. Cytogenet Cell Genet 48:66–67

Darlington CD (1926) Chromosome studies in the Scilleae. J Genet 16:237–251

Darlington CD (1936) The analysis of chromosome movements in *L. Podophyllum* vers pelle. Cytologta 7:242–247

Davidian AG, Dyomin AG, Galkina SA, Makarova NE, Dmitriev SE, Gaginskaya ER (2021) 45S rDNA repeats of turtles and crocodiles harbor a functional 5S rRNA gene specifically expressed in oocytes. Mol Biol Evol 39(1). https://doi.org/10.1093/molbev/msab324

Deakin JE, Potter S, O'Neill R, Ruiz-Herrera A, Cioffi MB, Eldridge MD, Fukui K, Marshall Graves JA, Griffin D, Grutzner F, Kratochvíl L (2019) Chromosomics: bridging the gap between genomes and chromosomes. Genes 10(8):627

Dewitte W, Murray JA (2003) The plant cell cycle. Annu Rev Plant Biol 54(1):235–264

Dubrovsky J, Ivanov V (2003) Celebrating 50 years of the cell cycle. Nature 426:759

Farmer JB, Moore JE (1905) On the maiotic phase (reduction divisions) in animals and plants. J Cell Sci 2(192):489–558

Fukagawa T, Earnshaw WC (2014) The centromere: chromatin foundation for the kinetochore machinery. Dev Cell 30(5):496–508

Fukui K (1996) Plant chromosomes at mitosis. In: Fukui K, Nakayama S (eds) Plant chromosomes: laboratory methods. CRC Press Inc., Boca Raton, pp 1–17

Garcia S, Garnatje T, Kovařík A (2012) Plant rDNA database: ribosomal DNA loci information goes online. Chromosoma 121(4):389–394

Guerra M (2000) Patterns of heterochromatin distribution in plant chromosomes. Genet Mol Biol 23:1029–1041

Hayashi K, Hasegawa J, Matsunaga S (2013) The boundary of the meristematic and elongation zones in roots: Endoreduplication precedes rapid cell expansion. Sci Rep 3(1):2723

Heitz E (1928) Das Heterochromatin der Moose. J Jahrb Wisenschs Bot 69:762–818

Heitz E (1931a) Die Ursache der gcsetzml1ssigen Zahl, Lage, Form Und Grbsse pflanzhcher Nucleolen. Planta 12:774–844

Heitz E (1931b) Nuklölen und Chromosomen in der Gattung Vicla. Planta 15:495–505

Heliot L, Kaplan H, Lucas L, Klein C, Beorchia A, Doco-Fenzy M, Menager M, Thiry M, O'Donohue MF, Ploton D (1997) Electron tomography of metaphase nucleolar organizer regions: evidence for a twisted-loop organization. Mol Biol Cell 8(11):2199–2216

Hennig W (1999) Heterochromatin. Chromosoma 108:1–9

Heslop-Harrison JS, Schwarzacher T (2011) The plant genome: An evolutionary view on structure and function; organisation of the plant genome in chromosomes. Plant J 66:18–33

Hizume M, Shibata F, Matsusaki Y, Garajova Z (2002) Chromosome identification and comparative karyotypic analyses of four *Pinus* species. Theor Appl Genet 105(4):491–497

Ibraimov A (2024) Heterochromatin and heterochromatic regions. Int J Clin Res Rep 3(3):1–10

Ivanov VB, Dubrovsky JG (1997) Estimation of the cell-cycle duration in the root apical meristem: a model of linkage between cell-cycle duration, rate of cell production, and rate of root growth. Int J Plant Sci 158(6):757–763

Jha TB, Bhowmick B (2021) Unravelling the genetic diversity and phylogenetic relationships of Indian *capsicums* through fluorescent banding. Genet Resour Crop Evol 68(1):205–225

Jha TB, Halder M (2023) Evaluation of karyotype diversity in Indian traditional aromatic rice cultivars through EMA-based non-fluorescent Giemsa and fluorescent DAPI staining. Genet Resour Crop Evol 71(4):1379–1400

Jha TB, Halder M (2025) Unlocking karyotype diversity in some valuable non-aromatic and aromatic indian rice cultivars through EMA-based Giemsa and DAPI staining. Genet Resour Crop Evol 72:6751–6769

Jha TB, Bhowmick B, Roy P (2021) Analysis of CMA-DAPI bands and preparation of fluorescent karyotypes in thirty Indian cultivars of *Lens culinaris*. Caryologia 74(2):65–77

Jha TB, Halder M, Bhowmick BK (2024) Giemsa-based chromosome staining and comparative fluorescent banding pattern in five valuable Indian plant species. Caryologia 77(3):37–45

John B (1988) The biology of heterochromatin. In: Verma RS (ed) Heterochromatin: molecular and structural aspects. Cambridge University Press, Cambridge, New York, New Rochelle, Melburn, Sydney. pp 1–147

Kornberg RD (1974) Chromatin structure: a repeating unit of histones and DNA: chromatin structure is based on a repeating unit of eight histone molecules and about 200 DNA base pairs. Science 184(4139):868–871

Lei B, Berger F (2019) H2A variants in *Arabidopsis*: versatile regulators of genome activity. Plant Commun 1(1):100015

Li LC, Okino ST, Zhao Z, Pookot D, Place RF, Urakami S, Enokida H, Dahiya R (2006) Small dsRNAs induce transcriptional activation in human cells. Proc Natl Acad Sci USA 103(46):17337–17342

Lima-de-Faria A (1976) The chromosome field I. Prediction of the location of ribosomal cistrons. Hereditas 83(1):1–22

Lima-de-Faria A (2010) One hundred years of chromosome research and what remains to be learned. Kluwer Academic Publishers

McClintock B (1934) The relatioship of a particular chromosomal element to the development of the nucleoli in *Zea mays*. Zeit Zellforsch Mik Anat 21:294–328

McStay B (2016) Nucleolar organizer regions: genomic 'dark matter'requiring illumination. Genes Dev 30(14):1598–1610

Mizianty M (1984a) Banding patterns in plant chromosomes. II. Bibliography (1970-1980) - Anthophyta. Acta Soc Bot Pol 53(Suppl. 1):485–487

Mizianty M (1984b) Banding patterns in plant chromosomes. II. Bibliography (1970-1980) - Anthophyta. Acta Soc Bot Pol 53:111–136

Mizianty M (1985) Banding patterns in plant chromosomes. II. Bibliography (1970-1980) - Anthophyta. Acta Soc Bot Poloniae 54(Suppl. 2):193–194

Mohannath G, Pontvianne F, Pikaard CS (2016) Selective nucleolus organizer inactivation in *Arabidopsis* is a chromosome position-effect phenomenon. Proc Natl Acad Sci USA 113(47):13426–13431

Morgan DO (2007) The cell cycle: principles of control. New Science Press Ltd, p 297

Morrison JH, Lea LV, Wilson GB (1959) Combined staining of nucleolus and chromosomes in squash preparations of plant and animal somatic tissues. Trans Am Microsc Soc 78(4):358–362

Moscone EA, Lambrou M, Ehrendorfer F (1996) Fluorescent chromosome banding in cultivated species of *capsicum* (Solanaceae). Plant Syst Evol 202:37–63

Murray BG, Bennett MD, Hammett KRW (1992) Secondary constrictions and NORs of *Lathyrus* investigated by silver staining and in -situ hybridization. Heredity 68(5):473–478

National Geographic Society (2023). Available at https://education.nationalgeographic.org

Pardue ML, Gall JG (1970) Chromosomal localization of mouse satellite DNA chromosomal localization of mouse satellite DNA. Science 168(3937):1356–1358

Penagos-Puig A, Furlan-Magaril M (2020) Heterochromatin as an important driver of genome organization. Front Cell Dev Biol 8:579137

Peska V, Garcia S (2020) Origin, diversity, and evolution of telomere sequences in plants. Front Plant Sci 11:117

Rattenbury JA (1952) Specific staining of nucleolar substance with aceto-carmine. Stain Technol 27(2):113–120

Sáez-Vásquez J, Medina FJ (2008) The plant nucleolus. In: Kader JC, Delseny M (eds) Advances in botanical research, vol 47. Elsevier, San Diego, pp 1–46

Samanta T, Jha TB, Ray S, Jha S (2023) Comparative cytogenetics and fluorescent chromosome banding in five indian species of *Dipcadi* Medik. Plants 12(13):2534

Santra I, Biswas D, Ghosh B (2024) Chromosomal characterization mediated by karyomorphological analysis and differential banding pattern in fenugreek (*Trigonella foenum-graecum* L.): a neglected legume. Caryologia 76(3):63–70

Schneeweiss HW, Schneeweiss GM (2013) Karyotype diversity and evolutionary trends in Angiosperms. In: Greilhuber J, Doležel J, Wendel JF (eds) Plant genome diversity, vol 2. Springer, Vienna, pp 209–230

Schrader F (1939) Kinetic regions in chromosomes. Nature 143:122

Schweizer D (1974) An improved giemsa C-banding procedure for plant chromosomes. Experientia 30(5):570–571

Schweizer D (1976a) DAPI fluorescence of plant chromosomes pre-stained with actinomycin D. Exp Cell Res 102(2):408–413

Schweizer D (1976b) Reverse fluorescent chromosome banding with chromomycin and DAPI. Chromosoma 58(4):307–324

Sharma AK, Sharma A (1980) Chromosome techniques: theory and practice, 3rd edn. Butterworths & Co. Ltd., London

Sheikh SA, Katsuhiko Kondo K (1996) Comparative C-banding and fluorescent- banding analysis of seven species of *Drosera* (Droseraceae). Cytologia 61:383–394

Sirri V, Urcuqui-Inchima S, Roussel P, Hernandez-Verdun D (2008) Nucleolus: the fascinating nuclear body. Histochem Cell Biol 129(1):13–31

Spinner NB, Ferguson-Smith MA (2019) Cytogenetic analysis. In: Pyeritz RE, Korf BR, Grody WW (eds) Emery and Rimoin's principles and practice of medical genetics and genomics, clinical principles and applications, 7th edn. Academic Press

Sumner AT, Evans HJ, Buckland RA (1971) New technique for distinguishing between human chromosomes. Nat New Biol 232:31–32

Vincent WS (1955) Structure and chemistry of nucleoli. Int Rev Cytol 4:269–298

Waldeyer HWG (1888) Über Karyokinese und ihre Beziehungen zu den Befruchtungsvorgängen. Arch Mikrosk Anat Entwickl 31:1–122

Watson J, Crick F (1953) Molecular structure of nucleic acids: a structure for Deoxyribose Nucleic Acid. Nature 171(4356):737–738

Yamamoto M, Takada N, Hirabayashi T, Kubo T, Tominaga S (2010) Fluorescent staining analysis of chromosomes in Pear (*Pyrus* spp.). J Japan Soc Hort Sci 79(1):23–26

Yamamoto M, Moriya S, Yamamoto T (2016) Detection of ribosomal RNA genes in apple (*Malus × domestica*) using fluorescence in situ hybridization. Chromosome Sci 19:33–36

Zavodnık M, Fajkus P, Franek M, Kopecky D, Garcia S, Dodsworth S, Orejuela A, Kilar A, Patacek J, Matl M, Hyskov A, Fajkus J, Vratislav Peska V (2023) Telomerase RNA gene paralogs in plants—the usual pathway to unusual telomeres. New Phytol 239:2353–2366

3 Preparation of Somatic Chromosomes by Classical Cytogenetics

Abstract

This chapter provides a comprehensive overview of the preparation of somatic chromosomes using classical cytogenetic techniques, emphasizing the importance of optimal processing for accurate chromosome analysis. The fundamental goals of plant cytogenetics involve examining chromosome number and distinct morphological configurations, particularly centromere and constriction positions. The introductory section outlines the necessary steps, which begin with the careful collection of healthy meristematic tissues, such as root tips, followed by a series of procedures including pretreatment, fixation, staining, and slide preparation. The focus is primarily on the preparation of somatic chromosomes. Key methods for germination and collection are detailed, providing practical guidance on how to obtain high-quality root tips from various plant species, including *Lens, Nigella, Vicia*, and Trigonella seeds; bulbous *Allium cepa, Allium sativum*, and *Albuca virens*; and non-bulbous *Aloe vera*. Notably, the impact of plant health, flowering timing, and collection conditions on chromosome viability and quality is discussed. Illustrated with student-friendly diagrams, this chapter is designed to inspire and facilitate the development of cytogenetic skills among students and researchers. It combines conceptual knowledge with hands-on methodologies, making it accessible for those new to cytogenetics. By elaborating on repeatable protocols and offering flowcharts for different plant species, it aims to enrich educational experiences and foster a deeper understanding of plant genomic structures, setting the groundwork for future research in cytogenetics.

T. B. Jha, M. Halder, *Classical and Molecular Plant Cytogenetics*,
https://doi.org/10.1007/978-981-95-8265-5_3

3.1 Introduction

The foundational objectives of classical plant cytogenetics are to study chromosome number and the species-specific morphological configuration of chromosomes, i.e., the position of primary constrictions (centromeres) and secondary constrictions (Lima-De-Faria 2010). Visual morphology, the sole determinant of subsequent cytogenetic parameters, depends primarily on how the chromosomes are processed before slide preparation. The process of somatic chromosome preparation starts with the collection of a healthy meristematic plant part, which is then gradually processed through pretreatment, fixation, staining, and slide preparation to achieve scientific chromosome analysis (Fig. 3.1). While meristematic tissue is located at the apical parts of roots, shoots, and leaves, root tips are preferred for somatic chromosome analysis because they are easy to access and process, structurally simple, and have a high mitotic index. Unlike leaves and shoots, roots also lack chlorophyll and other pigments and contain minimal ergastic materials that reduce the possibility of interference of these molecules during staining or create visual noise under the microscope. Learning and developing skills in somatic (mitotic) chromosome processing is a prerequisite for any student or researcher interested in cytogenetics (Jha and Ghosh 2016; Singh 2017).

In contrast, gametic (meiotic) chromosomes are observed within two specialized cell types, pollen mother cells (PMCs) and megaspore mother cells (MMCs), during meiotic cell division. Both cell types are developed from sporogenous tissues, but in distinct anatomical locations. PMCs, or microsporocytes, are found in the anther and undergo microsporogenesis to produce haploid pollen grains. MMCs, or megasporocytes, are located in the ovary and undergo megasporogenesis to form the embryo sac. PMCs are primarily considered for the preparation of gametic (meiotic) chromosomes because they are produced in large quantities inside the anther lobes of stamens, and the rupture of appropriate anthers results in the release of a large number of PMCs. On the other hand, megaspore mother cells (MMCs), although they make a minimal number of haploid cells, are difficult to analyze due to their deeply seated position within ovules. In this case, pretreatment is not needed as they are free of pectic substances and cellulosic cell walls. They are loose clumps of cells lacking cell walls, enclosed by callose boundaries, and are single cells undergoing reductional cell division (meiosis) to produce microspores or pollen. Freshly collected, unopened flower buds are the source material for gametic chromosome analysis. Collected flower buds can also be fixed instantly at division by immersing them in Carnoy's fixatives. However, the healthy growth of the plant, the timing of flowering, the size of the unopened buds, and the time of collection are all significant factors (Kianian and Kianian 2016; Windham et al. 2020).

This chapter, featuring a large number of illustrations in a student-friendly manner, presents the conceptual information and preparation methods for all the necessary chemicals required for somatic and gametic chromosomes. The chapter outlines the processing steps for preparing somatic chromosomes. Repeatable chromosome preparation methods have been elaborated using various classic materials and are ready to be applied to any plant material with minor modifications. A skill-developing flowchart with metaphase plates and additional information was prepared for each species. Classical cytogenetics is the foundation for revealing the amazing genomic structures under a microscope. It is expected to inspire and enrich our teachers and researchers, offering hands-on training to their students.

To begin with, somatic chromosome preparation requires familiarity with the parameters described in detail below.

3.2 Steps of Somatic Chromosome Preparation

3.2.1 Induction and Collection of Healthy Root Apical Meristems

It is possible to prepare somatic chromosomes from any active meristem of a plant, such as root tips, shoot tips, or leaf tips. Root tips are usually preferred to shoot or leaf tips because they grow in abundance and are less cytoplasm-heavy than shoots or leaves. To prepare somatic chromosomes, fresh and healthy root tips with their meristematic regions are collected. Depending on availability and feasibility, healthy root tips of any plant species may be collected from different sources. For class teaching materials, we usually collect root tips by germinating seeds (for *Lens, Nigella, Trigonella*, and *Vicia*), using roots from bulbs (for onion—*Allium cepa*, garlic—*Allium sativum*), or collecting roots from field-grown plants (for *Aloe vera* and *Albuca virens*). The length of roots, harvesting time, and proper cleaning should be considered while processing. Root length should not exceed one inch for seeds and bulbs. Roots collected from sand, sawdust, or soil need careful cleaning in water with the help of a fine brush.

- *Procedure for Germination of Seeds* (*Lens, Nigella, Trigonella,* and *Vicia*)
  To initiate the germination process, begin by taking 15–20 fresh seeds and submerging them in water overnight. Then, place the seeds on moist filter paper inside a container or Petri dish, or on moist sawdust. It is essential to keep the seeds in the dark and maintain a temperature between 20 °C and 25 °C. Keep in mind that the time required for germination may differ depending on the species.
- *Procedure for Induction of Roots From Bulbs* (*Allium cepa* and *Allium sativum*)
  To initiate the process, the dry bulbs are first thoroughly cleaned. Subsequently, any roots present on the bulb disc are carefully scraped with a blade and left to soak in water for 1–2 hours. Once the bulbs are ready, they are placed in damp sand inside a container and covered. Next, they are incubated in a dark place at a temperature ranging from 20 to 25 °C. After 48 hours, you can examine if the roots have been induced.
- *Collection of Healthy Roots From Potted or Field-Grown Plants* (*Aloe vera* and *Albuca virens*)
  Aloe vera and *Albuca virens* plants are typically grown in pots or fields with loose, well-draining soil. During the harvesting process, fresh and healthy roots are carefully removed from the soil to avoid damage to the root tips. Old and heavily thick roots are generally discarded.

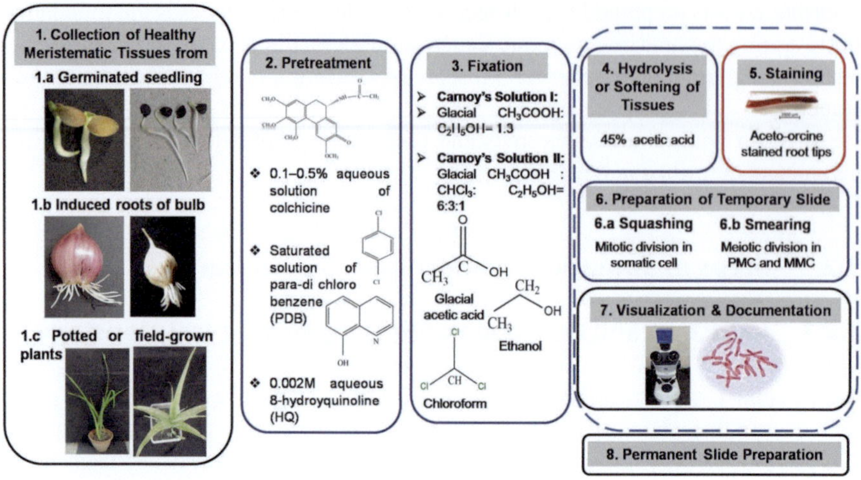

Fig. 3.1 Overview of steps of somatic chromosome preparation by classical method

3.2.2 Pretreatment or Prefixation

In the field of plant cytogenetics, the preparation of plant metaphase chromosomes involves an initial chemical process called pretreatment or prefixation. This process is crucial as it leads to the creation of high-quality metaphase plates for studying chromosome number and morphology. The pretreatment process helps to arrest cell division in metaphase and achieve chromosomal condensation, which is necessary for the subsequent analysis.

In plants, cell division is asynchronous, meaning that not all cells will undergo cell division or enter the cell cycle at the same time. Thus, not all cells will be in the same divisional stage at any given time. During classical cytogenetic studies of somatic chromosome, stages other than metaphase are not considered important, as mitotic metaphase is the ideal stage for counting chromosome number and morphology in the somatic cells of any plant. In this phase, maximum chromosome condensation occurs, and they become visible and countable under a microscope. Therefore, the probability of finding a scattered, countable plate with its morphological features greatly positively correlated with the increase frequency of metaphase. The main objective of pretreating meristematic tissues (root tips, shoot tips, and leaf tips) is to maximize the frequency of metaphase in the treated cell population by arresting cells at metaphase. In contrast, pretreatment step is not conducted during determination of mitotic index study because it takes all divisional phases into account to determine the frequency of cell division in a given sample.

Spindle fibers are attached to chromosomes and are responsible for the separation of chromosomes or chromatids during cell division. Spindle fibers or microtubules are microscopic tube-like structures made up of α- and β-tubulin. It has two ends, a "growing" or "positive (+)" end, and a "shrinking" or "negative (−)" end.

The growth and shrinking of spindle fibers are controlled by the rate of tubulin dimer polymerization at the positive end and the rate of depolymerization at the negative end. When certain chemicals are applied to the meristematic tissues, they cause a pause in the cell's division process during metaphase. The chemicals interfere with the dynamics of spindle fiber polymerization and depolymerization either by changing the viscosity of the cytoplasm or directly interacting with the microtubules. The increase in cellular viscosity caused by the chemical pretreatment process damages the spindle mechanism and results in the disorganization of the spindle apparatus. This disorganization releases all binding forces that hold the chromosomes within the cell, causing them to scatter throughout the cell. Pretreatment improves the clarity of primary and secondary constrictions by inducing differential hydration in different parts of the chromosome. Moreover, it causes chromosomal condensation (compactness and shortening) and straightening of the chromosome arms, which makes chromosome counts simpler and improves the visibility of constriction sites. Pretreatment also facilitates better penetration of fixatives at subsequent stages of chromosome preparation by removing undesirable deposits from the tissue surface.

Objectives of Pretreatment

(i) Arresting metaphase stages
(ii) Causing chromosomal condensation (compactness and shortening) and straitening of the chromosome arms
(iii) Improving the clarity of primary and secondary constrictions
(iv) Facilitating better penetration of fixative by removing undesirable deposits from the surface of the tissue
(v) Making cytoplasmic background more transparent

The process of pretreatment is heavily influenced by the chemicals used, their duration, and temperature. Many pretreating chemicals are now available. However, considering the diversity of plant kingdoms, it is challenging to generalize one pretreatment chemical for all plant species. In general, trials and errors are needed to determine the best pre-treatment chemical, duration of pretreatment, and pretreatment temperature for each plant species. The following paragraphs will provide basic knowledge about some frequently used pre-treating agents or chemicals, such as colchicine, para-dichlorobenzene (PDB), hydroxyquinoline (HQ), and aesculine and their preparation procedures.

Commonly Used Pretreatment Agents

- 0.1–0.5% aqueous solution of colchicine
- Saturated solution of para-dichloro benzene
- 0.002 M aqueous solution of 8-hydroyquinoline
- Ice-cold water (0 °C–2 °C and or 8–10 °C).

(a) *Colchicine*

Colchicine is a chemical compound with the molecular formula $C_{22}H_{25}O_6N$ (Fig. 3.2) and a molecular weight of 399.4 g mol^{-1}. It is a toxic and carcinogenic natural product that has been used for many years in chromosome research as a pre-treatment chemical. Originally, it was extracted from the roots and corms of the *Colchicum autumnale* plant, also known as meadow saffron (Singh 2017). However, it is now reported to be present in other plants, such as *Gloriosa superba* (Fig. 3.3), *Merendera persica, Androcymbium gramineum,* and *Sandersonia auranticata* (Paul et al. 2013).

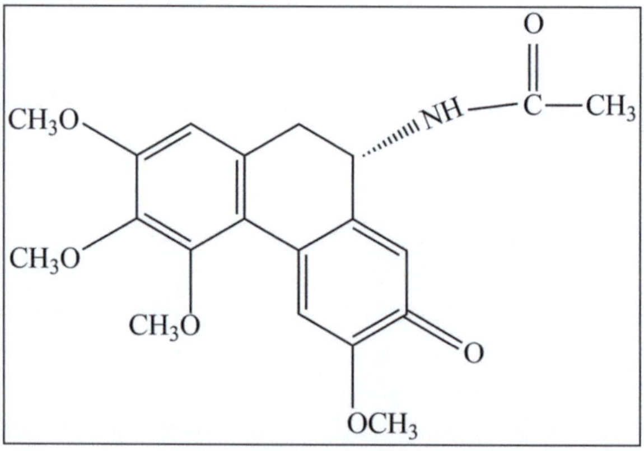

Fig. 3.2 Chemical structure of colchicine

Fig. 3.3 Flower of *Gloriosa superba*

In 2009, the Food and Drug Administration (FDA) approved colchicine as a drug for treating gout. The pure form of colchicine is a fine, pale-yellow powder that is sensitive to light exposure and can cause photoisomerization. Synthetic colchicine is now readily available. Colchicine is known as a mitotic blocker or "mitotic poison," as it causes the arrest of cells in metaphase stages and inhibits the progression of the cell cycle by preventing the formation of new microtubules (spindles). It directly interacts and binds to tubulin. In cytogenetics, colchicine-arrested metaphases are often referred to as C-mitosis or C-metaphase. It allows better visualization and analysis of somatic chromosomes. Colchicine is soluble in water. At low concentrations (0.1–0.5%), it is recommended for chromosome pretreatment (for a duration of 1–4 hours at 12–16 °C) for species having long- and medium-sized chromosomes. On the contrary, higher concentrations may induce polyploidy (colchiploidy) by doubling the chromosome number (endomitosis), whereas slightly prolonged treatment may cause clear splitting of sister chromatids and separation of the centromere. Since colchicine is highly carcinogenic and toxic, necessary precautions should be taken during its application. Colcemid, a synthetic derivative of the natural alkaloid colchicine, is sometimes preferentially used for metaphase arrest during cell division due to its lower toxicity and reversible interaction with the tubules.

Colchicine ($C_{22}H_{25}O_6N$)

- Natural plant alkaloid
- Source: *Colchicum autumnale* and *Gloriosa superba* (Colchicaceae)
- Recommended for long and medium sized chromosomes
- Higher concentration may induce polyploidy (colchiploid)
- Overtreatment may cause clear splitting of sister chromatids and splitting of the centromere

- *Preparation of 0.05% Colchicine Solution*
  - *Requirement:* (i) Colchicine, (ii) double distilled water (ddH_2O), (iii) an electronic balance, (iv) a measuring cylinder, and (v) a reagent bottle.
  - *Procedure:*
    - (i) 0.25 g colchicine is dissolved in 50 mL distilled water at room temperature for preparation of 0.5% stock solution (Fig. 3.4).
    - (ii) A 0.5% stock solution of colchicine is diluted 1:9 (v: v) with distilled water to get the 0.05% pretreatment solution (Fig. 3.4).
    - (iii) Since solutions are photosensitive, the solution should be kept in an amber-colored or dark-colored bottle in the refrigerator or at room temperature.

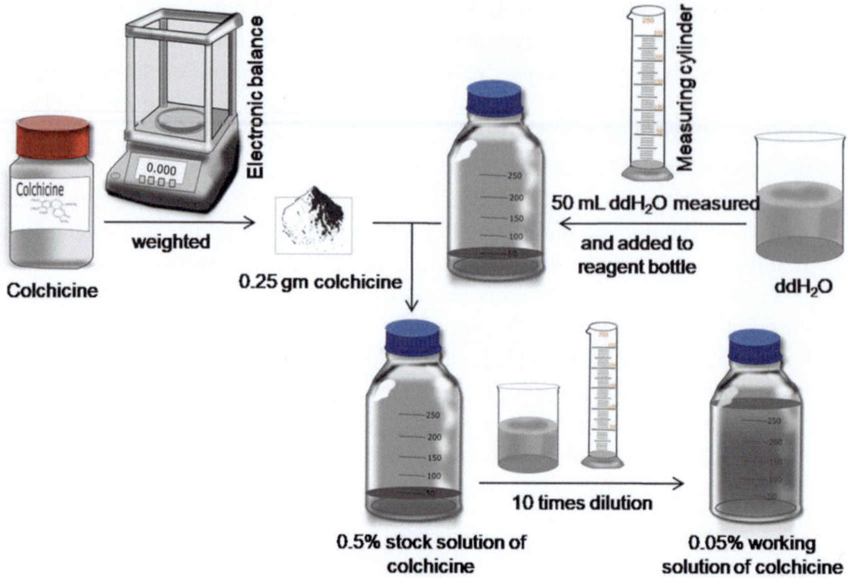

Fig. 3.4 Schematic representation of 0.05% colchicine solution preparation steps

(b) *Para-dichlorobenzene (PDB)*

Para-dichlorobenzene (PDB) is a colorless synthetic organic compound with the molecular formula $C_6H_4C_{l2}$ (Fig. 3.5) and a molecular weight of 147 g mol^{-1}. It has a strong odor and low water solubility. It is commonly used as a disinfectant, insecticide, and deodorant. It is also used in the production of mothballs. An oversaturated aqueous solution of PDB is an effective pretreatment chemical for plant tissues. It is recommended for the cytogenetic study of plant species that have both large and small chromosomes. This solution helps prevent cellular degradation and maintains the structural integrity of the chromosomes. In PDB, roots can be treated for 3–5 hours at 10–14 °C.

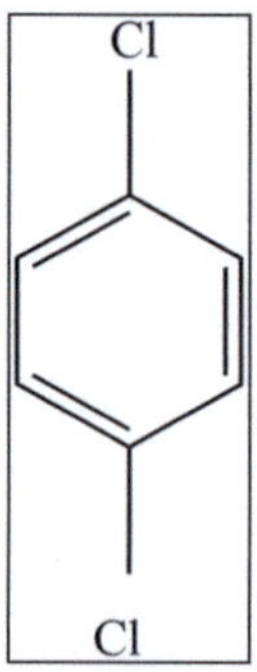

Fig. 3.5 Chemical structure of para-dichlorobenzene

Para-Dichlorobenzene ($C_6H_4C_{l2}$)

- Low solubility in water
- Super saturated aqueous solution of PDB is used
- Recommended for both large and small chromosomes

- *Preparation of Saturated Aqueous Solution of Para-dichlorobenzene*
  - *Requirement:* (i) Para-dichlorobenzene, (ii) ddH_2O, (iii) an electronic balance, (iv) a measuring cylinder, and (v) a reagent bottle.
  - *Procedure*:
    (i) Add 3 g of PDB to 100 mL of distilled water to make a supersaturated aqueous solution of PDB (Fig. 3.6). The bottle shook a few times, and some PDB crystals remained undissolved.
    (ii) Saturated aqueous solution of para-dichlorobenzene solution is stored at room temperature.
    (iii) The bottle should be shaken before use, and crystals should be avoided while used as a pretreating solution.

* It should be used with caution as it has carcinogenic properties.

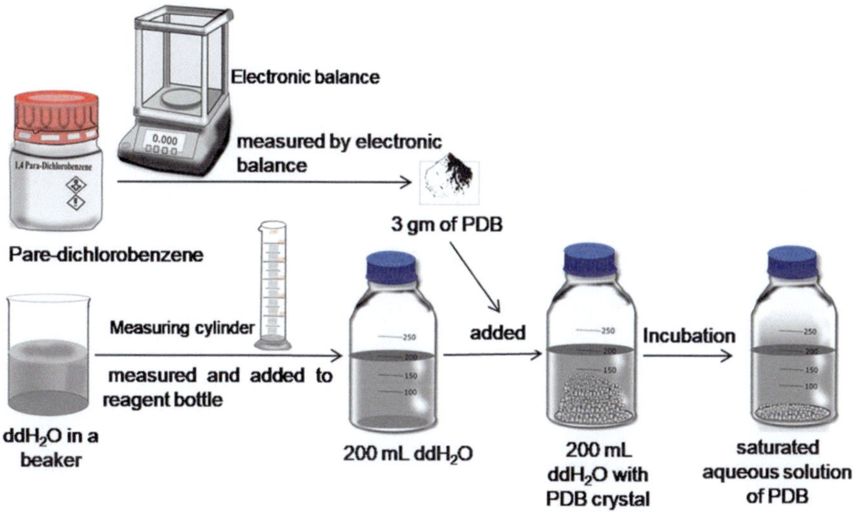

Fig. 3.6 Schematic representation of the supersaturated aqueous solution of PDB preparation steps

(c) *8-Hydroxyquinoline (8-HQ)*

8-Hydroxyquinoline is a heterocycle quinoline derivative with the molecular formula C_9H_7NO (Fig. 3.7) and a molecular weight of 145.16 g mol^{-1}. It has antiseptic properties and is used as an antibacterial agent. It can be found in

Cortinarius subtortus and *Allium stipitatum*. It is a white, off-white, or faintly yellow crystalline powder and is soluble in acetone, ethanol, chloroform, benzene, but is slowly soluble in water. A commonly used solution for studying the morphology of small chromosomes is a 2 mM or 0.002 M aqueous solution of 8-HQ. This solution should be kept at room temperature or refrigerator. Root tips are incubated for 3–4 hours at temperatures ranging from 10 °C to 20 °C for effective pretreatment. However, if the solution turns brown due to gradual deterioration, it should be discarded.

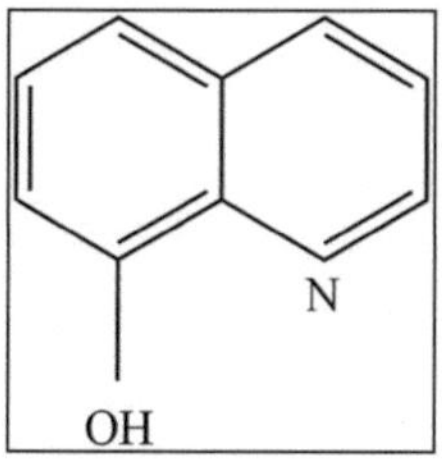

Fig. 3.7 Chemical structure of 8-hydroxyquinoline

8-Hydroyquinoline (C_9H_7NO)

- Heterocycle quinoline derivative
- Synthetic chemicals are used
- Source: *Cortinarius subtortus* and *Allium stipitatum*
- 2mM aqueous solution of 8-HQ is commonly used
- Very effective for morphological study of small chromosomes

- *Preparation of 8-Hydroxyquinoline (8-HQ) Solution*
  - *Requirement:* (i) 8-hydroxyquinoline, (ii) double-distilled water, (iii) an electronic balance, (iv) a measuring cylinder, and (v) a reagent bottle.
  - *Procedure*:
    (i) Add 29 mg of 8-hydroyquinoline crystals to 100 mL of double-distilled water (Fig. 3.8).
    (ii) 8-HQ crystals dissolve gradually in ddH_2O and develop pale yellow coloration.
    (iii) Kept the solution in amber-colored bottle overnight without shaking.
    (iv) The solution can be stored in the refrigerator or at room temperature.

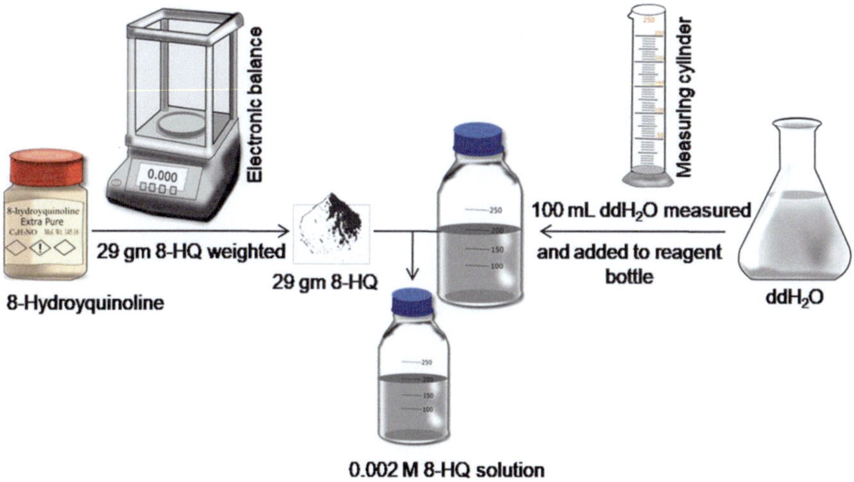

Fig. 3.8 Schematic representation of 0.002 M aqueous solution of 8-hydroxyquinoline preparation steps

(d) *Aesculin*

Aesculin is a natural glucoside with the molecular formula $C_{15}H_{16}O_9$ (Fig. 3.9) and a molecular weight of 340.28 g mol^{-1}. It is derived from the roots of *Aesculus hippocastanum, Aesculus californica*, and *Daphne mezereum*. It is water-soluble and is typically used as a pretreatment solution for 1 to 2 hours at a temperature of 4–10 °C. This solution works well for plants with small chromosomes and can also be used alone or in combination with PDB.

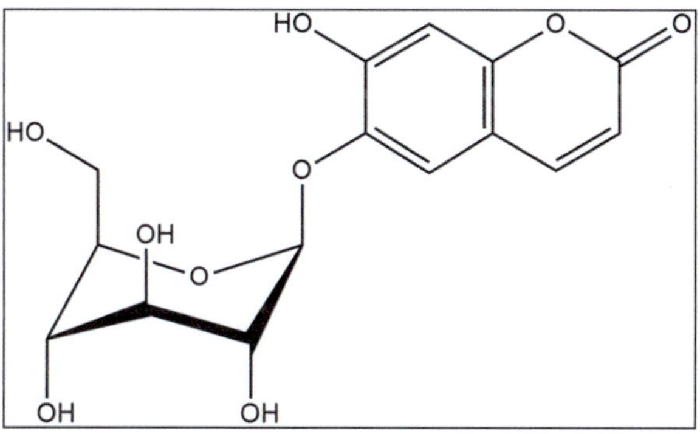

Fig. 3.9 Chemical structure of aesculin

Aesculin ($C_{15}H_{16}O_9$)

- Source- roots of *Aesculus californica, A. hippocastanum*, etc.
- Saturated aqueous solution is used separately or in association with PDB
- Recommended for small chromosomes

(e) *Ice-Cold Water*

The use of ice-cold water as a pre-treatment technique has been reported as the most affordable, cost-effective, and safe method (Singh 2017). This technique involves transferring 1.5 to 2 cm long root tips to a container that is two-thirds filled with ice-cold water. The container is then incubated in a refrigerator for 12 to 24 hours, depending on the materials. Ice-cold water is effective in arresting cell division at metaphase, just like other pretreating chemicals. Water alone can also cause spindle arrest by increasing cellular viscosity and differential hydration of chromosomes to bring about clear visualization. This is the reason why most of the other pre-treating agents mentioned above are used as aqueous solutions.

3.2.3 Fixation

During the process of chromosome processing, the fixation step is done after the pre-treatment is complete. The pretreated roots are washed with water before being placed in a fixative solution made up of various chemicals. Almost a century ago, La Cour (1929) emphasized the importance of using a fixative that quickly penetrates the roots without shrinking the cytoplasm, spreads the chromosomes for counting, and provides clear identification for constrictions, for the study of somatic chromosomes. Fixation is the most critical step in chromosome studies, as it selectively preserves the morphological organization and chemical composition of the chromosomes that need to be studied under a microscope. It kills cells without causing any cellular or chromosomal distortion, degradation, swelling, or shrinkage. Proper fixation maintains cellular and karyomorphological features, the structural integrity of the nucleic acid, and proteins of the chromosomes, and renders the cell contents insoluble by coagulation or precipitation of protein matter. The process of visualization of morphological features, including primary and secondary constrictions, depends on the accurate fixation of the cells. Proper fixation also helps accurately diagnose numerical and structural abnormalities and karyotyping and prevents microbial contamination. Finally, properly fixed materials can be stored at 4 °C or −20 °C for several months.

Fixation

- After pre-treatment, cells are killed without causing any cellular or chromosomal distortion, degradation, swelling, or shrinking during fixation.
- Fixatives are applied to selective preservation of the morphological organization and chemical composition of chromosomes.
- Helps in accurate karyotyping by preserving actual morphological features like primary and secondary constrictions.
- Helps in detection of numerical and structural abnormalities.
- Improving the visibility and clarity of chromosomes and their morphological features.
- It prevents bacterial decomposition.

Fixatives are a mix of different chemicals that are used to preserve biological samples. It's important to choose the right fixative for a particular material. Over the decades, a variety of chemicals have been used as fixatives for plant cytogenetic studies. Fixatives can be categorized as nonmetallic fixative and metallic fixative, depending on the chemicals used. Ethyl alcohol (C_2H_5OH), methyl alcohol (CH_3OH), glacial acetic acid (CH_3COOH), formaldehyde (CH_2O), propionic acid ($C_3H_6O_2$), picric acid ($C_6H_3N_3O_7$), and chloroform ($CHCl_3$) are commonly used in nonmetallic fixative preparation (Fig. 3.10). Metallic fixatives (Fig. 3.10) include chemicals like chromic acid (H_2CrO_4), osmic acid (OsO_4), platinic chloride (H_2PtCl_6), mercuric chloride ($HgCl_2$), uranium nitrate ($UO_2(NO_3)_2$), and lanthanum acetate ($C_6H_9LaO_6$). Metallic chemicals are rarely used as fixatives. Various combinations of glacial acetic acid and alcohol are commonly used as fixatives for chromosome preparation. However, neither of them is used alone as a fixative because they have different effects on tissue. Alcohol alone will dehydrate the tissue, while acetic acid alone can cause significant swelling and damage to cells and tissues. By combining them in different proportions, they can effectively fix the chromosomes. (i) Rapid tissue penetration ability, (ii) maintenance of the structure and integrity of chromosomes, (iii) precipitate chromatin to render the chromosome visible, (iv) anti-bacterial action, and (v) increase the satiability of acidic stain by increasing basophilic activity of the chromosomes are the features of good fixatives. A comprehensive list of fixatives can be found in the book *Chromosome Technique* (Sharma and Sharma 1980). However, in the following paragraphs, we will only describe the frequently used fixatives in our classes and research studies.

Fixatives

- Fixatives are a mixture of various chemicals.
- Based on chemicals, fixatives are termed nonmetallic and metallic.
- Nonmetallic chemicals: ethyl alcohol, methyl alcohol, glacial acetic acid, formaldehyde, chloroform.
- Metallic chemicals: chromic acid, osmic acid, platinic chloride, mercuric chloride, uranium nitrate.

Ethanol (C_2H_5OH; Mol. Wt. 46.07 g mol^{-1}) Ethanol (Fig. 3.10b) is a crucial component of several commonly used fixatives. It is preferred for its ability to quickly penetrate tissues and denature proteins irreversibly. Ethanol also dissolves lipids, precipitates proteins, and hardens the tissue by shrinking it. However, it can cause undesired hardening effects due to tissue dehydration and enhance the penetration of other fixative components. Ethanol can be combined with other acids, such as acetic acid, formaldehyde, and chloroform in varying ratios. Additionally, it helps reduce formaldehyde-induced artifacts.

Glacial acetic acid (CH_3COOH; Mol. Wt. 60.05 g mol^{-1}) Glacial acetic acid (Fig. 3.10d) is a commonly used component in a fixative solution. When used in combination with either ethanol or methanol, it helps to keep the chromosomal structure undisturbed and free of distortion. Although it does not fix cytoplasmic proteins, it does result in nucleoprotein precipitation, damage to the Golgi apparatus, and swelling of the mitochondria. Additionally, it dissolves histone proteins and causes softening of tissue. When combined with alcohol, acetic acid prevents tissue from hardening. It is also used as an effective solvent for both carmine and orcein dyes.

Propionic acid (C_2H_5COOH; Mol. Wt. 74.08 g mol^{-1}) Propionic acid (Fig. 3.10e) is a clear liquid with a strong, pungent, and unpleasant smell. It is an excellent substitute for glacial acetic acid.

Chloroform ($CHCl_3$; Mol. Wt. 119.37 g mol^{-1}) Chloroform (Fig. 3.10f) is a colorless liquid with toxic properties. As a component of fixative, it is used to prepare Carnoy's Solution II. It categorizes a hazardous chemical and needs careful handling.

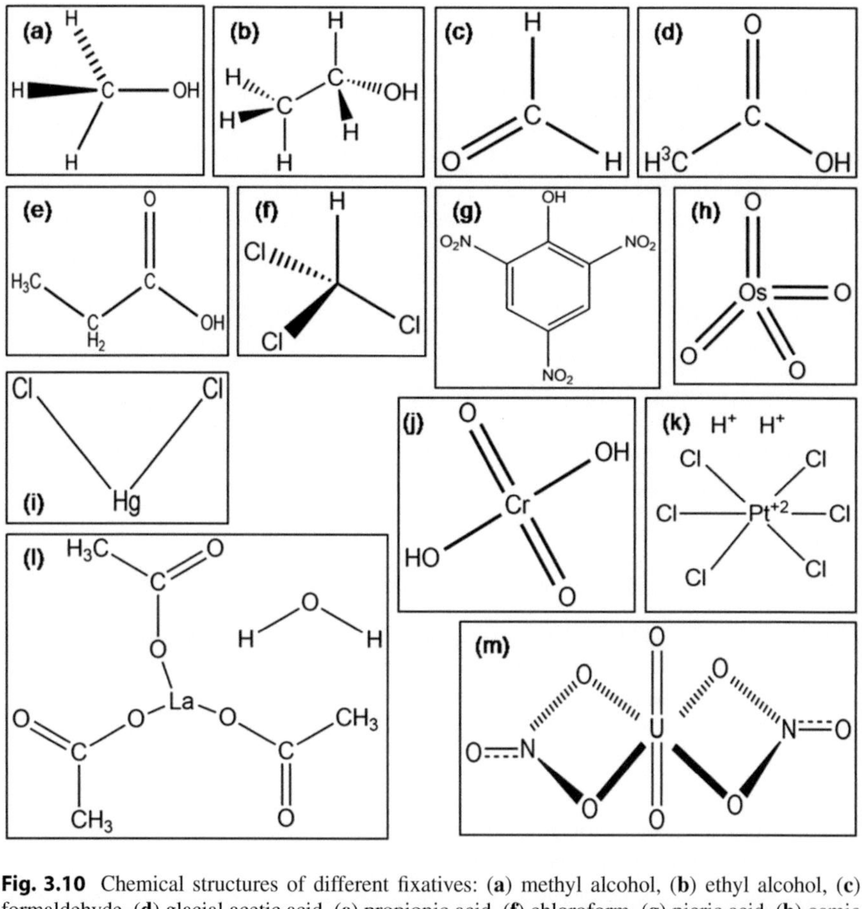

Fig. 3.10 Chemical structures of different fixatives: (**a**) methyl alcohol, (**b**) ethyl alcohol, (**c**) formaldehyde, (**d**) glacial acetic acid, (**e**) propionic acid, (**f**) chloroform, (**g**) picric acid, (**h**) osmic acid, (**i**) mercuric chloride, (**j**) chromic acid, (**k**) platinic chloride, (**l**) lanthanum acetate, and (**m**) uranium nitrate

3.2.3.1 Fixative Solutions

(a) *Carnoy's Solution I (combination of glacial acetic acid and 95% or absolute ethanol)*
By keeping the amount of glacial acetic acid constant, the proportion of 95% or absolute ethanol can be increased to create fixatives in ratios of 1:1, 1:2, and 1:3. Carnoy's Solution I is an excellent choice for fixing somatic and gametic cells. Freshly prepared fixatives can be stored at room temperature or in a refrigerator.

(b) *Carnoy's Solution II (combination of glacial acetic acid, chloroform, and ethanol)*
Carnoy's solution II is a fixative used for preserving plant materials with fatty and waxy substances, especially for species with small chromosomes. It is prepared by mixing six parts of ethanol with three parts of chloroform, and one part of glacial acetic acid (6:3:1). The presence of chloroform in Carnoy's solution II makes it an excellent solvent for fatty and waxy substances, resulting in better preservation of the specimen.

(c) *Propino alcohol*
Combination of concentrated propionic acid and 95% or absolute ethanol in a ratio of 1: 3 is used as fixative. This fixative gives better results in plants having small chromosomes.

(d) *Aceto-methanol*
One part of glacial acetic acid is added with three parts of methanol (1: 3) to prepare this fixative. It is one substitute for acetic alcohol.

Long-term preservation of fixed root tips can be achieved by directly transferring them in acetic alcohol, acetic methanol, or 70% ethyl alcohol at −20 °C.

3.2.4 Softening of Tissues

Tissue softening, or maceration, is another step in somatic chromosome preparation and analysis. Root tips are typically rigid and structured with multiple layers, where the middle lamella functions as intercellular cement that binds plant cells together. Tissue softening can be achieved by acid hydrolysis (1 N HCl) or treatment with 45% acetic acid. When roots are incubated in either 45% acetic acid or 1 N HCl for a specified period, these chemicals facilitate the breakdown of the middle lamella. During squashing, applying pressure to the coverslip spreads the softened tissue into a uniform monolayer, minimizing cell overlap and ensuring accurate determination of chromosome numbers and karyomorphological features. Without this important step, even optimally pretreated chromosomes can remain trapped within a dense, three-dimensional cell mass. Additionally, softening promotes slight cytoplasmic expansion, which helps to separate chromosomes into a flat, horizontal plane, enhancing both visibility and stain permeability. It is important to standardize treatment duration across species. Moreover, 45% acetic acid not only softens the tissues but also induces cell swelling.

3.2.5 Stain and Staining

In general, most cellular components are colorless and transparent with low contrast, making differentiation difficult under a microscope. To overcome this problem, different coloring agents, including dyes and stains, are regularly used in both conventional and molecular cytogenetic research. They improve the contrast between different cellular and extracellular components, making it easier to visualize and differentiate them under a microscope.

Why staining is used in bioscience experiments?
Colorless and transparent cellular components with *low relative contrast*

Staining

↓

Increasing relative contrast for better visualization and differentiation of cellular components

Dyes are colouring agents that contain impurities (Table 3.1). They are commonly used in industries to color clothes and to stain organic molecules in biological laboratories. Dyes bind with cellular components or molecules, giving them a specific color. Chemical modifications can improve the quality of dyes. Some common biological dyes are pyronin G, aniline blue, methyl green, and Orange G. Stains, on the other hand, are a mixture of chemically modified and almost impurity-free synthetic dyes. They are produced with great attention and precautions and used to stain tissues, cells, sub-cellular components, and molecules. Like dyes, stains increase contrast to improve visibility under a microscope.

Many stains used in staining techniques are derived from aniline or nitrobenzene and consist of three components: the benzene ring, chromophore, and auxochrome. The benzene ring is a colorless structural component of the stain or dye. The effectiveness of a particular stain or dye depends on the interaction of its specific chemical configurations that determine its chemical affinities toward cellular components and physical adsorption properties. The chromophore, such as a quinonoid ring, is responsible for the characteristic color of the dye by strongly absorbing and transmitting light of a specific wavelength. In contrast, auxochromes are functional

Table 3.1 Difference between dyes and stains

Dye	Stain
Dye is a crude form of natural coloring agent	Stain is a mixture of chemically modified synthetic dyes
More impure than stain	Stains have almost no impurity
Natural compounds	Commercially produced
Used for coloring clothes in industries and for staining organic molecules	Used for staining tissues, cells, sub-cellular components, and molecules

groups, such as amino and hydroxyl groups that provide the ionic properties of dye or stain. The presence of these chemical configurations is necessary for a dye molecule to impart color or adhere to cellular components.

Based on the chemical nature and behavior of chromogen, dyes or stains are usually classified as acidic, basic, amphoteric, or neutral. Acidic or anionic dyes or stains, such as acid fuchsin, congo red, methyl blue, aniline blue, lacmoid, Orange G, Nigrosine, and India ink (Table 3.2), carry a negatively charged chromophore, which stains positively charged basic (acidophilus) structures of the cell. In contrast, basic or cationic dyes/stains, such as basic fuchsin, methyl green, methylene blue, malachite green, pyronin G, crystal violet, toluidine blue, carmine, orcein, and safranin, have a positively charged chromophore, which stains negatively charged acidic (basophilic) structures, such as nucleic acids. Some stains are amphoteric, which means they are neither acidic nor basic and can react with both acidic and basic components of the cell. Acridine orange is an example of an amphoteric dye. Neutral dyes are a mixture of acidic and basic dyes where both anionic and cationic parts are colored, such as Giemsa.

There are three types of stains: vital, nonvital, and supravital. Vital stains are safe, water-soluble dyes that are used on living cells or tissues. However, they are not very effective for staining chromosomes. One example of a vital stain is methylene blue. Nonvital stains are the most effective for staining chromosomes. They can be toxic or non-toxic and are used on fixed or non-living cells or tissues that have already been killed by certain chemicals. Orcein, Carmine, Feulgen, and Giemsa are good examples of nonvital stains. Giemsa stain was originally used for animal systems but is now commonly used for plant chromosomes. Nonvital stains used in conventional chromosome staining are also called non-fluorescent stains.

Table 3.2 Difference among acidic, basic, and amphoteric dyes/stains

Properties	Acidic or anionic dye/stains	Basic or cationic dyes/stains	Amphoteric dyes/stains
Charge of chromophore	Negative	Positive	Neither positive nor negative
Interact with	Positively charged basic (acidophilus) components of the cell	Negatively charged acidic (basophilic) components of the cell	Both acidic and basic components of the cell
Example	Acid fuchsin, Congo red, aniline blue, lacmoid, Orange G, Bigrosine	Basic fuchsin, crystal violet, toluidine blue, carmine, orcein, safranin	Acridine orange

Occasionally, some chemicals like ferric chloride and ferric acetate are added to the stain to increase the color intensity and stability of the stain, called mordants. They form a chemical bond between the dye and the cellular constituents to enhance the affinity of the dye. The application of mordants may speed up and strengthen the staining process, making it more resistant to fading or washing and improving the contrast between the stained cellular components and their background for prolonged periods. However, the use of excessive mordants can result in excess staining and precipitation of the dye.

Mordants are chemicals occasionally added to the stain that forms chemical bonds between the stain and the cellular constituents to increase the color intensity and stability of the stain. Examples: ferric chloride and ferric acetate.

The final step in processing chromosomes before visualization is staining. Staining creates differential contrast between the chromosomes and their cytoplasm, so that they are easier to see under the microscope. The stain applied interacts with the different cellular components of the chromosomes, such as proteins and DNA. A proper staining protocol is crucial for precise chromosome counting and analysis of karyomorphometric features. It also helps in identifying structural and numerical abnormalities in the karyotype, such as loops and deletions. There are several dyes or stains available for this purpose, but the following paragraphs focus on the commonly used stains in plant chromosome studies.

(a) *Carmine*

Carminic acid is a natural red dye that is obtained from the shells of female cochineal bugs, specifically the *Dactylopius coccus* species. These bugs are small, flat, and oval-shaped with gray-white bodies. They gather in large clusters on prickly pear cacti, such as the *Coccus cacti* or *Opuntia coccinellifera* varieties. To protect themselves from predators, female insects produce and secrete acidic substances on their bodies. The collected insects are dried and boiled to extract cochineal, which is then converted into carminic acid. It's worth noting that while carmine dye is commonly used for chromosome staining, it is a "semi-synthetic" dye. To create carmine, the dye is first treated with lead acetate to produce lead carminate, which is then treated with sulfuric acid. The result is crimson red-colored carmine, which is produced by mixing alum with carminic acid. Carmine has been widely used for many decades to stain meiotic and somatic chromosomes in plant cytogenetics. The chemical formula of the dye is $C_{22}H_{20}O_{13}$ (Fig. 3.11).

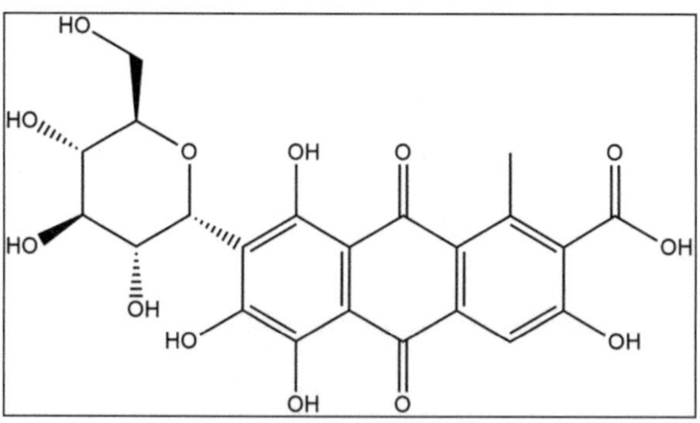

Fig. 3.11 Chemical structure of carmine

(b) *Orcein*

Orcein is a type of dye that has a deep purple color. Its chemical formula is $C_{28}H_{24}N_2O_7$ (Fig. 3.12) and it has a molecular weight of 500.499 gm mol^{-1}. Orcein is derived from a colorless compound called orcinol 3, 5-di-hydroxytoluene. Natural and synthetic versions of orcein are available on the market. Natural orcein is obtained from two types of lichens—*Rocella tinctoria* and *Lecanora porella*. Acetic-orcein was introduced by La Cour in 1941 as a chromosome stain.

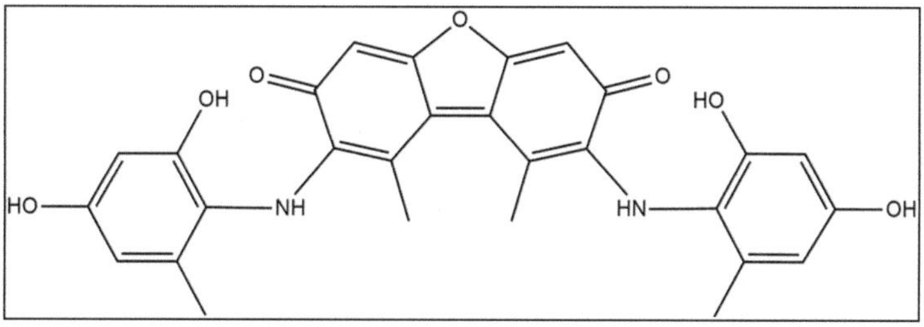

Fig. 3.12 Chemical structure of orcein

- *Preparation of 1 or 2% Aceto-carmine and Aceto-orcein*
  - *Requirement:* (i) Glacial acetic acid, (ii) double distilled water, (iii) measuring cylinder, (iv) beaker or conical flask, (v) carmine or orcein powder, (vi) electronic balance, (vii) glass rod, (viii) heating apparatus, and (ix) filter paper.
  - *Procedure*:
    (i) To prepare a 100 mL 45% acetic acid solution, take 55 mL of distilled water and 45 mL of glacial acetic acid in a measuring cylinder and mix them in a 200 mL beaker or conical flask (Fig. 3.13).
    (ii) Next, use a digital balance to weigh out 1 gm or 2 gm of carmine or orcein powder.
    (iii) Cover the mouth of the beaker or conical flask containing the 100 mL 45% acetic acid solution with cellophane paper to prevent the acetic acid from evaporating.
    (iv) Place the beaker or flask on a heating apparatus and heat it for 10 minutes at a simmering point to avoid rapid boiling.
    (v) Slowly add and dissolve 1 or 2 gm of carmine or orcein powder in 100 mL of 45% acetic acid at a simmering point with occasional stirring using a glass rod (Fig. 3.13).
    (vi) After adding the carmine powder, keep the solution for 5–10 minutes at a simmering point until the color becomes dark red (Fig. 3.13).

(vii) Filter the mixture after cooling it down to room temperature and keep it in a colored bottle. Make up the final volume with 45% acetic acid, if some amount of stain volume is lost.

(viii) The filtered 1% or 2% carmine or orcein solution can be kept for a long time at room temperature but may require filtration if not used regularly.

*Note: For softening of the tissue and separation of cells during tissue smearing, 1 N HCl solution may be added to the aceto-orcein solution in a ratio of 1:9 (v: v). 1 N HCl dissolves the middle lamella by interacting with calcium and magnesium pectate.

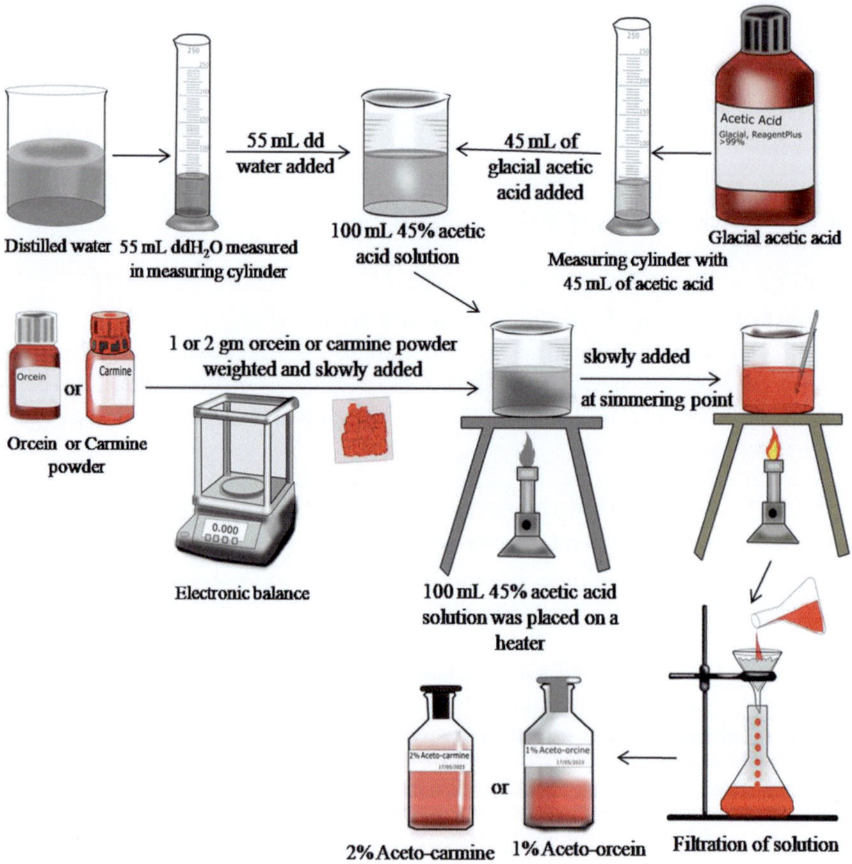

Fig. 3.13 Steps of aceto-orcine or aceto-carmine stain preparation

(c) *Basic Fuchsin and Feulgen/Schiff's Reagent*

Basic fuchsin is a magenta-red color dye with chemical formula $C_{19}H_{17}N_3$ (Fig. 3.14a) and molecular weight 323.82 gm mol^{-1}. It is soluble in both water and alcohol. It is an important ingredient in the Feulgen reagent, which is also known as Schiff's reagent or fuchsin sulphurous acid. Basic fuchsin comes in a powdered form, while Feulgen reagent is only available in liquid form on the market or can be prepared in a laboratory through a chemical reaction between basic fuchsin and sulfurous acid.

In 1924, Robert Feulgen and Heinrich Rossenbeck discovered Feulgen stain and used it as one of the most common, oldest, and specific reagents in histochemistry. Feulgen reaction is considered to be the most effective about chromosome staining (Sharma and Sharma 1980). The Feulgen solution is prepared through a process that involves the conversion of β-rosaniline chloride to leucosulphinic acid by sulfurous acid. This acid is obtained by reacting HCl with potassium metabisulphite. The excess SO_2 produced in this reaction then combines with leucosulphinic acid to produce bi-N-aminosulphinic acid, which is commonly referred to as Schiff's reagent (Fig. 3.14b). Different methods of Feulgen stain preparation are available (Fukui 1996). However, the following method is based on the principles of Sharma and Sharma (1980) with required modifications.

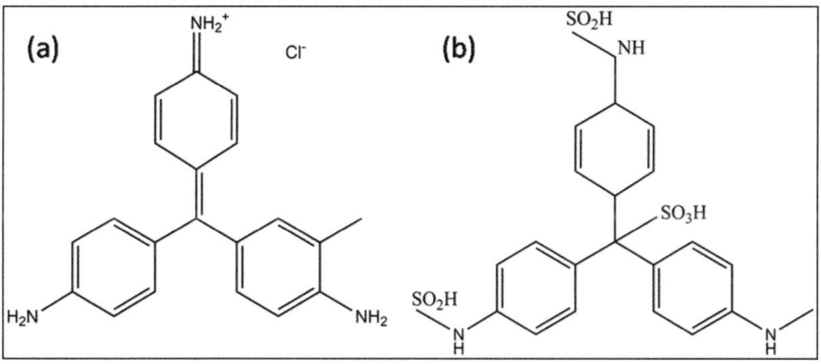

Fig. 3.14 Chemical structure of (**a**) basic fuchsin and (**b**) Feulgen reagent

- *Preparation of Feulgen / Schiff's Reagent*
  - *Requirement:* (i) Certified basic fuchsin powder, (ii) double distilled water, (iii) 1 N HCl, (iv) potassium metabisulfite ($K_2S_2O_5$), (v) activated charcoal, (vi) conical flask, (vii) heating apparatus, (viii) digital balance, (ix) filtration unit, (x) measuring cylinder, (xi) parafilm, (xii) stopper, (xiii) aluminum foil, (xiv) amber-colored bottle, and (xv) glass rod.
  - *Procedure*:
    (i) Slowly dissolve 0.5 gm basic fuchsin in 90 mL of boiling ddH_2O (Fig. 3.15).
    (ii) Thoroughly shake the flask and let the temperature of the solution drop to 50–58 °C. Then, filter the solution and cool the filtrate down to room temperature.
    (iii) Subsequently, add 10 mL 1 N HCl and 0.5 gm potassium metabisulphite ($K_2S_2O_5$) into the filtrate at room temperature. Carefully shake the flask repeatedly (Fig. 3.15).
    (iv) Seal the mouth of the container with a rubber cork stopper and paraffin. Additionally, wrap the container with black paper or aluminum foil to prevent light exposure (Fig. 3.15). Incubate the stained container in a dark chamber at room temperature for 24 hours or until a light straw color develops. Always seal the stopper with paraffin whenever the flask is opened, and never remove the black paper to prevent exposure to oxygen and light, which can cause the color to return.

    [Note that within 24 hours of preparation, the solution will turn straw-colored due to the bleaching effect of SO_2, which is released by the action of HCl on potassium metabisulphite. The amount of potassium metabisulphite to be added depends on its freshness. If enough SO_2 is not released, the bleaching will be incomplete, and the solution will not turn straw-colored, which will make subsequent staining ineffective. This solution is called Schiff's reagent and is a clear, colorless liquid produced by preparing leuco fuchsin. Diamond fuchsin is recommended for the best results.]

    (v) If the solution does not turn straw-colored or shows a very faint straw color after 24 hours, 0.5 gm activated charcoal powder can be added to the solution (Fig. 3.15). The flask should be adequately shaken, covered, and kept overnight in a refrigerator (4 °C). The charcoal powder makes the solution colorless by absorbing the coloring particles.
    (vi) Finally, the transparent solution can be filtered in an amber-colored bottle and stored at a refrigerated temperature for future use (Fig. 3.15).

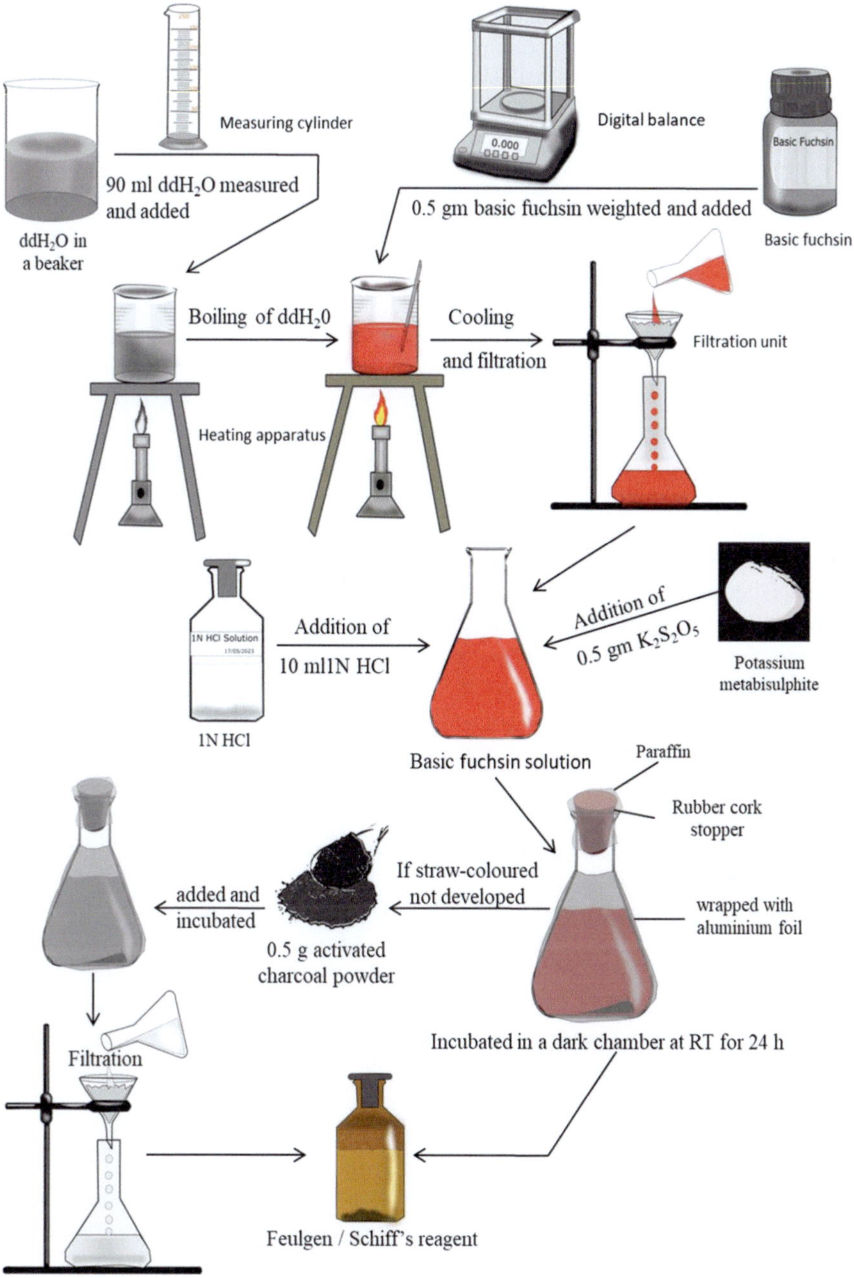

Fig. 3.15 Steps of Feulgen stain preparation

The above-mentioned stains are commonly used for staining somatic chromosomes. However, the primary staining solution used conventionally is a mixture of 2% aceto-orcein and HCl (9:1). One limitation of this stain is that it stains not only chromosomal DNA and protein but also cytoplasm. In contrast, Feulgen stain is a specific stain for chromosomal DNA and does not interact with protein.

3.2.6 Preparation of Temporary Cytological Slides by Squash Method

Flowering plants have two types of cells primarily used for chromosome preparation: diploid (2n) somatic meristematic cells in root tips and haploid (n) gametic cells in anthers. The terms "squash" and "smear" are commonly used for the preparation of somatic and gametic chromosomes, respectively.

To prepare somatic chromosomes, pre-treatment and fixation are mandatory. Finger pressure is applied to spread the somatic cells containing metaphase chromosomes evenly. This process is known as squash preparation which helps to obtain scattered metaphase chromosomes to count the chromosome number and visualize their detailed morphology. On the other hand, gametic chromosomes are prepared from unopened flower buds using the smearing process. No pre-treatment is required for this method. Fresh flower buds can be directly used for gametic chromosome preparation. However, fixed (in any fixative) and subsequently preserved (in 70% alcohol) flower buds are alternative sources for meiotic study so that buds remain usable for a few more months. This is especially useful for seasonally growing flowers and flowers from remote phytogeographical locations.

- *Requisition Requirement:* (i) Stained root tips, (ii) grease-free slide, (iii) preferably square cover slips (22 mm), (iv) blade, (v) 45% acetic acid in dropping bottle, (vi) blotting paper, (vii) spirit lamp, (viii) paraffin block or wax, (ix) marker, (x) tapper, (xi) scalpel, (xii) brush, and (xiii) needle.
- *Procedure*:

  To prepare a temporary sealed cytological slide by the squash method using a stained root sample for microscopy, follow these steps:

  (i) Place the root at the center of a clean slide and add a drop of 45% acetic acid using a brush (Fig. 3.16).

  (ii) Excise tip of the root with a blade and discard the remainder.

  (iii) Carefully mount the root tip with a coverslip using a needle. Position the edge of the coverslip on the slide so that it makes contact with the edge of the mounting medium, then gently lower the coverslip to minimize the formation of air bubbles.

(iv) To create a temporary slide for microscopic observation, start by placing a fold of blotting paper on a flat surface. Apply moderate and even pressure on the cover slip with your thumb or a tapper to spread the cells and chromosomes into a single plane (Fig. 3.16).

(v) Begin your preliminary observation with a low-power objective lens to assess whether the root tip cells are well scattered and if there are cells containing metaphase chromosome plates. If not, add a few drops of 45% acetic acid, gently heat it over a spirit lamp, and tap it to improve the spread.

(vi) If necessary, perform a 45% acetic acid wash to eliminate any excess stain in the background. Once finished, seal the edges of the cover slip with paraffin or nail polish, and the slide will be ready for immedi.ate microscopic observation (Fig. 3.16).

A properly sealed temporary slide can be used for observations over the course of 2–7 days if stored in a refrigerator. However, for long-term use, it is advisable to prepare a permanent slide.

3.2.7 Preparation of Permanent Slides

Long-term preservation of a temporary seal chromosome slide prepared by the squash and smear method following classical stains, such as orcein, Feulgen, and carmine, requires processing for permanent mounts. Slides that are processed for permanent mounts can be preserved for extended periods that allow visualization of same slide under a compound microscope in future. Choosing the right mounting medium is essential for ensuring the long-term preservation of a slide. An effective mounting medium should have a refractive index that closely matches that of glass to minimize refraction, harden quickly when exposed to air, and prevent the decolorization of tissues, cells, or chromosomes. Proper mounting achieves several objectives, including rendering the tissue transparent, enhancing visibility under a microscope, securely holding the tissue in place with the coverslip, and preserving the slides for an extended duration.

There are several methods available for creating these permanent mounts, and some of these methods are outlined below.

- *Permanent Slide Preparation by Tertiary Butyl-Alcohol Schedule*
  - *Requirement:* (i) Temporarily prepared slide, (ii) blade, (iii) glass rod, (iv) petridish, (v) acetic-ethanol solution, (vi) slide, (vii) cover glass, (viii) ethanol: butanol, (ix) n-butanol grade I, (x) n-butanol grade II, and (xi) entelan or euparal.

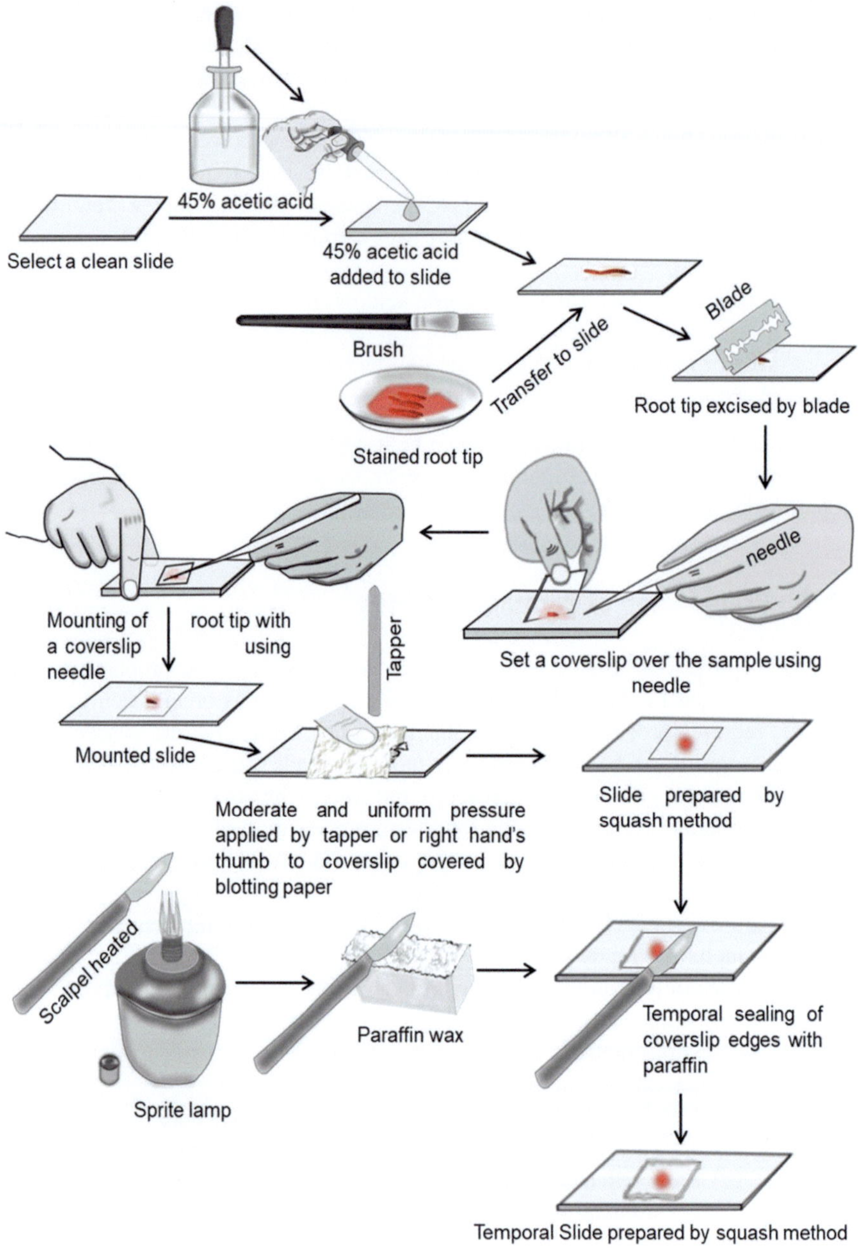

Fig. 3.16 Steps of somatic slide preparation by the squash method

- *Procedure*:
  (i) Carefully remove the temporary paraffin seal using a blade, taking care not to disturb the cover glass (Fig. 3.17).
  (ii) Place the slide in an inverted position on a glass rod within a petri dish containing a 1:1 mixture of glacial acetic acid and ethanol until the cover glass separates (Fig. 3.17).
  (iii) Transfer both the slide and cover glass to a second petri dish filled with a 1:1 solution of ethanol and butanol for 10 to 15 minutes (Fig. 3.17).
  (iv) Subsequently, move the slide and cover glass to n-butanol grade I and then to n-butanol grade II solutions, allowing for 10 to 15 minutes in each (Fig. 3.17).
  (v) Before mounting with a drop of entelan or euparal, allow the slide and coverslip to air dry separately, creating two distinct slides (Fig. 3.17). Let them dry on a hot plate overnight. The used solutions can be reused multiple times.

- *Permanent Slide Preparation Using Freshly Prepared Slides**
  (i) Add 45% acetic acid to one side of the slide and use blotting paper on the opposite side to absorb any excess stain.
  (ii) Place the slide, along with a coverslip, in a covered container containing alcohol-soaked blotting paper. The alcohol vapors will gradually replace the acetic acid beneath the coverslip. Allow it to sit for 4 to 5 days.
  (iii) Afterward, add a few drops of euparal or entelan (both soluble in alcohol) to two sides of the slide inside the container, and let it rest for an additional day. This will enable the entelan or euparal to permeate between the slide and the coverslip.
  (iv) Finally, allow the slide to dry in open air.
  *Please note that this process is not very familiar and can be quite time-consuming

- *Permanent Slide Preparation Using Dry Ice Method*
  This process of making temporary preparations permanent is both speedy and convenient. It involves the following steps:
  (i) Place the slide on a flat piece of dry ice in such a manner that the cover glass faces upward.
  (ii) Allow the slide to remain on the dry ice for one minute.
  (iii) After this time, gently apply pressure to detach the cover glass from the slide.
  (iv) Immediately immerse both the slide and cover glass in 95% ethanol.
  (v) Perform the mounting using either entelan or euparal.

3.2.8 Visualization and Documentation

To visualize somatic and gametic chromosomes, it is essential to utilize a compound binocular or trinocular microscope equipped with objective lenses of 10×, 20×, 40×, and 100×, along with 10× or 15× eyepieces. Proper focusing and the ability to locate well-scattered metaphase figures are critical for accurately counting chromosome numbers and analyzing their morphology.

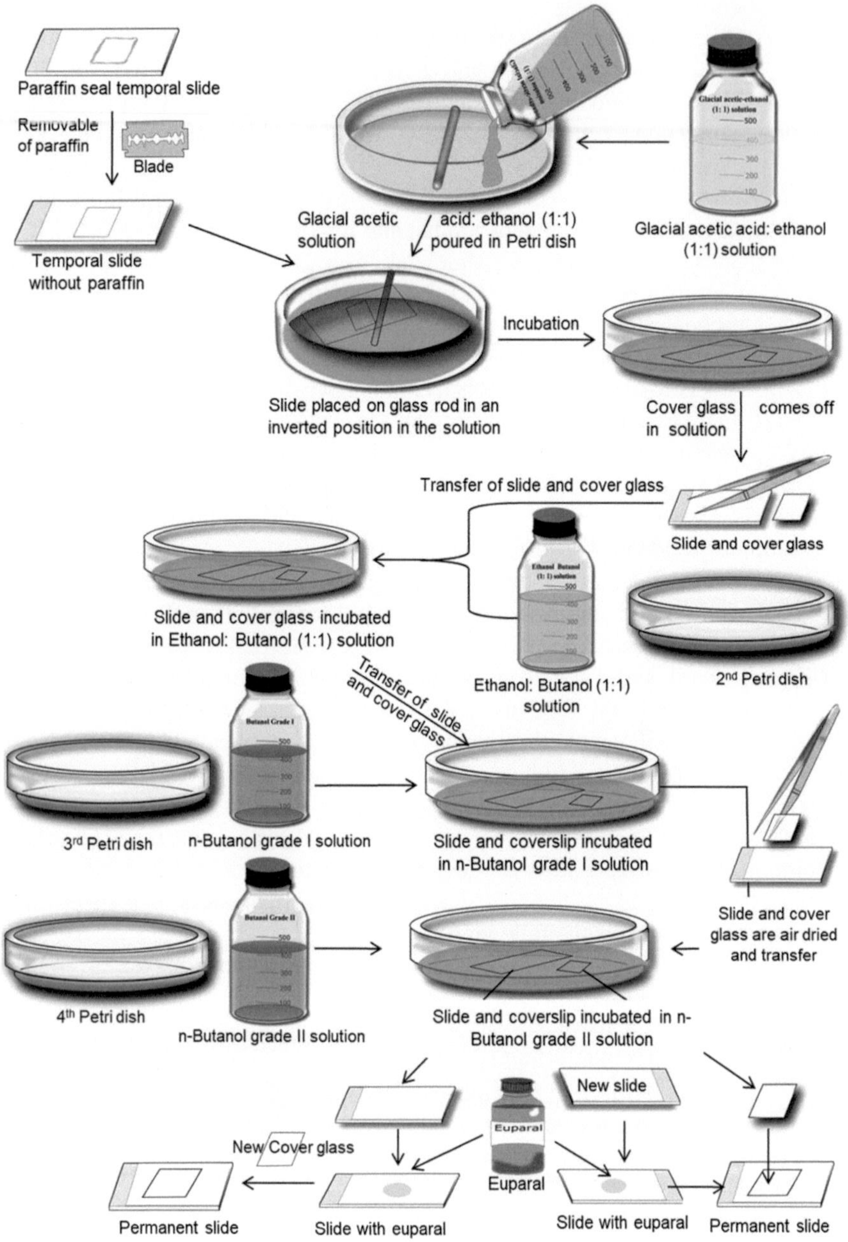

Fig. 3.17 Steps of preparation of permanent slides

Metaphase divisional phases can be documented by either manually drawing a plate using a camera lucida or employing digital documentation methods based on the available microscope facilities. Additionally, you can determine the magnification of a specific stage manually or by using a digital platform with specialized software. Documentation with a microscope and karyotype analysis is discussed in detail in Chapters 5 and 12.

3.3 Case Studies

The following texts present a practical training protocol and a methodology for chromosome preparation. This includes essential documentation and key karyomorphological features, all supported by species-specific chromosome plates that we developed through our instructional work on the following valuable and classic plant materials.

(i) *Allium cepa* L. (Onion), $2n = 16$ chromosomes (Fig. 3.18)
(ii) *Allium sativum* L. (Garlic), $2n = 16$ chromosomes (Fig. 3.19)
(iii) *Trigonella foenum-graecum* L (Methi), $2n = 16$ chromosomes (Fig. 3.20)
(iv) *Lens culinaris* Medik. (Musur), $2n = 14$ chromosomes (Fig. 3.21)
(v) *Aloe vera* Burm. f., $2n = 14$ chromosomes (Fig. 3.22)
(vi) *Nigella sativa* L. (Kalo jeera), $2n = 12$ chromosomes (Fig. 3.23)
(vii) *Vicia faba* L. (Broad bean), $2n = 12$ chromosomes (Fig. 3.24)
(viii) *Ornithogalum virens/Albuca virens,* (Lindl.) J.C. Manning & Goldblatt, $2n = 6$ chromosomes (Fig. 3.25)

3.3.1 Species: *Allium cepa* L. (Onion)

3.3.1.1 Chromosome Preparation Protocol (Fig. 3.18)

(i) Bulbs and seeds of *Allium* showed effectively root induction and growth at lower temperature range of 22–25 °C within 2–3 days.
(ii) Healthy roots, about 1–2 cm long, are usually collected between 10 am and 11 am.
(iii) Before pretreatment, make sure the roots are no longer than 1 cm. Carefully clean the roots using a fine-tip brush, ensuring that the root apical meristems are not damaged.
(iv) For pretreatment, place the roots in a glass vial containing pretreatment chemicals. Use either a saturated PDB solution or a 0.5% colchicine solution for 4.5 to 5 hours at 12–16 °C for effective chromosome condensation. However, variations may occur depending on the specific cultivar.

(v) After pretreatment, pour off the pretreatment chemical and rinse the roots with water. Remove the water, then add a fixative solution (1 part acetic acid to 3 parts alcohol or methanol), and leave the roots in it for one hour to overnight.
(vi) Transfer the roots from the fixative into 45% acetic acid and incubate for 20–25 minutes for softening tissues or hydrolysis using 1 N HCl for 10–15 minutes.
(vii) For staining, place the roots in a mix of 2% aceto-orcein and 1 N HCl (9:1 ratio) and heat gently for a few seconds. Leave the root tips in this solution for at least an hour to stain them well.
(viii) For squashing, put the root tip on a clean, dry slide with a drop of 45% acetic acid. Cover with a glass cover slip and apply uniform pressure.

3.3.1.2 Key Karyomorphological Features

Observations of the metaphase plates under the microscope revealed a diploid chromosome number of $2n = 16$ (Fig. 3.18). The karyotype comprises five metacentric and three submetacentric pairs of long chromosomes (Fig. 3.18). Out of three submetacentric pairs of chromosomes, one submetacentric pair (the sixth pair) contains a terminal secondary constriction (Fig. 3.18). Total chromatin length is 241.09 ± 9.85 μm. The model karyotype formula is 10 m + 4sm + 2sm.t.

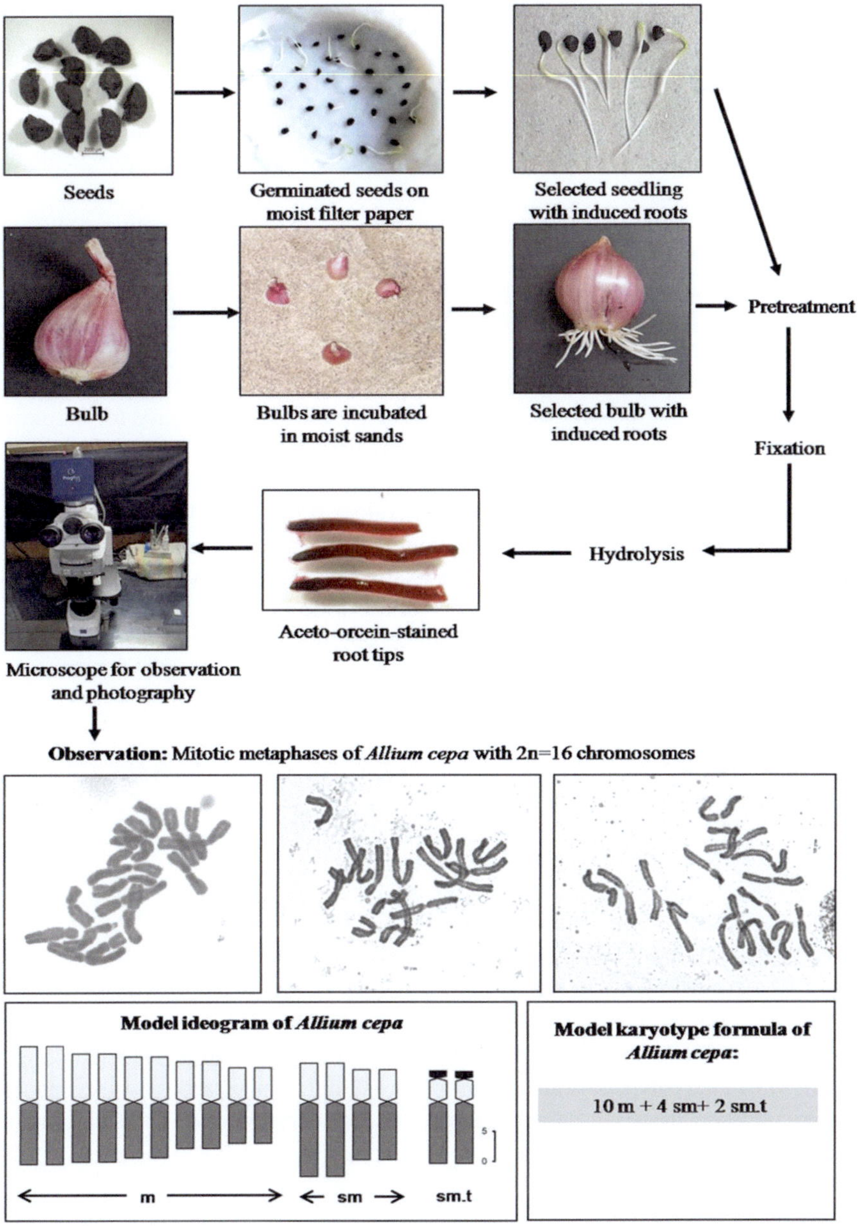

Fig. 3.18 Various stages of the classical methodology used for preparing, staining, and observing somatic chromosomes in *Allium cepa*. This process resulted in well-scattered metaphase plates containing 2n = 16 chromosomes. Model ideogram and karyotype formulas are also included

3.3.2 Species: *Allium sativum* L. (Garlic)

3.3.2.1 Chromosome Preparation Protocol (Fig. 3.19)

(i) Bulbs of *Allium sativum* serve as the source material for root induction. Lower temperature ranges of 22–25 °C promote root development within 2–3 days in *Allium sativum*.
(ii) Healthy roots, typically 1–2 cm in length, are harvested between 10 am and 11 am.
(iii) At pretreatment, root length should not exceed 1 cm. Roots are cleaned with a fine-tip brush to avoid damaging the root apical meristems.
(iv) For pretreatment, roots are placed in a glass vial containing the appropriate chemicals. Both a saturated solution of PDB for 4.5 to 5 hours and a 0.5% colchicine solution for 4.5 to 5 hours at 12–16 °C are effective for chromosome condensation, although results may vary depending on the cultivar.
(v) During fixation, the pretreatment chemical is decanted and roots are rinsed with water. After water removal, a fixative solution (1:3 acetic alcohol or acetic methanol) is added for one hour to overnight.
(vi) Roots are then transferred from the fixative to 45% acetic acid for 20–25 minutes to soften tissues or to 1 N HCl for 10–15 minutes for hydrolysis.
(vii) In the staining process, roots are immersed in a mixture of 2% aceto-orcein and 1 N HCl (9:1 ratio) and gently heated for a few seconds. Root tips remain in this solution for over an hour to ensure thorough staining.
(viii) For squashing, a clean, dry slide is prepared and only the root tip is placed in a drop of 45% acetic acid. The sample is covered with a glass cover slip and uniform pressure is applied.

3.3.2.2 Key Karyomorphological Features

Observations of the metaphase plates under the microscope revealed a diploid chromosome number of $2n = 16$ (Fig. 3.19). The karyotype shows almost all chromosomes uniformly long, with five metacentric pairs and three submetacentric pairs (Fig. 3.19). Position of secondary constriction is not terminal but interstitial (Fig. 3.19). The number of secondary constricted chromosomes may vary within a karyotype from two, three, and four in different root apical meristems (Jha et al. 2024). Model karyotype formula: 10 m + 4sm + 2sm.t, which carries two secondary constrictions on the sixth pair of submetacentric chromosomes. Total chromatin length is 224.18 ± 5.85 μm.

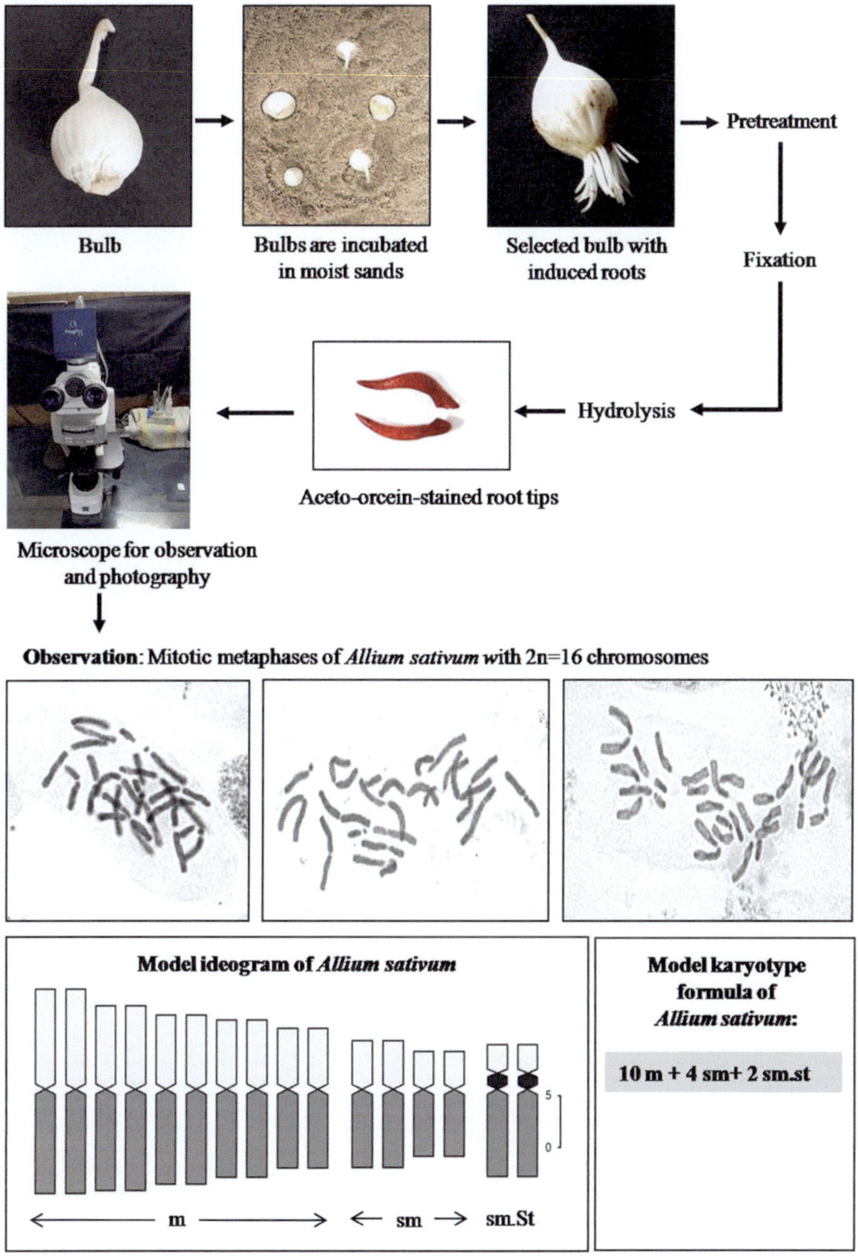

Fig. 3.19 Various stages of the classical methodology used for preparing, staining, and observing somatic chromosomes in *Allium sativum*. This process resulted in well-scattered metaphase plates containing $2n = 16$ chromosomes. Model ideogram and karyotype formulas are also included

3.3.3 Species: *Trigonella foenum-graecum* L

3.3.3.1 Chromosome Preparation Protocol (Fig. 3.20)

(i) Seeds are used for root induction for *Trigonella*. The species prefers a moderate temperature range of 22–27 °C for seed germination and root development within 2–3 days.
(ii) Healthy roots, about 1–2 cm long, are usually harvested between 10 am and 11 am.
(iii) At the time of pretreatment, the root size should remain within 1 cm. Carefully wash the roots using a fine-tip small brush, ensuring that the root tip is not damage during the process.
(iv) For pretreatment, place the roots in a glass vial containing pretreatment chemicals. A saturated solution of PDB with a pinch of aesculin, used for 4.5 to 5 hours, for effective chromosome condensation. However, the time may change depending on the specific cultivar.
(v) During fixation, decant the pretreatment chemical and rinse the roots with water. After removing the water, add a fixative solution (1:3 acetic alcohol or acetic methanol) and leave the roots in it for one hour to overnight.
(vi) Move the roots from the fixative to 45% acetic acid for 15 to 20 minutes to soften the tissues, or use 1 N HCl for 10–15 minutes.
(vii) For staining, the roots are transferred in a mixture of 2% aceto-orcein and 1 N HCl (9:1) and gently heated for a few seconds. Leave the root tips in this mixture for over an hour so they get fully stained.
(viii) For squash preparation, put just the root tip on a clean, dry slide in a drop of 45% acetic acid. Cover with a glass slip and press down evenly.

3.3.3.2 Key Karyomorphological Features

Observations of the metaphase plates under the microscope revealed a diploid chromosome number of $2n = 16$ (Fig. 3.20). The karyotype shows three pairs of nearly metacentric (m), two pairs of submetacentric (sm), and three pairs of subtelocentric (st) chromosomes. Interstitial secondary constriction was located on the two pairs of nearly metacentric chromosomes at submetacentric (second and fourth pair) positions (Fig. 3.20). Total chromatin length was 165.52 ± 18.42 μm, and the model karyotype formula is 6st + 4sm + 2m + 4m.sm.

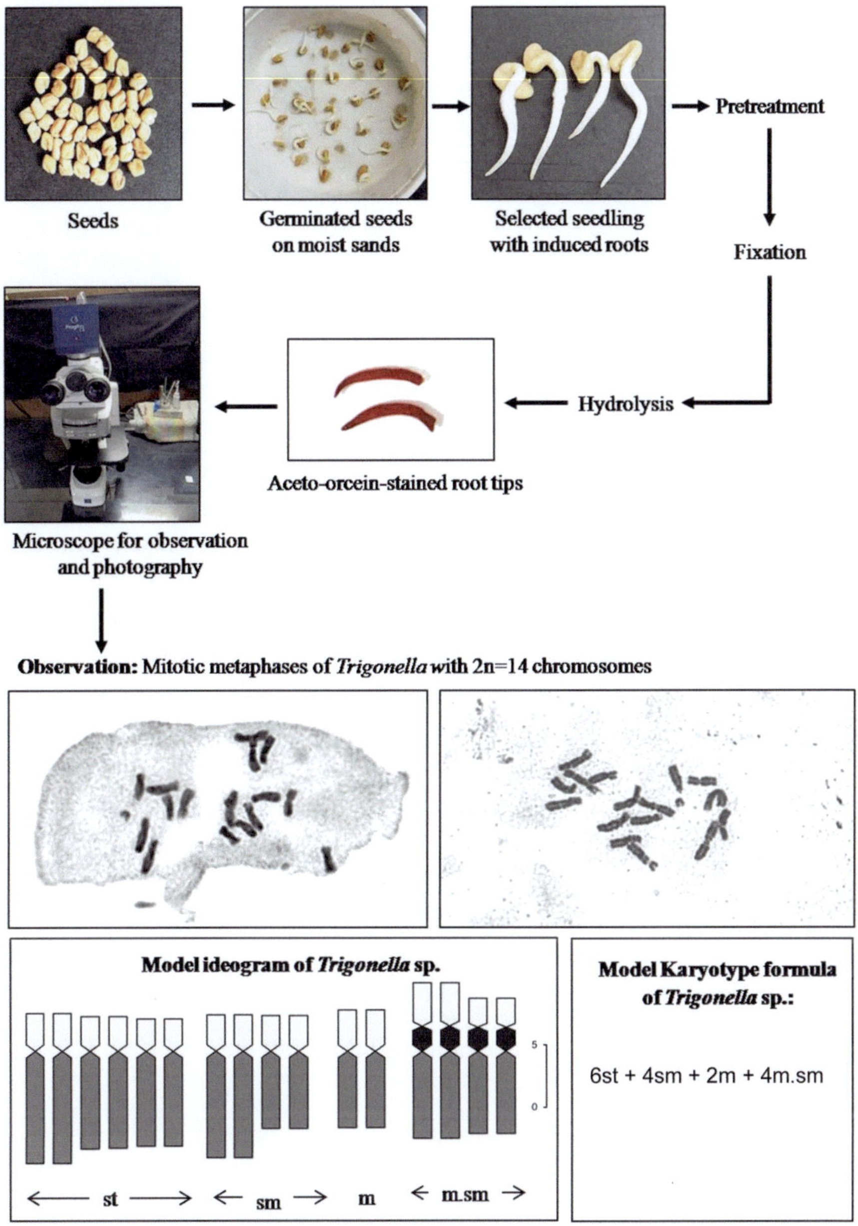

Fig. 3.20 Various stages of the classical methodology used for preparing, staining, and observing somatic chromosomes in *Trigonella foenum-graecum*. This process resulted in well-scattered metaphase plates containing $2n = 16$ chromosomes. Model ideogram and karyotype formulas are also included

3.3.4 Species: *Aloe vera* Burm. f

3.3.4.1 Chromosome Preparation Protocol (Fig. 3.21)

(i) Healthy roots with lengths of 1–2 cm were directly collected from potted *A. vera* plants between 10 am and 11 am. Roots were gently washed with a fine-tip brush to protect the root apical meristems.
(ii) For pretreatment, roots were placed in a glass vial with either a saturated PDB solution for 5 hours or a 0.5% colchicine solution for 4 hours at 12–16 °C. Both methods help with chromosome condensation, though results may vary by cultivar.
(iii) For fixation, the pretreatment solution was poured off, and roots were rinsed with water. After removing the water, a fixative (1:3 acetic alcohol or acetic methanol) was added and incubated for one hour to overnight.
(iv) Roots were then transferred to 45% acetic acid for 15–20 minutes to soften the tissue, or to 1 N HCl for 10–15 minutes for hydrolysis.
(v) For staining, roots were placed in a mix of 2% aceto-orcein and 1 N HCl (9:1 ratio) and gently heated for a few seconds. The root tips stayed in this solution for over an hour to ensure good staining.
(vi) For squashing, a clean, dry slide was used. Only the root tip was placed in a drop of 45% acetic acid, covered with a glass cover slip, and even pressure was applied.

3.3.4.2 Key Karyomorphological Features (Fig. 3.21)

Observations of the metaphase plates under the microscope revealed a diploid chromosome number of $2n = 14$ (Fig. 3.21). A distinct bimodal karyotype shows three pairs of small submetacentric (sm), and four pairs of long submetacentric (sm) to subterminal chromosomes (st) (Fig. 3.21). Terminal secondary constriction was located on the one pair of the long submetacentric chromosomes. Total chromatin length was 295.86 ± 10.72 μm and the model karyotype formula is 6st + 6sm + 2sm. t.

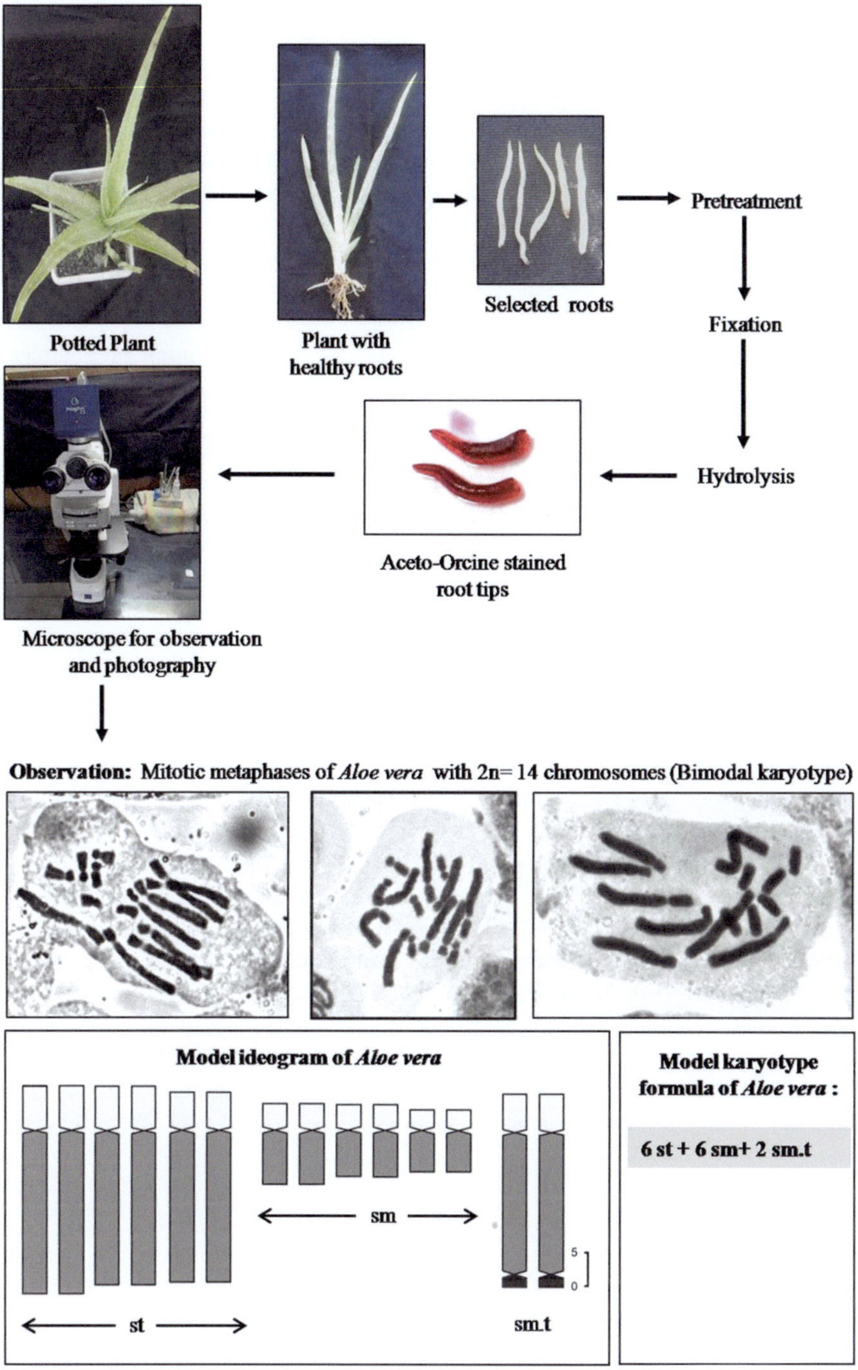

Fig. 3.21 Various stages of the classical methodology used for preparing, staining, and observing somatic chromosomes in *Aloe vera*. This process resulted in well-scattered metaphase plates containing $2n = 14$ chromosomes. Model ideogram and karyotype formulas are also included. (Metaphase plates are taken from Das et al. 2010. Nucleus 53:89–94. With permission)

3.3.5 Species: *Lens culinaris* Medik

3.3.5.1 Chromosome Preparation Protocol (Fig. 3.22)

(i) Seeds are used for the germination of roots. The species prefers a moderate temperature range of 22–27 °C to promote root development within 2–3 days.
(ii) Healthy roots about 1–2 cm in length are typically harvested between 10 am and 11 am.
(iii) At the time of pretreatment, the root size should remain within 1 cm. Collected roots were carefully washed with water and cleaned using a fine-tip brush, without damaging the root apical meristems.
(iv) For pretreatment, place the roots in a glass vial containing pretreatment chemicals. A saturated solution of PDB for 5 hours at 12–16 °C is effective for chromosome condensation. However, variations may occur depending on the specific cultivar.
(v) During fixation, pretreatment chemical was decanted and the roots were rinsed with water. After removing the water, add a fixative solution (1:3 acetic alcohol or acetic methanol) for a duration ranging from one hour to overnight.
(vi) Transfer the roots from the fixative into 45% acetic acid for 15–20 minutes for softening tissues or hydrolysis using 1 N HCl for 10–15 minutes.
(vii) In the staining process, roots are placed in a mixture of 2% aceto-orcein and 1 N HCl (in a 9:1 ratio) and gently heated for a few seconds. The root tips were incubated in this solution for over an hour to achieve appropriate staining.
(viii) For squashing, take a clean, dry slide and place only the root tip in a drop of 45% acetic acid. Cover it with a glass cover slip and apply uniform pressure.

3.3.5.2 Key Karyomorphological Features (Fig. 3.22)

Observations of the metaphase plates under the microscope revealed a diploid chromosome number of $2n = 14$ (Fig. 3.22). Karyotype shows four pairs of nearly metacentric (m), two pairs of submetacentric (sm), and one pair of subterminal chromosomes (st) (Fig. 3.22). The position of secondary constriction in this species is not terminal but interstitial. Interstitial secondary constriction was located on the 4th nearly metacentric chromosome pair, which is difficult to identify through the classical method (Fig. 3.22). Model karyotype formula in this germplasm is 6m + 4sm + 2st + 2 m.m.

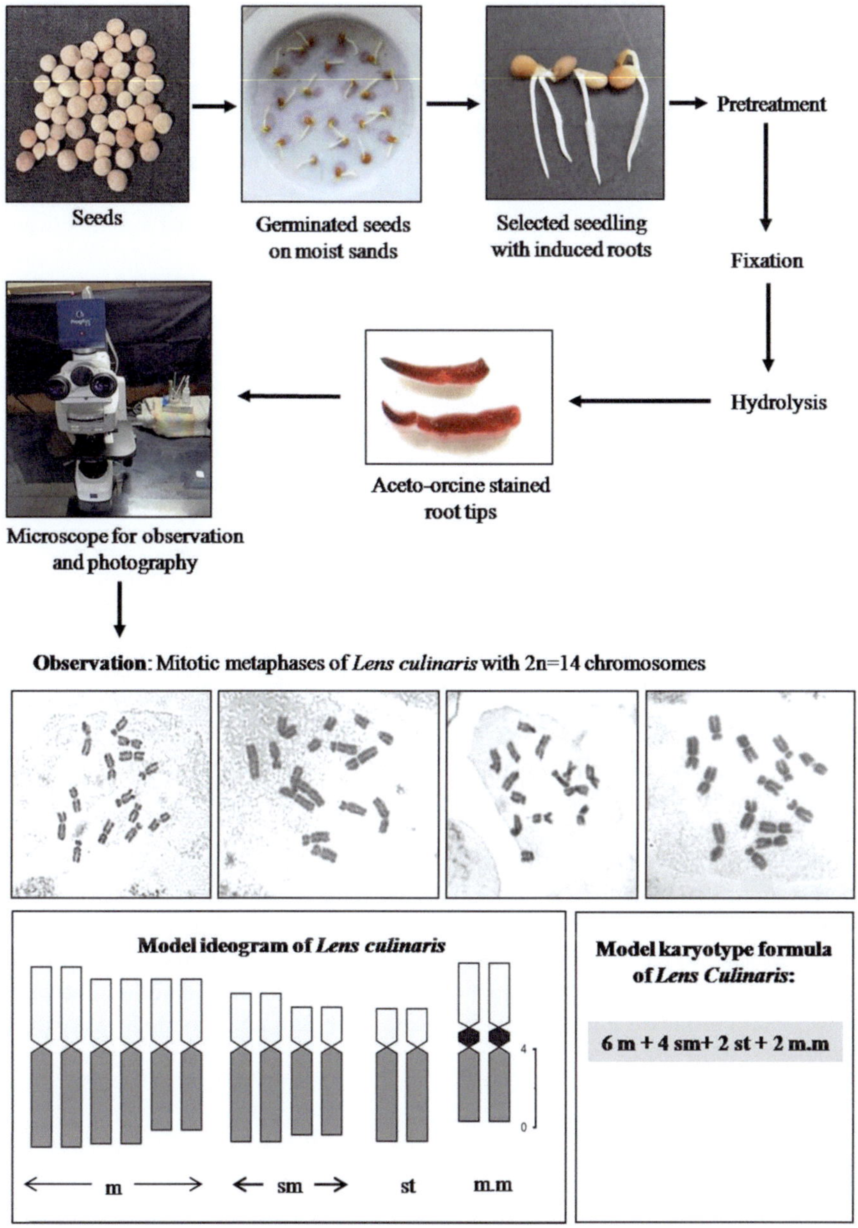

Fig. 3.22 Various stages of the classical methodology used for preparing, staining, and observing somatic chromosomes in *Lens culinaris*. This process resulted in well-scattered metaphase plates containing $2n = 14$ chromosomes. Model ideogram and karyotype formulas are also included

3.3.6 Species: *Nigella sativa* L

3.3.6.1 Chromosome Preparation Protocol (Fig. 3.23)

(i) Seeds are used for the germination of roots. The species prefers a moderate temperature range of 23–25 °C to encourage root development within 8–10 days.
(ii) Healthy roots, measuring 1–2 cm in length, are typically harvested between 10 am and 11 am.
(iii) At the time of pretreatment, the root size should remain within 1 cm. Carefully wash the roots using a fine-tip brush, ensuring that you do not damage the root apical meristems.
(iv) For pretreatment, place the roots in a glass vial containing pretreatment chemicals. A 0.5% Colchicine solution for 4 to 4.5 hours at 12–16 °C is effective for chromosome condensation; however, variations may occur depending on the specific cultivar.
(v) During fixation, decant the pretreatment chemical and rinse the roots with water. After removing the water, add a fixative solution (1:3 acetic alcohol/ acetic methanol) for a duration ranging from one hour to overnight.
(vi) Transfer the roots from the fixative into 45% acetic acid for 15–20 minutes for softening tissues or hydrolysis, with 1 N HCl for 10–15 minutes.
(vii) In the staining process, roots are placed in a mixture of 2% aceto-orcein and 1 N HCl (in a 9:1 ratio) and gently heated for a few seconds. The root tips should remain in this solution for over an hour to achieve thorough staining.
(viii) For squashing, take a clean, dry slide and place only the root tip in a drop of 45% acetic acid. Cover it with a glass cover slip and apply uniform pressure.

3.3.6.2 Key Karyomorphological Features (Fig. 3.23)

Observations of the metaphase plates under the microscope revealed a diploid chromosome number of $2n = 12$. Karyotype shows one pair of short subterminal chromosomes (st) and five pairs of long metacentric chromosomes (m). Terminal secondary constrictions were located on the short subterminal chromosome pairs (6th). Total chromatin length was 159.0.9 ± 18.06 μm, and the model karyotype formula is 10m + 2st.t.

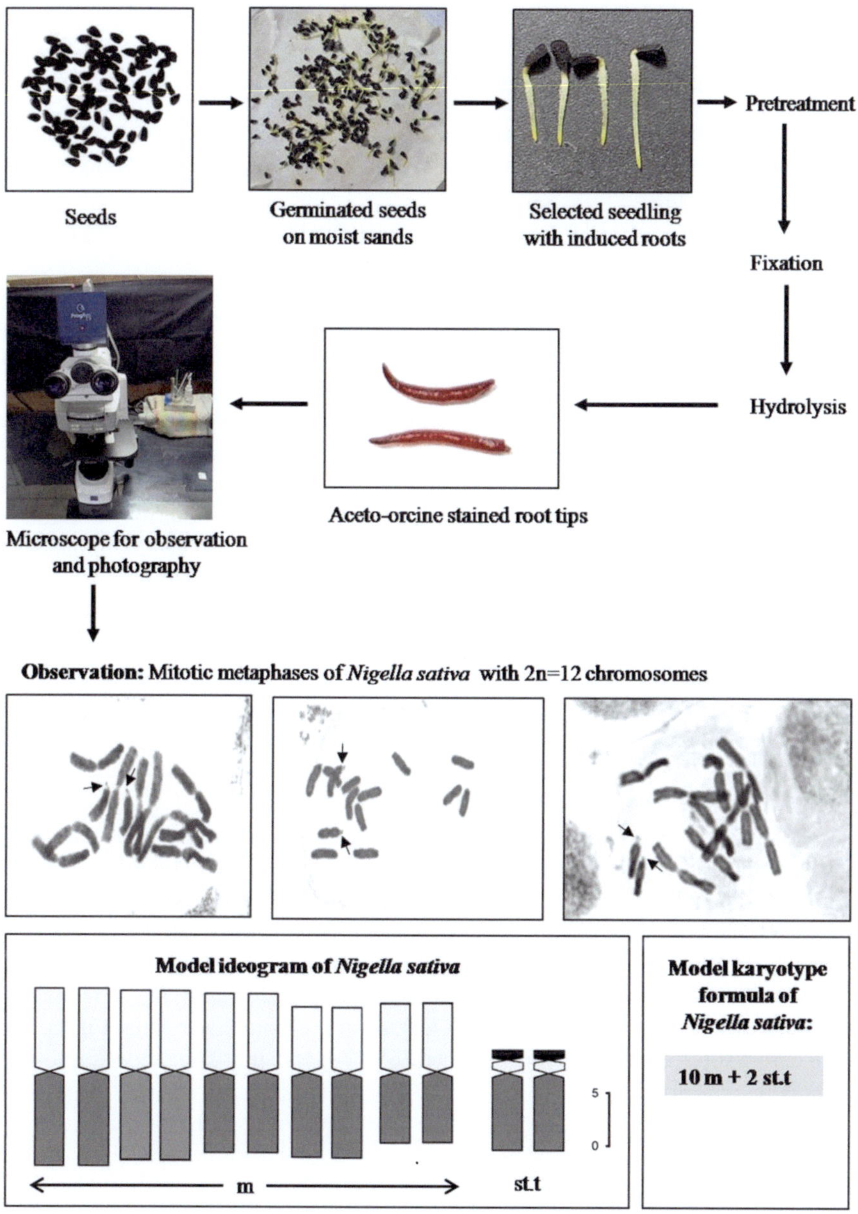

Fig. 3.23 The various stages of the classical methodology used for preparing, staining, and observing somatic chromosomes in *Nigella sativa*. This process resulted in well-scattered metaphase plates containing $2n = 12$ chromosomes. Model ideogram and karyotype formulas are also included

3.3.7 Species: *Vicia faba* L.

3.3.7.1 Chromosome Preparation Protocol (Fig. 3.24)

(i) Seeds are germinated on water-soaked blotting paper at a colder temperature range of 23–25 °C within 3–5 days. This temperature encourages root development.
(ii) Healthy roots of 1–2 cm in length are typically harvested between 11 am and 11.30 am.
(iii) For pretreatment, ensure roots are no longer than 1 cm. Gently wash roots with a fine-tip brush, taking care not to damage the apical meristems.
(iv) For pretreatment, place the roots in a glass vial containing pretreatment chemicals. Treat with 0.5% colchicine solution for 4 hours at 12–16 °C, which is effective for chromosome condensation; however, variations may occur depending on the specific cultivar.
(v) For fixation, decant the pretreatment solution, rinse roots with water, then add a fixative (1:3 acetic alcohol or acetic methanol) for one hour to overnight.
(vi) Transfer roots from the fixative to 45% acetic acid for 15–20 minutes to soften tissues, or use 1 N HCl for 10–15 minutes for hydrolysis.
(vii) For staining, immerse roots in a 2% aceto-orcein and 1 N HCl mixture (9:1 ratio), gently heat for a few seconds, and leave the root tips in the solution for at least one hour.
(viii) To squash, place the root tip on a clean, dry slide in a drop of 45% acetic acid, cover with a glass cover slip, and apply even pressure.

3.3.7.2 Key Karyomorphological Features

Observations of the metaphase plates under the microscope revealed a diploid chromosome number of $2n = 12$ (Fig. 3.24). The karyotype distinctly revealed one pair of long chromosomes and five pairs of comparatively short chromosomes (Fig. 3.24). Karyotype includes two pairs of subterminal (st), three pairs of nearly terminal (t), and one pair of nearly metacentric (m) chromosomes. Position of secondary constriction in this germplasm is not terminal but interstitial (Fig. 3.24). It is located below the small arm of the long nearly metacentric chromosome pair (Fig. 3.24). Total chromatin length was estimated to be 165.47 ± 4.85 μm. The model karyotype formula is 4st + 6t + 2m.st.

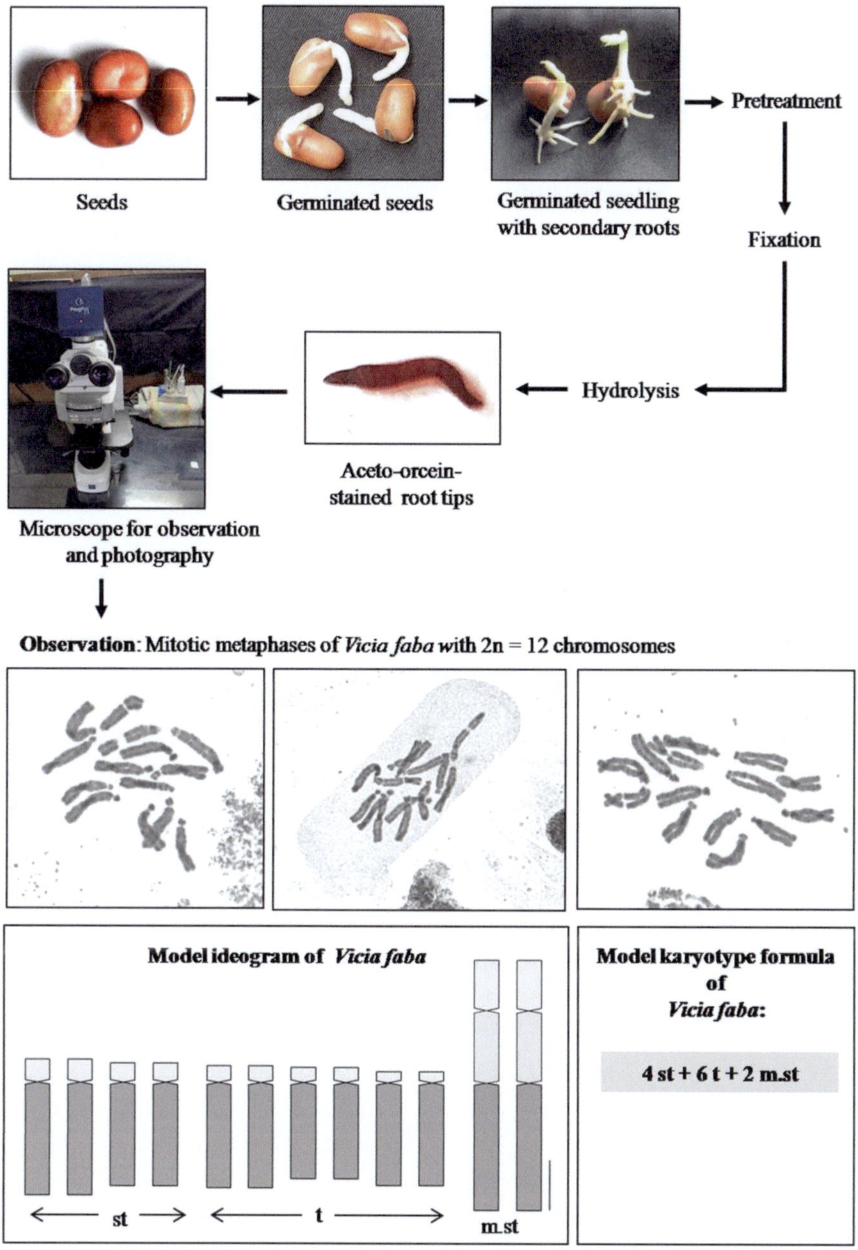

Fig. 3.24 Various stages of the classical methodology used for preparing, staining, and observing somatic chromosomes in *Vicia faba*. This process resulted in well-scattered metaphase plates containing $2n = 12$ chromosomes. Model ideogram and karyotype formulas are also included

3.3.8 Species: *Ornithogalum virens/ Albuca virens* (Lindl.) J.C. Manning & Goldblatt

3.3.8.1 Chromosome Preparation Protocol (Fig. 3.25)

(i) Healthy roots of 1–2 cm in length were harvested directly from potted *Albuca virens* plants between 11:00 am and 11:45 am.
(ii) At the time of pretreatment, root size was maintained within 1 cm. Roots were carefully cleaned with a fine-tip brush and wash using water taking care of no damage of root apical meristems.
(iii) For pretreatment, roots were placed in a glass vial containing the appropriate pretreatment chemicals. A 0.5% colchicine solution was applied for 4 hours at 12–16 °C to induce chromosome condensation, although the optimal conditions may vary depending on the cultivar.
(iv) During fixation, the pretreatment chemical was decanted and roots were rinsed with water. After water removal, a fixative solution (1:3 acetic alcohol or acetic methanol) was added for one hour to overnight.
(v) Roots were then transferred from the fixative to 45% acetic acid for 20–25 minutes to soften tissues or to 1 N HCl for 10–15 minutes for hydrolysis.
(vi) For staining, roots were placed in a mixture of 2% aceto-orcein and 1 N HCl (9:1 ratio) and gently heated for a few seconds. Root tips remained in this solution for over an hour to ensure thorough staining.
(vii) For squashing, a clean, dry slide was prepared and only the root tip was placed in a drop of 45% acetic acid. A glass cover slip was applied and uniform pressure was exerted.

3.3.8.2 Key Karyomorphological Features (Fig. 3.25)

Observations of the metaphase plates under the microscope revealed a diploid chromosome number of $2n = 6$ (Fig. 3.25). No distinct size variation. The position of primary constriction in one pair is nearly terminal (t), while in the other two pairs it is subterminal (st) (Fig. 3.25). Terminal secondary constriction was located on one pair of subterminal chromosomes (Fig. 3.25). Total chromatin length was 39.93 ± 4.02 μm, and the model karyotype formula is 2t +2st + 2st.t.

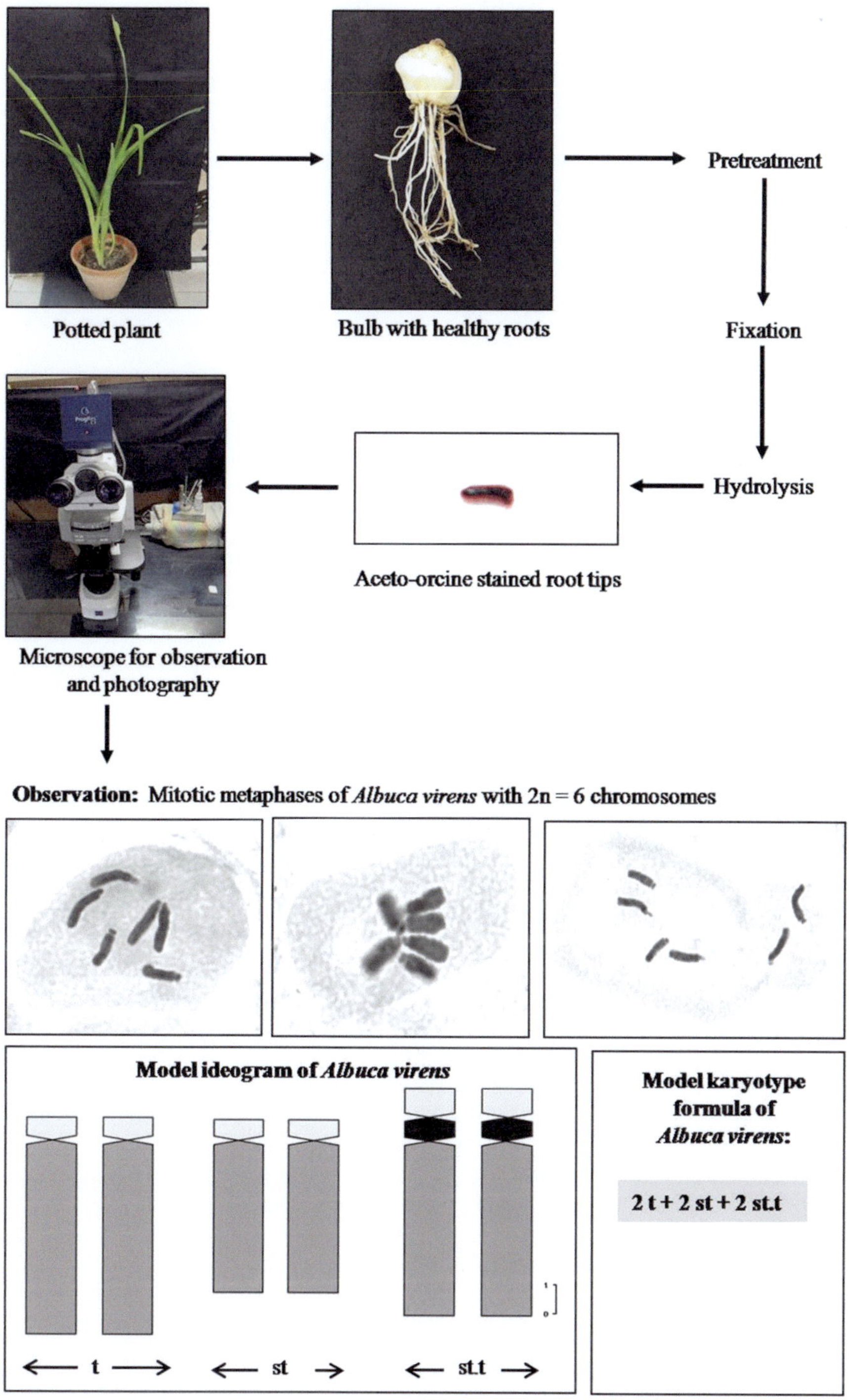

Fig. 3.25 Various stages of the classical methodology used for preparing, staining, and observing somatic chromosomes in *Albuca virens*. This process resulted in well-scattered metaphase plates containing $2n = 6$ chromosomes. Model ideogram and karyotype formulas are also included

References

Das A, Mukherjee P, Ghorai A, Jha TB (2010) Comparative karyomorphological analyses of in vitro and in vivo grown plants of *Aloe vera* L. BURM. f. Nucleus 53(3):89–94. https://doi.org/10.1007/s13237-010-0001-1

Fukui K (1996) Plant chromosomes at mitosis. In: Fukui K, Nakayama S (eds) Plant chromosomes: laboratory methods. CRC Press Inc., Boca Raton, pp 1–17

Jha TB, Ghosh B (2016) Plant tissue culture basis and applied. Revised 2nd edn. Platinum Publishers, Kolkata

Jha TB, Halder M, Bhowmick BK (2024) Giemsa-based chromosome staining and comparative fluorescent banding pattern in five valuable Indian plant species. Caryologia 77(3):37–45

Kianian SF, Kianian PMA (2016) Plant cytogenetics methods and protocols. Springer Science Business Media, New York

Lima-De-Faria A (2010) One hundred years of chromosome research and what remains to be learned. Kluwer Academic Publishers

Paul P, Halder M, Jha S (2013) Alkaloids derived from tyrosine: penethylisoquinoline (autumnaline, colchicine). In: Ramawat KG, Mérillon JM (eds) Natural products. Springer, Berlin, Heidelberg, pp 461–478

Sharma AK, Sharma A (1980) Chromosome techniques: theory and practice, 3rd edn. Butterworths & Co. Ltd., London

Singh RJ (2017) Plant cytogenetics, 3rd edn. CRC Press Taylor & Francis Group

Windham MD, Pryer KM, Poindexter DB, Li FW, Rothfels CJ, Beck JB (2020) A step-by-step protocol for meiotic chromosome counts in flowering plants: a powerful and economical technique revisited. Appl Plant Sci 8(4):e11342

Preparation of Somatic Chromosomes by EMA-Based Molecular Cytogenetics

4

Abstract

This chapter delves into advancements in somatic chromosome preparation in molecular cytogenetics, specifically the enzymatic maceration and air drying (EMA) method. Traditional plant cytogenetics has faced challenges due to the thick cell walls and cytoplasm of plant cells, hindering high-quality chromosome preparation and molecular analysis. The EMA technique, developed since the 1970s, overcomes these limitations by removing cytoplasmic debris to free the chromosomes, thereby allowing detailed examination and molecular insights. Grounded in the foundational work of Emil Heitz, the chapter highlights the significance of heterochromatin and euchromatin in understanding chromosome morphology, further enhanced by advancements in DNA analysis. The repetitive DNA sequences identified within plant genomes serve as essential molecular markers for studying genetic diversity and genome evolution. This chapter also underscores the dynamic nature of centromeres and their varied DNA organization across species. By utilizing EMA-based preparations with Giemsa staining and fluorescent dyes, researchers can visualize distinct morphological and molecular characteristics of chromosomes, including the identification of heterochromatic regions and nucleolar organizing regions (NOR). The chapter concludes with a detailed EMA-based protocol tailored for specific species, offering valuable resources for educational purposes and advancing the field of plant molecular cytogenetics. Through this comprehensive overview, the chapter aims not only to enhance understanding but also to facilitate further research and exploration in chromosome analysis and cytogenetic techniques.

T. B. Jha, M. Halder, *Classical and Molecular Plant Cytogenetics*,
https://doi.org/10.1007/978-981-95-8265-5_4

4.1 Introduction

With advancements in our understanding of genes, chromatin, and the unique genomic structures of chromosomes, classical plant cytogenetics has recognized its limitations and started adopting modern chromosome preparation and analysis methodologies. Researchers have identified thick cell walls and cytoplasm as primary constraints in obtaining high-quality chromosome preparations and performing advanced analyses. Furthermore, procedural limitations make these preparations unsuitable for further experiments. In contrast, while human chromosome analysis began later, it evolved rapidly because human cells lack cell walls. By the early 1970s, the preparation of plant chromosomes had incorporated enzymatic maceration, yielding promising results. Gradually, plant chromosome preparation and analysis began to integrate with a basic molecular cytogenetics approach known as the enzymatic maceration and air drying (EMA) method (Schweizer et al. 1978). This evolution enabled the preparation of high-quality chromosomes, free from cell walls and cytoplasm, thereby facilitating a detailed examination of chromosome morphology as well as their analysis at the molecular level. Since 1970, the EMA method for plant chromosome preparation with Giemsa staining has become fundamental to molecular cytogenetic techniques. Its development has made significant strides, particularly in the use of non-fluorescent and fluorescent stains, fluorescent in situ hybridization (FISH), genomic in situ hybridization, and spectral karyotyping.

Molecular cytogenetics benefited tremendously from the work of Emil Heitz, a German-Swiss botanist and geneticist. In 1928 and 1929, in the first half of the twentieth century, he contributed important conceptual ideas about chromosome morphology long before the work of others. Through his improved chromosomal staining techniques, Heitz observed that certain regions of mitotic chromosomes stained more densely than others, which led him to introduce the terms "heterochromatin" and "euchromatin." He proposed that heterochromatin consists of chromosome regions that do not carry genes or are associated with genetically inactive areas (Heitz 1928, 1929). This concept remains foundational in today's understanding of heterochromatin. Heitz also recognized the connection between satellite chromosomes and nucleolus formation. His observations are now classified as constitutive and facultative heterochromatin. The discovery of DNA and advancements in molecular biology have shed light on the molecular nature of heterochromatin, revealing that it is primarily composed of repetitive DNA sequences. Research has shown that many plant genomes comprise these repetitive sequences (Biscotti et al. 2015). These sequences evolve more rapidly than coding sequences and serve as useful molecular markers for analyzing genetic diversity and studying genome evolution (Kalendar and Schulman 2006). Repetitive DNA sequences have been employed as cytogenetic probes to investigate chromosomal organization in plants (Cuadrado et al. 2008), as karyotyping markers (Han et al. 2008), and for cytogenetic comparative analysis among closely related species (Zhang et al. 2016). The chromosomal locations of repetitive DNA sequences found in pericentromeric or subtelomeric regions or dispersed throughout chromosome regions (Kubis et al.

1998; Setiawan et al. 2019) are reported to be species-specific. The centromere in every eukaryotic chromosome is recognized as a multifunctional dynamic complex (Heslop-Harrison and Schwarzacher 2013). It is reported that despite mediating the evolutionarily conserved function of directing chromosome segregation, they exhibit surprising diversity in their DNA sequence organization (Tong et al. 2019; Barbosa et al. 2022). DNA sequence analysis in rice, maize, and *Arabidopsis* has shown that centromeres are highly repetitive DNA sequences with considerable size variation and sequence divergence (Ma et al. 2007). Plant molecular cytogenetics, utilizing EMA-based chromosome preparations followed by Giemsa staining and DNA base-specific fluorescent dyes, has opened a new window for revealing the distinct morphological and molecular features directly on the individual chromosomes (Yamamoto and Mukai 1989; Wang et al. 2023; Jha et al. 2024a, b). This approach helps to identify the DNA nature of constitutive heterochromatic regions throughout the chromosome length, in addition to the marked primary and secondary constricted areas. This is the only direct visualization method through which anyone can see GC-rich, AT-rich, or GC-AT neutral heterochromatic regions on chromosomes under a microscope. This advanced method can successfully confirm the landmark position and molecular nature of secondary constrictions in the nucleolar organizing regions (NOR) within any plant karyotype.

This chapter provides an EMA-based species-specific chromosome preparation protocol, along with conceptual information and preparation methods for Giemsa and other commonly used fluorescent dyes and buffers. The application of the protocol has been elaborated with several class teaching materials and is ready to be applied to any plant material with minor modification. The results of our experiments are expected to inspire teachers and researchers, enriching their work, and to offer hands-on training to their students in the field of molecular cytogenetics. Our efforts are to reconfirm that the age-old discipline of plant cytogenetics has evolved from a classical to a molecular foundation.

4.2 Enzymatic Maceration and Air Drying (EMA) Method

4.2.1 Preparation of Enzyme Cocktail

(i) Prepare 100 mL of enzyme mixture by mixing 2 gm of Cellulase Onozuka RS, 1.5 gm of Macerozyme R-10, and 0.3 gm of Pectolyase Y-23.
(ii) Prepare a stock solution of 200 mM Ethylenediaminetetraacetic acid (EDTA) by dissolving 745 mg of EDTA in 10 mL of distilled water.
(iii) Add 500 μL of the EDTA stock solution to the 100 mL enzyme mixture to prepare a working enzyme cocktail solution to use in the EMA-based chromosome preparation technique.
(iv) Finally, adjust the pH of the enzyme mixture to 4, and then filter it.
(v) Distribute the mixture into 1.5 mL Eppendorf tubes for storage at −20 °C.

4.2.2 Steps of the EMA Method

The primary goal of plant cytogenetics is to achieve high-quality chromosome preparation. The EMA technique is a contemporary method that facilitates the removal of cell walls and cytoplasm. For somatic chromosome preparation, young and healthy roots are regarded as the most suitable material. Roots can be obtained through seed germination, induction from bulbs, or from plants cultivated in pots. The following are important steps of EMA-based chromosome preparation.

(i) *Germination or Induction of Roots:* Begin by soaking fresh seeds or discs from bulbs in water, then place them on moist filter paper, sawdust, or sand. To enhance germination, keep these materials in a dark environment.

(ii) *Germination Time:* The duration required for germination or root induction varies depending on the specific species and environmental conditions. However, using a temperature-controlled incubator can facilitate year-round root development.

(iii) *Pretreatment Process:* Collect healthy root tips and clean them thoroughly with water. Subsequently, treat the tips with suitable pretreatment chemicals at the specified temperature for the designated duration (see Chap. 3 for details).

(iv) *Fixation Process:* Rinse the roots in water, and then transfer them to a freshly prepared fixative solution, which is a 1:3 mixture of acetic acid and methanol or acetic acid and alcohol. This fixation should occur overnight. After fixation, proceed with chromosome preparation or store the roots at −20 °C for future use.

(v) *Cold Treatment:* Place the fixed root tips in water and store them in a refrigerator at 4 °C for at least 3–5 hours, preferably overnight, depending on the species.

(vi) *Excising Root Tips:* After the cold treatment, carefully remove the roots from the water. Using fine forceps and a razor blade on a glass slide, excise the tip portion (1–2 mm) of each root. Place the excised tips into an Eppendorf tube containing 100 μL of an enzyme mixture, maintaining a 1:1 ratio of enzyme to distilled water.

(vii) *Enzymatic Digestion:* Proceed with the enzymatic digestion of the micro root tips at 37 °C, which is a critical step in the process. The duration of digestion may vary depending on the species, ranging from 40 to 120 min. It is also important to consider that the required digestion time can depend on the age of the enzyme mixture.

(viii) *Stopping Enzymatic Digestion:* Once the enzymatic digestion is complete, carefully add a few drops of water to the enzyme mixture. Using a pasture pipette, transfer the root tips into a petri dish containing water, where they should be incubated for 15–20 min.

(ix) *Maceration of Root Tips:* A pasture pipette is used to place a single root tip onto a clean slide and gently blot away any excess water that remains. Small fine-tipped forceps or tweezers and a freshly prepared fixative solution, made of one-part acetic acid and three parts methanol, are then used to uniformly spread the enzyme digested root tip on the glass slide. Proper maceration ensures that the cells become invisible on the side.

Alternatively, add a drop of fixative onto the root tip, cover it with a cover glass, and gently spread the root cells using a squash technique.

(x) *Air drying of slide:* For slides without cover glasses, allow them to air dry completely for at least one hour to overnight.

For slides with cover glasses, an additional step is required to detach the cover glass. Place the slide on dry ice for a few minutes, and then carefully remove the cover glass with a blade, taking care not to lose any cells that may be attached to it.

4.2.3 Advantages of the EMA Method

(i) In botanical studies, thick cell walls can impede the preparation of high-quality chromosome samples. The EMA method effectively breaks down these cell walls and removes other cytoplasmic debris, facilitating the direct spreading of chromosomes onto a glass slide, with or without a cover slip.
(ii) This technique allows for the observation of the detailed structures of chromosomes against a background devoid of cytoplasm, making sure the chromosomes are evenly spread across the glass slide.
(iii) The prepared slides can be used for a variety of staining techniques, including Giemsa staining, fluorescent banding, and FISH.
(iv) With the ability to identify individual chromosomes, we can now pinpoint DNA base-specific heterochromatin blocks along the entire length of the chromosomes.
(v) The enzymatic maceration method has yielded clearly defined chromosome samples, advancing chromosomal analysis from traditional methods to molecular techniques.

4.3 Buffer Solutions Used in Molecular Cytogenetics

Buffer solutions consist of weak acids and their salts or weak bases and their salts. They help maintain a nearly constant pH level, even when strong acids or bases are added. The concentration and pH of the buffer provide a stable environment for staining. If the buffer is too acidic, the stain will appear too red, and the nuclei will be too faint. Conversely, if the buffer is too basic, the stain will appear too blue, and the details of the cytoplasm will be indistinct.

4.3.1 Preparation of McIlvaine Buffer (pH 7)

(i) Dissolve 1.33 gm of citric acid ($H_3C_6H_5O_7$) in 50 mL of distilled deionized water to prepare 0.1 M citric acid solution.
(ii) Dissolve 7.33 gm of dibasic sodium phosphate (Na_2HPO_4) in 200 mL of distilled deionized water to prepare 0.2 M dibasic sodium phosphate solution.
(iii) In a separate container, mix 36 mL of the citric acid solution with 164 mL of the sodium phosphate solution to make a total of 200 mL of McIlvaine buffer.
(iv) Appropriately label the container containing the prepared buffer and store at −20 °C.

4.3.2 Preparation of McIlvaine Buffer with $MgCl_2$ (pH 7)

(i) Prepare a stock solution of 5 mM $MgCl_2$ or $MgSO_4{\cdot}7H_2O$.
(ii) Add 500 μL of 5 mM $MgCl_2$ or $MgSO_4{\cdot}7H_2O$ solution to 100 mL of McIlvaine buffer.
(iii) Store McIlvaine buffer plus $MgCl_2$ solution at −20 °C.

4.3.3 Preparation of Phosphate Buffer (pH 7)

Phosphate buffer is an essential solution for preparing diluted Giemsa stains. It comprises disodium hydrogen phosphate (Na_2HPO_4) and potassium dihydrogen phosphate (KH_2PO_4). The process for preparing 100 mL of phosphate buffer is outlined below.

(i) Add 2.39 gm of Na_2HPO_4 to 40 mL of double-distilled water and make up a volume of 50 mL.
(ii) In a separate container, add 0.9 gm of KH_2PO_4 to 40 mL of double-distilled water and make up a volume of 50 mL.
(iii) Prepare a stock solution of phosphate buffer by mixing both solutions to create 100 mL of phosphate buffer.
(iv) Adjust the pH of the solution to 7 and store it in the refrigerator.

To prepare a 2% Giemsa solution, combine 1:1 phosphate buffer and distilled water, then add the 2% Giemsa solution.

4.4 Stains and Counterstains Used in Molecular Cytogenetics

Giemsa, DAPI, CMA, Propidium Iodide, Distamycin (Counterstain), and Actinomycin D (Counterstain) are the most common and widely used non-fluorescent and fluorescent stains and counterstains for molecular cytogenetics. The following sections deal with these stains and the Table 4.1 showing important characteristic features of these stains.

Table 4.1 List of common fluorescent stains and counterstains with salient features applied in plant chromosome research

Sl. No.	Stain/ counter stain	Chemical nature	DNA binding chemistry	DNA base specificity	Absorption max. (nm)	Fluorescence max. (nm)	Emission color
1	DAPI	Indole group of antibiotics	Minor groove binder	AT	355	450	Blue
2	CMA	Aureolic acid group antibiotics	Minor groove binder	GC	430	570	Yellow-green
3	PI	Fluorescent stain	Intercalation binder	Total DNA	535	617	Red
4	Mithramycin	Aureolic acid group of antibiotics	Minor groove binder	GC	430	580	Yellow
5	TOTO-1	–	–	Total DNA	514	531	Red
6	Methyl green	Cationic aniline dye	Major groove binder	AT	638	Non-fluorescent	
7	Distamycin A (DA)	Oligopeptide antibiotic	Intercalation binder	AT	340	Non-fluorescent	
8	Actinomycin D (AMD)	Non-fluorescent anticancer drug	Intercalation binder	GC	455	±Non-fluorescent	

4.4.1 Non-fluorescent Stain: Giemsa

German scientist Gustav Giemsa (1904) introduced this stain in the early twentieth century to study the malaria parasite *Plasmodium*. It is a non-fluorescent, differential dye usually used in standard bright-field microscopy. Giemsa belongs to the Romanowsky group of neutral stains, which are composed of a mixture of acidic and basic dyes, primarily eosin Y (acidic) and methylene blue and azures (basic). Its molecular formula is $C_{14}H_{14}ClN_3S$ (Fig. 4.1). For many decades, the use of Giemsa stain was restricted to bacteria, yeast, and animal cells (Barcia 2007). However, since the 1970s, the stain has demonstrated remarkable staining properties for both animal and plant chromosomes. Giemsa staining marked a new era in plant chromosome research by significantly enhancing chromosome analysis and facilitating the identification of individual chromosomes. This technique laid the groundwork for advancements in molecular cytogenetics. In addition to routine chromosome staining, Giemsa can produce a specific banding pattern known as G-banding, which preferentially stains heterochromatic regions on chromosomes. This banding pattern was first observed in mouse chromosomes and later in human chromosomes (Pardue and Gall 1970; Seabright 1971). Vosa and Marchi (1972) were the first to report G-banding in plant chromosomes, and it has since become the most widely used non-fluorescent and stable banding method for chromosomes. Interestingly, Giemsa can also produce R- and C-bands when the chromosome processing protocol is modified. R-banding generates positive bands in GC-rich euchromatin sequences, while C-banding highlights the AT-rich centromeric regions, which consist of constitutive heterochromatin.

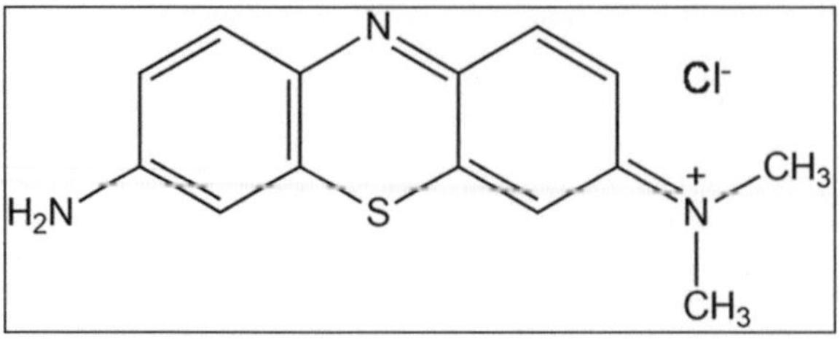

Fig. 4.1 Molecular structure of Giemsa

The stain can be prepared in the laboratory as described below, or 2% Giemsa solution is readily available from Merck, Germany, in the market.

4.4.1.1 Preparation of Giemsa Stain

There are various methods available for preparing Giemsa stain. The following procedure can be used for easier preparation.

- *Requisition:* (i) Giemsa powder, (ii) Glycerine, (iii) Methanol, (iv) conical flask, and (v) heater.
- *Preparation of Stock Solution:*
  (i) Dissolve 3.8 gm of Giemsa powder into 250 mL of methanol.
  (ii) Heat the solution up to ~60 °C.
  (iii) Slowly add 250 mL of glycerin to the solution at the time of heating and keep the solution for 24 hours at room temperature.

  Filter the solution using Whatman filter paper no.1 in an amber-colored bottle, label it, and store it in a cool, dark place with a firm stopper. The solution should be left to stand for about 1–2 months before use.

- *Preparation of Working Solution:*
  Add 10 mL of stock solution to 80 mL of distilled water and 10 mL of methanol.

4.4.1.2 Principle of Giemsa Stain

Giemsa is a neutral, non-fluorescent stain composed of methylene blue, azure, and eosin Y. The first two components are basic dyes that target the acidic nucleus and chromosomes, resulting in a blue-purple coloration. Eosin, an acidic dye, acts on the cytoplasm and cytoplasmic granules, producing a red-orange hue. This stain is particularly effective on slides containing air-dried chromosome preparations that have been fixed with methanol. This non-fluorescent stain can induce bands on the plant chromosomes when chromosomes are specially treated with trypsin and urea before adding Giemsa solution. This banding pattern is referred to as G-banding (Ruiyang et al. 1994). It specifically targets the phosphate groups in DNA and preferentially attaches to repetitive DNA sequences rich in adenine and thymine (A-T).

4.4.1.3 Steps of Giemsa Staining and Documentation

(i) Slides prepared by EMA methods are placed inside a coupling jar containing 2% of Giemsa solution for the required duration (which may vary by species, from 10 to 40 min) at room temperature.

(ii) Rinse the slides with ddH_2O for 4–5 times. Blot off excess water droplets using an air drier, and allow the slides to dry completely before observing them under a bright-field microscope.
(iii) Add 40–50 μL of xylene to the slide, and cover it with a rectangular cover glass measuring 24 mm × 32, 40, or 50 mm. Examine xylene-mounted slides under a standard branded microscope equipped with a CCD camera, computer, and software for chromosomal documentation and analysis.
(iv) Determine the 2n chromosome number by counting more than 20 well-scattered metaphase plates and analyzing karyotypes using a minimum of five to ten Giemsa-stained, cytoplasm-free, well-scattered metaphase plates with distinct chromosomal configurations.
(v) Prepare a data sheet to record the exact position of the metaphase plates, and take photographs as part of the documentation process.

4.4.2 Fluorescent Dyes or Fluorochromes

In 1871, Adolf von Baeyer first synthesized the fluorescent dye known as fluorescein (Wienert et al. 1993). Fluorescent dyes are organic molecules called fluorochromes that can undergo fluorescence. In fluorescent microscope, when these fluorochrome molecules are exposed to specific excitation wavelengths, they emit distinct excitation wavelengths in the form of fluorescence. The excitation wavelength indicates the amount of energy required to excite the fluorochrome. In contrast, the emission wavelength represents the energy of the photon emitted by the fluorochrome. Typically, the emission wavelength is longer than the excitation wavelength. Figure 4.2 illustrates the excitation and emission spectra of DAPI stain. All fluorescent dyes are categorized as nucleic acid dyes and are used to locate AT- or GC-rich heterochromatic and euchromatic sequences directly on chromosomes. The binding sites of common DNA-specific fluorochromes in plant chromosome research are displayed in Fig. 4.3. For example, propidium iodide, distamycin A, and actinomycin D act as intercalating dyes, interacting between two nitrogen bases. Whereas DAPI and CMA interact at the minor groove of the DNA helix.

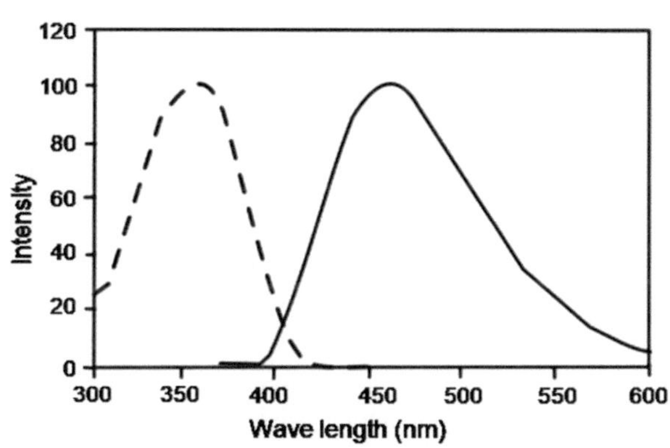

Fig. 4.2 Excitation and emission spectra of DAPI

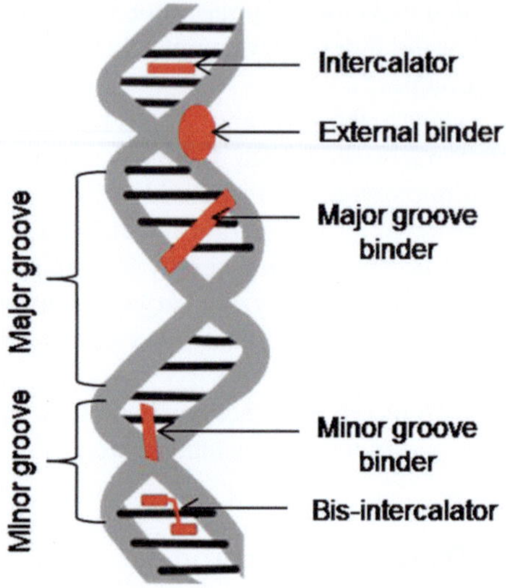

Fig. 4.3 Schematic diagram showing binding sites of common DNA-specific fluorochrome dyes and counterstains applied in plant cytogenetics

Fluorochrome banding was introduced by Caspersson and his colleagues in plant chromosomes in 1968 and was later successfully applied to human and other animal chromosomes with necessary modifications (Fedak and Kim 2008). By using quinacrine mustard as a banding agent, they achieved differential yellow fluorescence on chromosomes, which became known as Q-banding. Since 1970, chromosome banding has become more practical and useful for identifying specific chromatin regions. Fluorochromes can distinctly induce banding in constitutive heterochromatic regions (heterochromatic banding), throughout the euchromatic length of the chromosomes (euchromatic banding), and in the nucleolar organization region (NOR banding). Some commonly known fluorochromes used in plant chromosome analysis are discussed below.

4.4.2.1 4′,6-Diamidino-2-Phenylindole (DAPI)

DAPI is a synthetic photostable fluorescent DNA stain with the chemical formula of $C_{16}H_{15}N_5.2HCl$ (Fig. 4.4) and molecular weight of 350.25 gm mol^{-1}. The stain exhibits significant blue fluorescence at excitation and emission wavelengths of

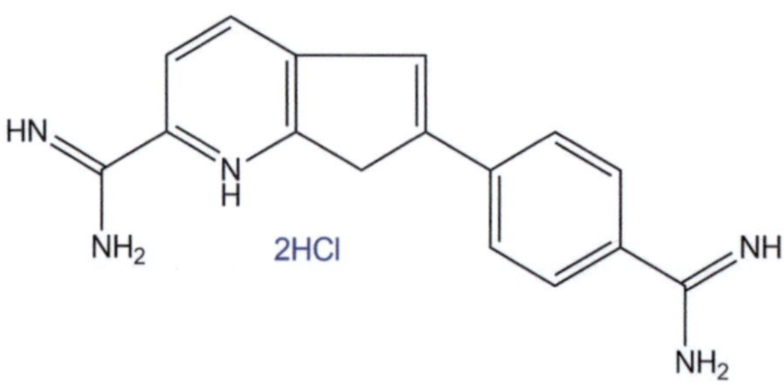

Fig. 4.4 Molecular structure of 4′,6-Diamidino-2-Phenylindole (DAPI)

358 nm and 463 nm, respectively. DAPI binds to the minor groove of AT-rich DNA sequences. It is also invisible to regular transmission microscopy. DAPI is effective for both fixed and living cells.

Preparation of DAPI Solution

To prepare a DAPI stock solution, you can use either double-distilled water or McIlvaine buffer. Begin by dissolving 1 mg of DAPI powder in 5 mL of double-distilled water or McIlvaine buffer. This will create a stock solution with a concentration of 100 μg mL^{-1}. Once prepared, aliquot the stock solution and store it at −20 °C.

For your working solution, dilute the stock to achieve a concentration of 0.1 to 0.5 μg mL^{-1}, and store this working solution at −20 °C.

Steps of DAPI Staining and Documentation

(i) The Giemsa-destained air-dried slides were utilized for DAPI staining.
(ii) The slides were incubated for 30 min in McIlvaine buffer at room temperature.
(iii) Each slide was transferred individually from the McIlvaine buffer, and excess water droplets were carefully blotted off using a blower in semi-dark conditions.
(iv) A 60–70 μL volume of a 0.1–2 μg mL^{-1} DAPI solution was added, and the slide was covered with a long rectangular cover glass, followed by incubation in the dark for 10–30 min (note that concentration and time may vary depending on species).
(v) The cover glass was gently removed in McIlvaine buffer and allowed to rest for 10 min.
(vi) Each slide was transferred individually from the McIlvaine buffer, and excess water droplets were carefully blotted off using a blower in semi-dark conditions.
(vii) A 60–70 μL volume of a 0.25 mg mL^{-1} solution of the counterstain Actinomycin D (AMD) was added, covered with a long rectangular cover glass, and kept in the dark for 5–20 min.
(viii) The cover glass was gently removed in McIlvaine buffer and allowed to rest for 10 min.
(ix) Each slide was transferred individually from the McIlvaine buffer, and excess water droplets were carefully blotted off using a blower in semi-dark conditions.
(x) A 60–70 μL volume of glycerol was added, covered with a cover glass, and the slide may be directly observed under a fluorescent microscope fitted with a DAPI filter or stored at 4 °C overnight for maturation.
(xi) The same metaphase plate captured with the Giemsa stain can be recorded using a CCD camera attached to the microscope and associated software.
(xii) After removing the cover glass, the slides are rinsed with water, excess water is blotted off with a blower, and then they are destained in 70% methanol or a 1:3 acetic alcohol solution for at least one hour.

4.4.2.2 Chromomycin A3 (CMA)

Chromomycin A3 (CMA) is a naturally extracted product derived from *Streptomyces griseus* and *Streptomyces cavourensis*. This antibiotic exhibits antibacterial, antifungal, and antitumor activities, and it is widely used as a fluorescent stain for DNA. The molecular formula of CMA is $C_{57}H_{82}O_{26}$ (Fig. 4.5) with a molecular weight of 1183.2 gm mol^{-1}. CMA is a GC-specific fluorescent stain, with an excitation wavelength of 445 nm and an emission wavelength of 575 nm.

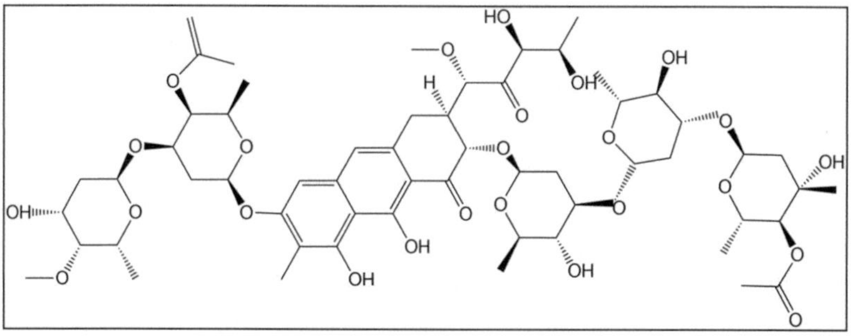

Fig. 4.5 Molecular structure of Chromomycin A3

Preparation of CMA Solution

(i) To prepare the stock solution, dissolve 5 mg of CMA in 5 mL of McIlvaine buffer containing $MgCl_2$.
(ii) Allow the solution to sit undisturbed at 4 °C overnight. Store the stock solution at −20 °C.
(iii) For the working solution, prepare a concentration between 0.1 and 0.5 mg mL^{-1} of CMA.
(iv) Aliquot the working solution and store it at −20 °C.

The time and concentration of CMA staining vary by species. CMA generally acts as a positive and confirmatory stain for GC-rich repetitive DNA sequences present in NOR regions, subtelomeric and other satellite DNA locations. On very few occasions, it stains the primary constriction regions (*Trigonella* sp.).

Steps of CMA Staining and Documentation

(i) Utilize the same DAPI-destained air-dried slides for the CMA staining procedure.
(ii) Incubate the slides in McIlvaine buffer at room temperature for 30 min.
(iii) Remove the buffer and place the slides back in a McIlvaine buffer containing 5 mM $MgCl_2$ for an additional 15–20 min at room temperature.
(iv) Remove each slide individually and gently blot off excess water droplets using a blower in semi-dark conditions.

(v) Add 60–70 μL of a CMA solution (0.1–0.5 mg L^{-1}), cover the slide with a long rectangular cover glass, and incubate in the dark for 40–100 min (note that the concentration and time may vary depending on the species).
(vi) Carefully take off the cover glass while submerged in McIlvaine buffer with 5 mM $MgCl_2$ and let the slide rest for 10 min.
(vii) Remove each slide individually and gently blot off excess water droplets using a blower in semi-dark conditions.
(viii) Add 60–70 μL of a counterstain solution, Distamycin A (DA) at 0.25 mg mL^{-1}, cover the slide with a long rectangular cover glass, and keep in the dark 20–30 min (this step is optional; timing may need to be optimized for each species).
(ix) Repeat step vi and then step iv again.
(x) Add 60–70 μL of glycerol, cover the slide with a cover glass, and store the slides in the dark at 4 °C for at least 48 h for maturation.
(xi) Capture the same metaphase plate previously obtained with Giemsa and DAPI staining using a fluorescent microscope equipped with a CMA filter cassette and a CCD camera connected to a computer.
(xii) After removing the cover glass, wash the slides with water, gently blot off excess water with a blower, and destain in 70% methanol or 1:3 acetic alcohol for a minimum of one hour.

4.4.2.3 Combined CMA-DAPI Staining and Documentation

(i) The Giemsa-destained air-dried slides were utilized for combined CMA-DAPI staining. The slides were incubated for 30 min in McIlvaine buffer at room temperature.
(ii) Each slide was removed individually, and excess water droplets were carefully blotted off using a blower in semi-dark conditions.
(iii) Add 50–60 μL of 0.1 mg mL^{-1} of Distamycin A solution in McIlvaine buffer on each slide, cover the slide with a long rectangular cover glass, and incubate in the dark for 10 min (note that the concentration and time may vary depending on the species).
(iv) Put the slides inside the Coplin jar containing McIlvaine buffer with 5 mM $MgCl_2$, carefully remove the cover glass, and let the slide rest for 10 min.
(v) Each slide was removed individually, and excess water droplets were carefully blotted off using a blower in semi-dark conditions.
(vi) Add 60–70 μL of 0.1–0.5 mg mL^{-1} of CMA in McIlvaine buffer with 5 mM $MgCl_2$. Cover the slide with a long rectangular cover glass, and incubate in the dark for 40–120 min (note that the concentration and time may vary depending on the species).
(vii) Step VI was repeated.
(viii) Add 60–70 μL of 0.1–2 μg mL^{-1} of DAPI in McIlvaine buffer, cover the slide with a long rectangular cover glass, and incubate in the dark for 10–30 min (note that the concentration and time may vary depending on the species).
(ix) The cover glass was gently removed in McIlvaine buffer and allowed to rest for 10 min.

(x) Put the slides inside the Coplin jar containing McIlvaine buffer with 5 mM $MgCl_2$, carefully remove the cover glass, and let the slide rest for 10 min.
(xi) Add 60–70 μL of glycerol, cover the slide with a cover glass, and store the slides in the dark at 4 °C for at least 48 h for maturation.
(xii) Capture the same metaphase plate previously captured through Giemsa staining using a fluorescent microscope equipped with a CMA filter cassette and a CCD camera connected to a computer. Use the DAPI filter first, and then move on to the CMA filter for capturing the photographs.
(xiii) After removing the cover glass, wash the slides with water, gently blot off excess water with a blower, and destain in 70% methanol or 1:3 acetic alcohol for a minimum of one hour.

4.4.2.4 Propidium Iodide (PI)

Propidium iodide (PI) is a light-emitting intercalating agent used for staining cells. It serves as a DNA stain in flow cytometry to assess cell viability and analyze DNA content during the cell cycle. Additionally, it can be employed in microscopy to observe the nucleus and other organelles that contain DNA. Since PI cannot cross the cell membrane, it effectively differentiates between necrotic, apoptotic, and healthy cells. Furthermore, PI is an RNA-binding protein, which necessitates treatment with nucleases to distinguish between DNA and RNA staining. It is a water-soluble compound. Prepare 100 μg mL^{-1} stock solutions in water and preserve them at −20 °C. A working solution before use is diluted to 0.1–5 μg mL^{-1} with 1× phosphate buffer. PI is not effective in live cells, but it is commonly used on dead cells or fixed tissues.

4.4.2.5 Combination of PI—DAPI (CPD) Staining and Documentation

(i) Preparation of the plant chromosomes for CPD staining requires passing through RNase A, pepsin, and formaldehyde treatment (She et al. 2006).
(ii) The slides were then stained with a mixture of PI and DAPI.
(iii) Stock solutions of 1 mg mL^{-1} DAPI and 1 mg mL^{-1} PI as described above were diluted with distilled water to 10% and 2% (v/v), respectively.
(iv) μL of each were added to 200 μL of Vectashield (Vector Laboratories, California, USA), to give a final concentration of approximately 3 μg mL^{-1} DAPI and 0.6 μg mL^{-1} PI (Andras et al. 2000).
(v) A mixture of PI and DAPI was added to the slide containing chromosome plates and covered with a cover glass.
(vi) After carefully bottling the excess fluid, the edges of the cover glass were sealed with rubber solution (Dunlop Adhesives, Birmingham, UK; Andras et al. 2000).
(vii) For observation and documentation, a fluorescence microscope equipped with a cooled CCD camera and stain-specific PI and DAPI filters can be used.

4.4.3 Counterstaining

It is a technique that is used to enhance or modify banding patterns with fluorochromes that bind and fluoresce uniformly throughout the chromosome. It is also used to strengthen banding patterns that lack high resolution. Counterstaining is performed with fluorescent or non-fluorescent chemicals after the application of a primary fluorescent stain. Distamycin A (DA) and Actinomycin D (AMD) are two widely used counterstains, besides others (Schweizer 1981).

4.4.3.1 Distamycin A

Distamycin A (DA) is an oligopeptide antibiotic biosynthesized by *Streptomyces distallicus*. The antibiotic targets the minor groove of DNA, preferentially binding to DNA sequences containing five consecutive A-T pairs. In fluorescent banding with CMA, DA is used as a counterstain to enhance the primary stain's staining intensity. DA is dissolved in Mcllavaine's buffer, and 0.1–0.2 mg mL^{-1} is used as the working solution. The stock solution needs to be stored at –20 °C.

4.4.3.2 Actinomycin D (AMD)

Actinomycin D, also known as Dactinomycin, is a potent antibiotic isolated from *Streptomyces* species and the oldest anticancer drug. It is a non-fluorescent DNA ligand that exhibits a base-pair binding affinity complementary to that of the primary stain (Schweizer 1976; Jorgenson et al. 1978; Sahar and Latt 1978). It is generally used as a counterstain with the fluorescent dye DAPI. Actinomycin D is either directly dissolved in Macilvaine buffer or dissolved in one to two drops of methanol and then added to the buffer. The working solution is 0.2–0.25 mg mL^{-1}. It can be kept in a refrigerator for several months.

4.5 Species-Specific Case Studies (Protocols, Key Features of Fluorescence and Photographic Plates)

The following pages present our findings from fluorescent banding studies conducted on the valuable and classic plant materials mentioned in Chap. 3 and listed below. These results, supported by species-specific chromosome plates stained with non-fluorescent Giemsa and the DNA base-specific fluorescent stains DAPI and CMA, were developed through our instructional endeavors. The general staining protocol for Giemsa, DAPI, and CMA is detailed on the previous pages. The following pages provide species-specific enzymatic digestion and fluorescent staining protocols (including time and concentration) for eight class teaching species. Additionally, key fluorescent features and fluorescent karyotype formulas, based on fluorescent stains, are provided for each species.

Designation of chromosome type based on fluorescent banding pattern has no standard protocol like Leaven classification in a non-fluorescent karyotype.

Researchers can designate a symbol based on their choice. We have designated chromosomes based on fluorescent banding types as follows: "A" for chromosomes with CMA-positive centromeric region, "B" for DAPI-positive centromeric region, "C" for chromosomes with CMA-positive terminal regions, "D" for chromosomes with CMA-positive secondary constriction regions, "E" for chromosomes with CMA-positive terminal and secondary constriction regions, CMA- and DAPI-neutral chromosomes are designated with "F," "G" for chromosomes with CMA-positive centromeric and secondary constriction regions, and "H" for chromosomes having DAPI-positive centromeric regions and CMA-positive secondary constriction regions.

We hope these species-specific protocols for EMA-based chromosome preparation and fluorescence staining will inspire and enrich our readers, encouraging them to apply this method to the listed materials and to other plant species.

(i) *Allium cepa* L. (Onion), $2n = 16$ chromosomes (Fig. 4.6).
(ii) *Allium sativum* L. (Garlic), $2n = 16$ chromosomes (Figs. 4.7, 4.8, and 4.9).
(iii) *Trigonella foenum-graecum* L (Methi), $2n = 16$ chromosomes (Figs. 4.10 and 4.11).
(iv) *Lens culinaris* Medik. (Musur), $2n = 14$ chromosomes (Figs. 4.12 and 4.13).
(v) *Aloe vera* L., $2n = 14$ chromosomes (Figs. 4.14 and 4.15).
(vi) *Nigella sativa* L. (Kalo jeera), $2n = 12$ chromosomes (Figs. 4.16 and 4.17).
(vii) *Vicia faba* (Broad bean), $2n = 12$ chromosomes (Fig. 4.18).
(viii) *Albuca virens/Ornithogalum virens,* $2n =6$ chromosomes (Fig. 4.19).

4.5.1 Species: *Allium cepa* L. (Fig. 4.6)

Protocol parameters	Range of concentration and duration of treatment	Optimum concentration and duration of treatment for this germplasm
Time of enzymatic digestion at 37 °C	65–75 min	70 min
Concentration of Giemsa stain	1.5–2%	2%
Duration of Giemsa staining	10–15 min	10 min
Concentration of DAPI stain	0.1–0.2 μg mL^{-1}	0.2 μg mL^{-1}
Duration of DAPI staining	10–15 min	10 min
Concentration of Actinomycin D (AMD) stain	0.1–0.25 mg L^{-1}	0.25 mg L^{-1}
Duration of (AMD) staining	5–10 min	6 min
Concentration of CMA stain	0.1–0.5 mg mL^{-1}	0.5 mg mL^{-1}
Duration of CMA staining	60–80 min	80 min
[a]Concentration of distamycin A (DA)	0.1–0.25 mg L^{-1}	Not used
[a]Duration of DA staining	10–30 min	Not used

[a]Optional

Key Fluorescent Feature (Fig. 4.6)

Analysis of more than 25 metaphase chromosome plates stained with contrasting DNA base-specific CMA and DAPI fluorochrome in this germplasm revealed the following fluorescent feature:

(i) The terminal region of all $2n$ = 16 chromosomes shows 32 CMA^{+ve} $DAPI^{-ve}$ signals.
(ii) Chromosome pair sixth confirms carrying a terminal secondary constriction with CMA $^{+ve}$ DAPI $^{-ve}$ signals.
(iii) No DAPI-positive signal was observed.
(iv) Based on CMA and DAPI banding pattern the model fluorescent karyotype remains 14A + 2B.

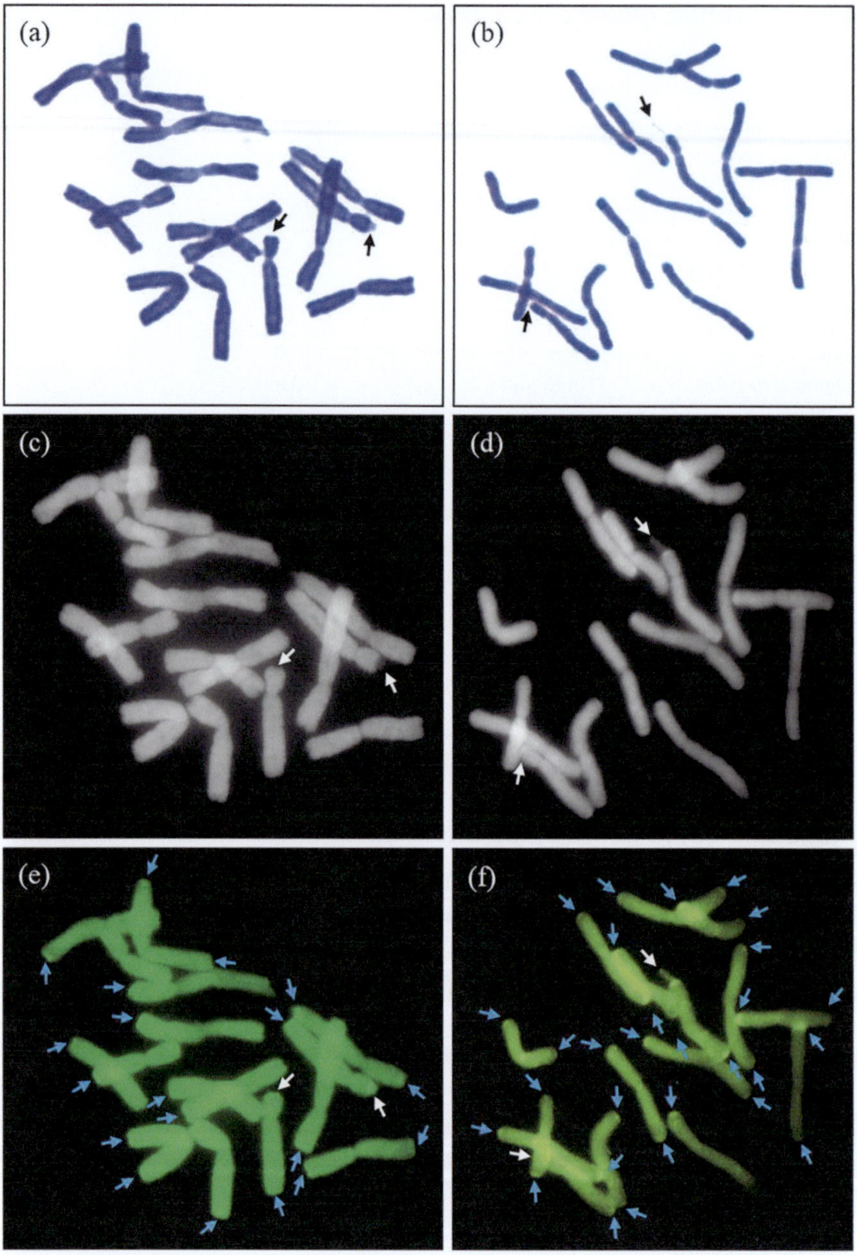

Fig. 4.6 EMA presents somatic chromosome preparations of *Allium cepa* ($2n = 16$ chromosomes) that have been stained using Giemsa, DAPI, and CMA methods. The cytoplasm-free Giemsa-stained chromosomes exhibit distinct morphological characteristics (**a**, **b**); DAPI-stained metaphases are shown in (**c**, **d**); and CMA-stained metaphases are illustrated in (**e**, **f**). Black and white arrows highlight the secondary constrictions located at the terminal regions of one chromosome pair, while blue arrows indicate CMA-positive bands found at both terminal and intercalary positions across all chromosomes, including one pair of SAT-bearing chromosomes

4.5.2 Species: *Allium sativum* L. (Figs. 4.7, 4.8, 4.9)

Protocol parameters	Range of concentration and duration of treatment	Optimum concentration and duration of treatment for this germplasm
Time of enzymatic digestion at 37 °C	65–75 min	70 min
Concentration of Giemsa stain	1.5–2%	2%
Duration of Giemsa staining	10–15 min	10 min
Concentration of DAPI stain	0.1–0.2 μg mL^{-1}	0.2 μg mL^{-1}
Duration of DAPI staining	10–15 min	10 min
Concentration of Actinomycin D (AMD) stain	0.1–0.25 mg L^{-1}	0.25 mg L^{-1}
Duration of (AMD) staining	5–10 min	6 min
Concentration of CMA stain	0.1–0.8 mg mL^{-1}	No positive results in any concentrations
Duration of CMA staining	60–120 min	120 min
[a]Concentration of distamycin A (DA)	0.1–0.25 mg L^{-1}	Not used
[a]Duration of DA staining	Not used	Not used

[a]Optional

Key Fluorescent Feature (Figs. 4.7, 4.8, and 4.9)

Analysis of more than 25 metaphase chromosome plates stained with contrasting DNA base-specific CMA and DAPI fluorochrome in this germplasm revealed the following fluorescent feature:

(i) Not a single chromosome ($2n = 16$) shows CMA^{+ve} signals.
(ii) Interstitial secondary constrictions are marked on the two chromosomes of the sixth pair, one in the eighth pair, and one in the first pair (2, 3, and 4 in numbers in different roots), behaving CMA-DAPI neutral.
(iii) No DAPI-positive signal was observed.
(iv) Based on the CMA and DAPI banding pattern, the model fluorescent karyotype remains 16C.

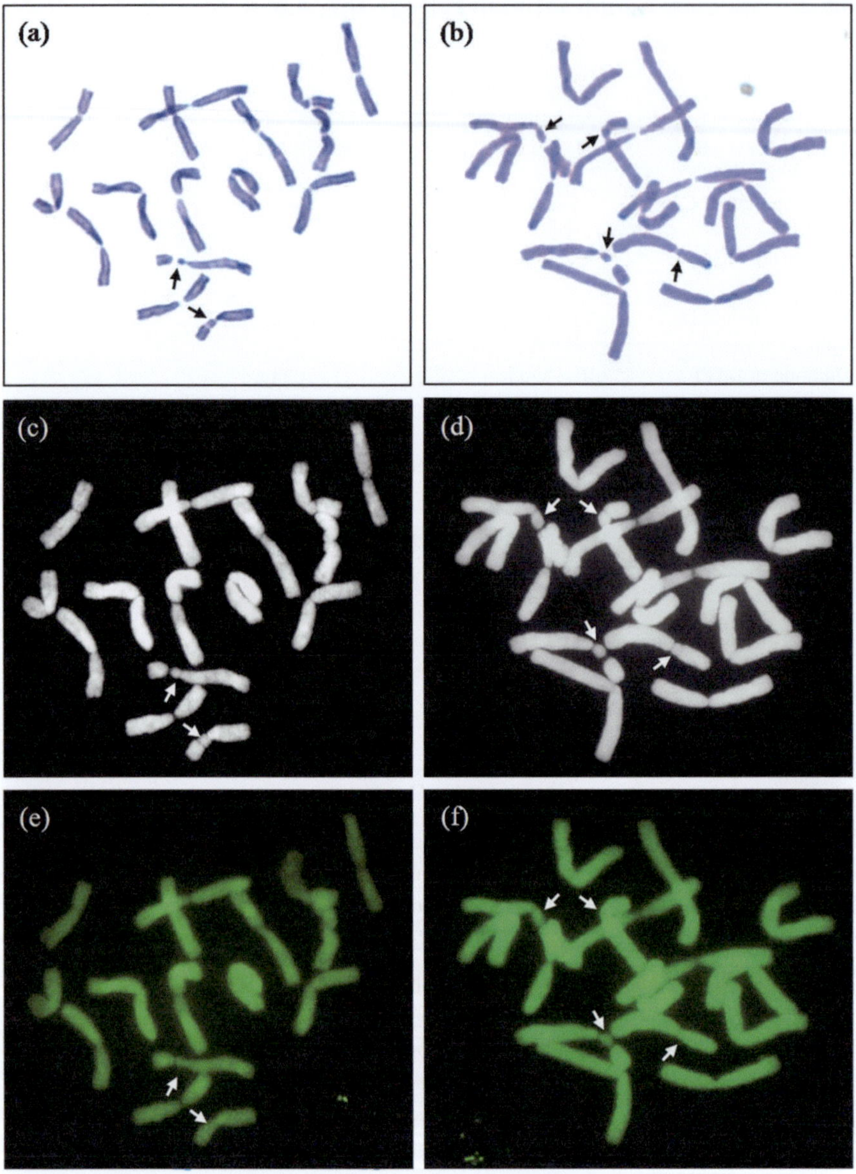

Fig. 4.7 EMA illustrates the preparation of somatic chromosomes based on Giemsa, DAPI, and CMA staining from metaphases of *Allium sativum* ($2n = 16$ chromosomes). The cytoplasm-free Giemsa-stained somatic chromosomes exhibit distinct morphology, featuring one pair (**a**) and two pairs (**b**) of intercalary secondary constrictions. The DAPI-stained metaphases are shown in (**c**) and (**d**), while the CMA-stained metaphases are presented in (**e**) and (**f**). Black and white arrows indicate the intercalary secondary constrictions on the respective pairs of chromosomes (**a**, **c**, **e** for one pair; **b**, **d**, **f** for two pairs)

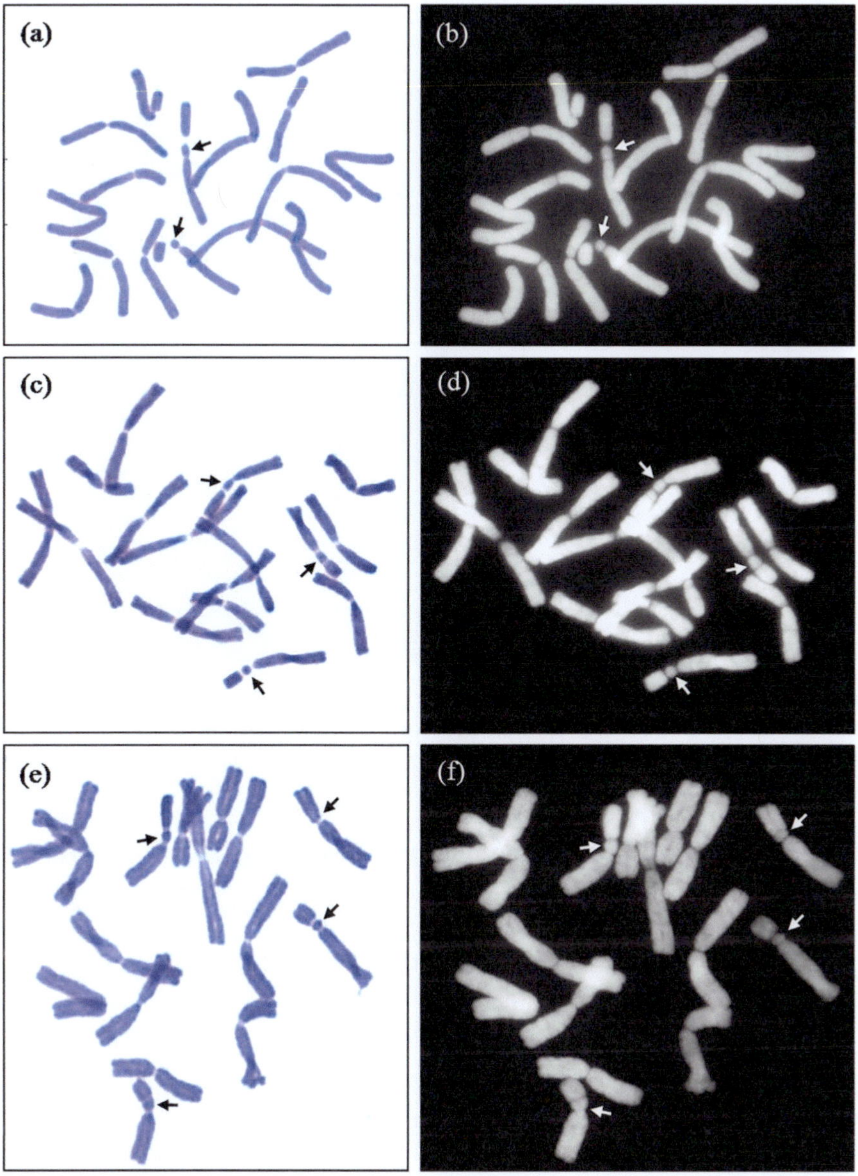

Fig. 4.8 EMA illustrates a somatic metaphase chromosome preparation of *Allium sativum* ($2n = 16$ chromosomes) stained with Giemsa and DAPI. The displayed cytoplasm-free Giemsa- and DAPI-stained somatic chromosomes reveal distinct morphologies, showcasing two (**a**, **b**), three (**c**, **d**), and four (**e**, **f**) intercalary secondary constrictions. Black and white arrows highlight the locations of the intercalary secondary constrictions

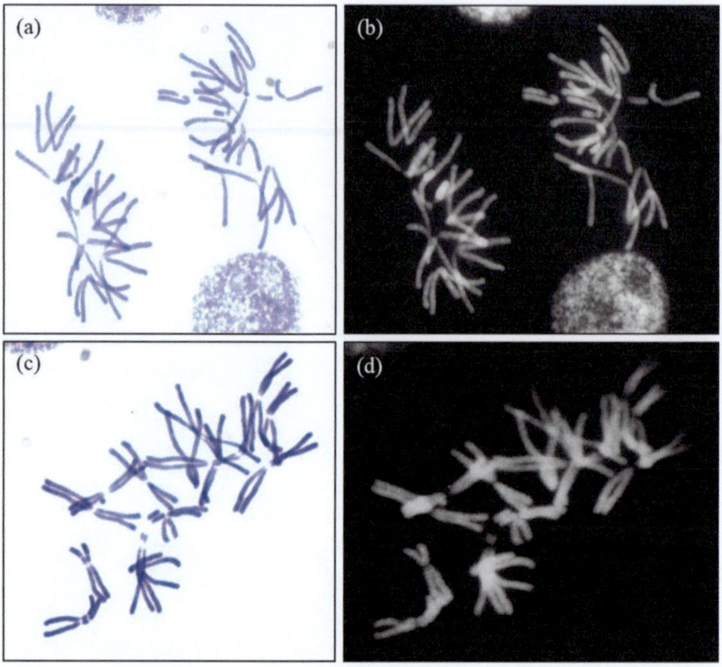

Fig. 4.9 EMA presents another somatic chromosome preparation of *Allium sativum* ($2n$ = 16 chromosomes), also stained with Giemsa and DAPI. This figure depicts the anaphasic separation of chromatids (**a**, **b**), as well as metaphase chromosomes exhibiting diplochromatic characteristics (**c**, **d**)

4.5.3 Species: *Trigonella foenum-graecum* L. (Figs. 4.10 and 4.11)

Protocol parameters	Range of concentration and duration of treatment	Optimum concentration and duration of treatment for this germplasm
Time of enzymatic digestion at 37 °C	65–75 min	75 min
Concentration of Giemsa stain	1.5–2%	2%
Duration of Giemsa staining	10–15 min	10 min
Concentration of DAPI stain	0.1–0.2 μg mL^{-1}	0.1 μg mL^{-1}
Duration of DAPI staining	10–15 min	12 min
Concentration of Actinomycin D (AMD) stain	0.1–0.25 mg L^{-1}	0.25 mg L^{-1}

Protocol parameters	Range of concentration and duration of treatment	Optimum concentration and duration of treatment for this germplasm
Duration of (AMD) staining	5–10 min	5 min
Concentration of CMA stain	0.2–0.5 mg mL^{-1}	0.5 mg mL^{-1}
Duration of CMA staining	60–80 min	90 min
[a]Concentration of distamycin A (DA)	0.1–0.25 mg L^{-1}	Not used
[a]Duration of DA staining	Not used	Not used

[a]Optional

Key Fluorescent Feature (Figs. 4.10 and 4.11)

Analysis of more than 25 metaphase chromosome plates stained with contrasting DNA base-specific CMA and DAPI fluorochrome in this germplasm revealed the following fluorescent feature:

(i) A distinct and intense four CMA^{+ve} / $DAPI^{-ve}$ signal on the interstitial region of the second and fourth pair of chromosomes confirmed the position of secondary constrictions in this germplasm.
(ii) Unique16 CMA^{+ve} / $DAPI^{-ve}$ bands were obtained in the centromeric regions of all the chromosomes, pointing to the possibility of GC-rich sequences in the centromeric region.
(iii) No DAPI-positive signal was observed.
(iv) Based on the CMA and DAPI banding pattern, the model fluorescent karyotype remains 12A + 4G.

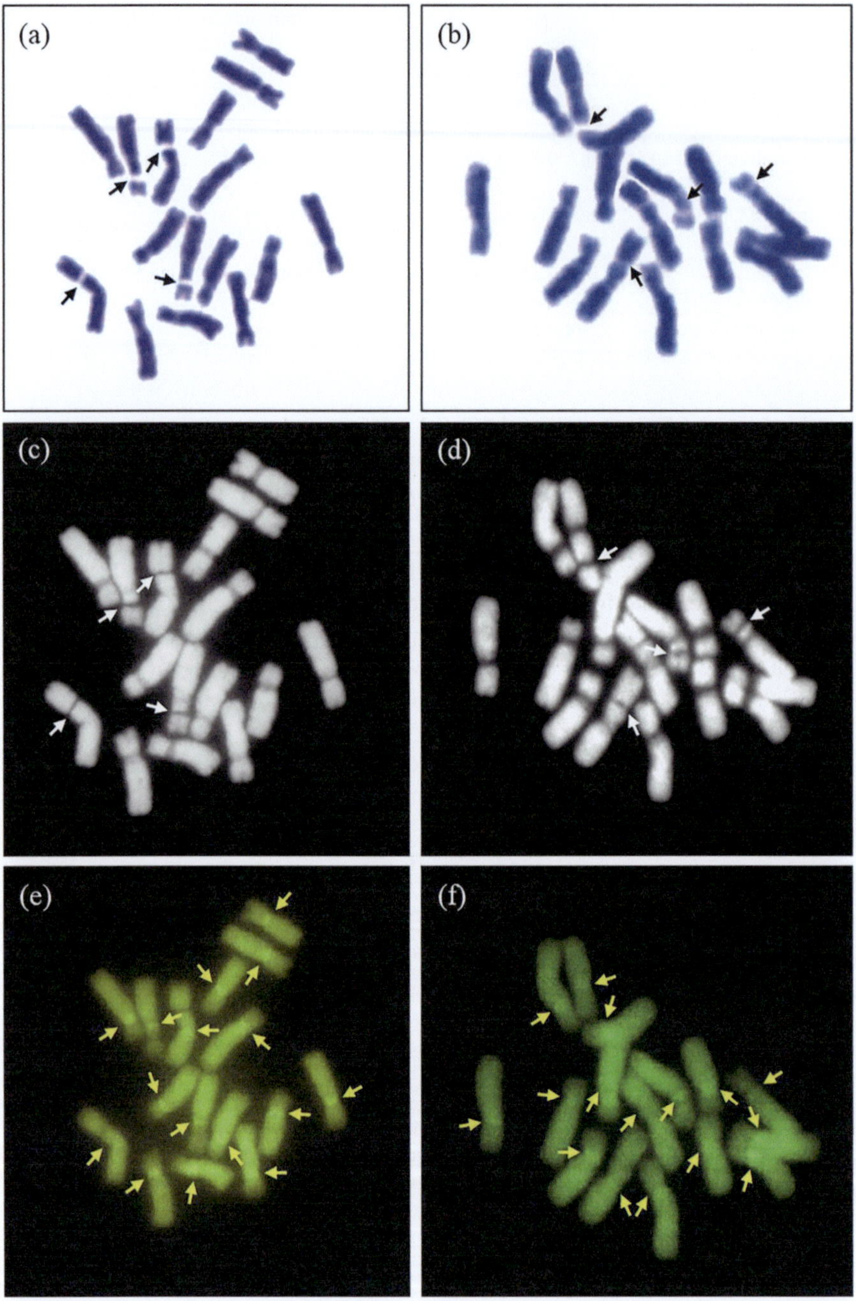

Fig. 4.10 EMA depicts somatic chromosome preparations of *Trigonella foenum-graecum*, utilizing Giemsa, DAPI, and CMA staining on the same metaphase plates. The images display well-separated chromosomes ($2n = 14$) with distinct morphology as observed through Giemsa staining (**a**, **b**), DAPI staining (**c**, **d**), and CMA staining (**e**–**f**). Black and white arrows indicate intercalary secondary constrictions on two pairs of chromosomes, while yellow arrows highlight the locations of CMA-positive bands

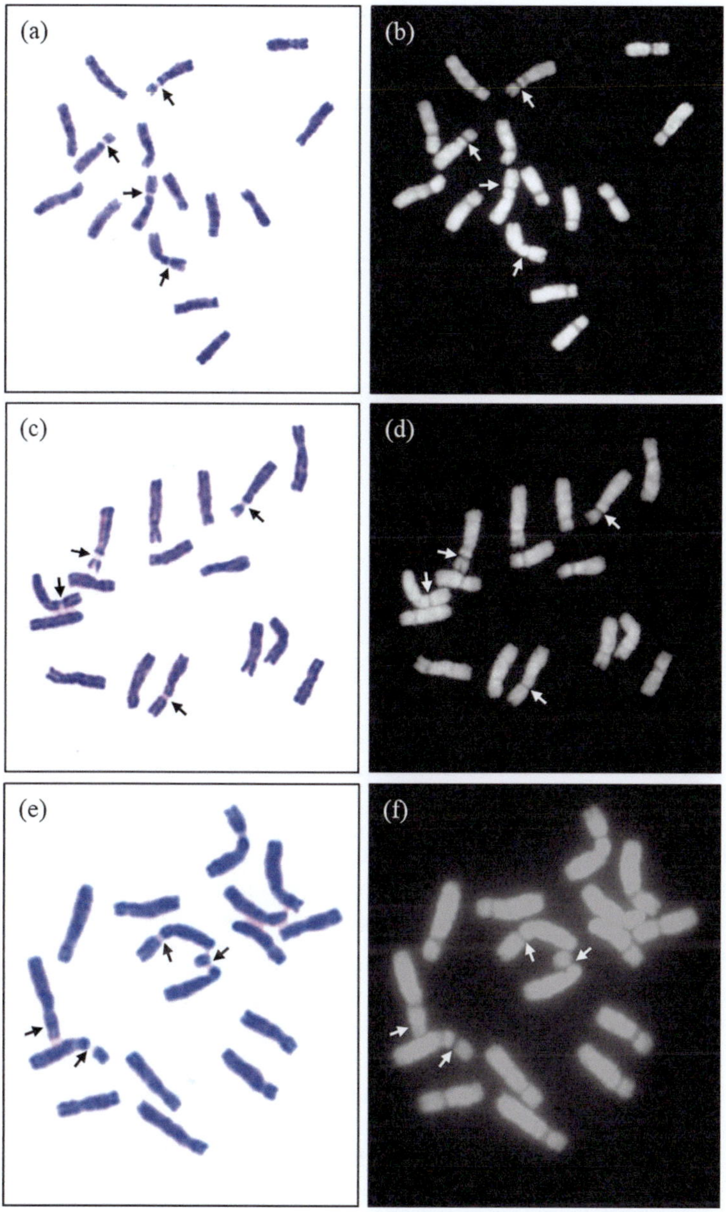

Fig. 4.11 EMA-based somatic chromosome preparations of *Trigonella foenum-graecum*, followed by Giemsa and DAPI staining on the same metaphase plates, reveal well-scattered, distinct chromosomes with a diploid number of $2n = 14$. The Giemsa-stained images are presented in (**a**, **c**, **e**), while the DAPI-stained images are shown in (**b**, **d**, **f**). Black and white arrows indicate the locations of intercalary secondary constrictions on the chromosomes

4.5.4 Species: *Aloe vera* Burm. f. (Figs. 4.12 and 4.13)

Protocol parameters	Range of concentration and duration of treatment	Optimum concentration and duration of treatment for this germplasm
Time of enzymatic digestion at 37 °C	65–75 min	68 and 70 min
Concentration of Giemsa stain	1.5–2%	2%
Duration of Giemsa staining	10–15 min	12 min
Concentration of DAPI stain	0.1–0.2 μg mL^{-1}	0.2 μg mL^{-1}
Duration of DAPI staining	10–15 min	10 min
Concentration of Actinomycin D (AMD) stain	0.1–0.25 mg L^{-1}	0.25 mg L^{-1}
Duration of (AMD) staining	5–10 min	6 min
Concentration of CMA stain	0.2–0.5 mg mL^{-1}	0.3 and 0.5 mg mL^{-1}
Duration of CMA staining	60–80 min	90 min
[a]Concentration of distamycin A (DA)	0.1–0.25 mg L^{-1}	Not used
[a]Duration of DA staining	Not used	Not used

[a]Optional

Key Fluorescent Feature (Figs. 4.12 and 4.13)

Analysis of more than 25 metaphase chromosome plates stained with contrasting DNA base-specific CMA and DAPI fluorochrome in this germplasm revealed the following fluorescent feature:

(i) Unique and distinct $DAPI^{+ve}$ signals were located at the centromeric region of all six small and four long chromosomes, one of which bears a secondary constriction.
(ii) Centromeric regions of another four long chromosomes, in the karyotype, remain CMA- DAPI neutral.
(iii) Terminal secondary constriction on the first and third pairs of the long chromosomes showed very faint CMA^{+ve} signals corresponding to $DAPI^{-ve}$ signals.
(iv) The fluorochrome banding pattern of *A. vera* may be 8B + 2H + 4F.

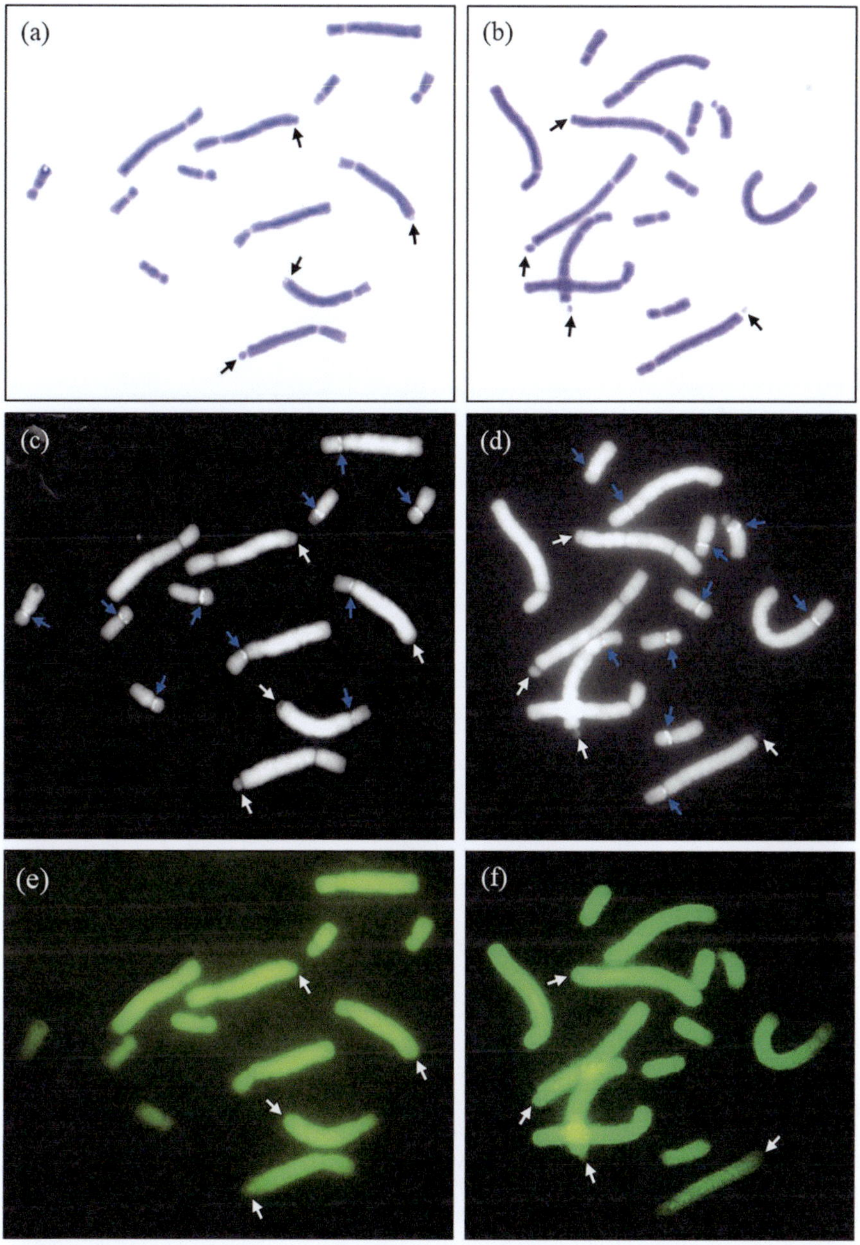

Fig. 4.12 EMA presents a somatic chromosome preparation based on Giemsa, DAPI, and CMA staining of metaphases in *Aloe vera* ($2n = 14$ chromosomes). The cytoplasm-free Giemsa-stained metaphases show a total of $2n = 14$ somatic chromosomes, which consist of eight long and six small morphologically distinct chromosomes (**a**, **b**). Additionally, DAPI-stained metaphases are shown in panels (**c**, **d**), along with CMA-stained metaphases in (**e**, **f**). Black and white arrows indicate the terminal secondary constrictions on two pairs of chromosomes, while blue arrows mark the locations of DAPI-positive bands on the chromosomes

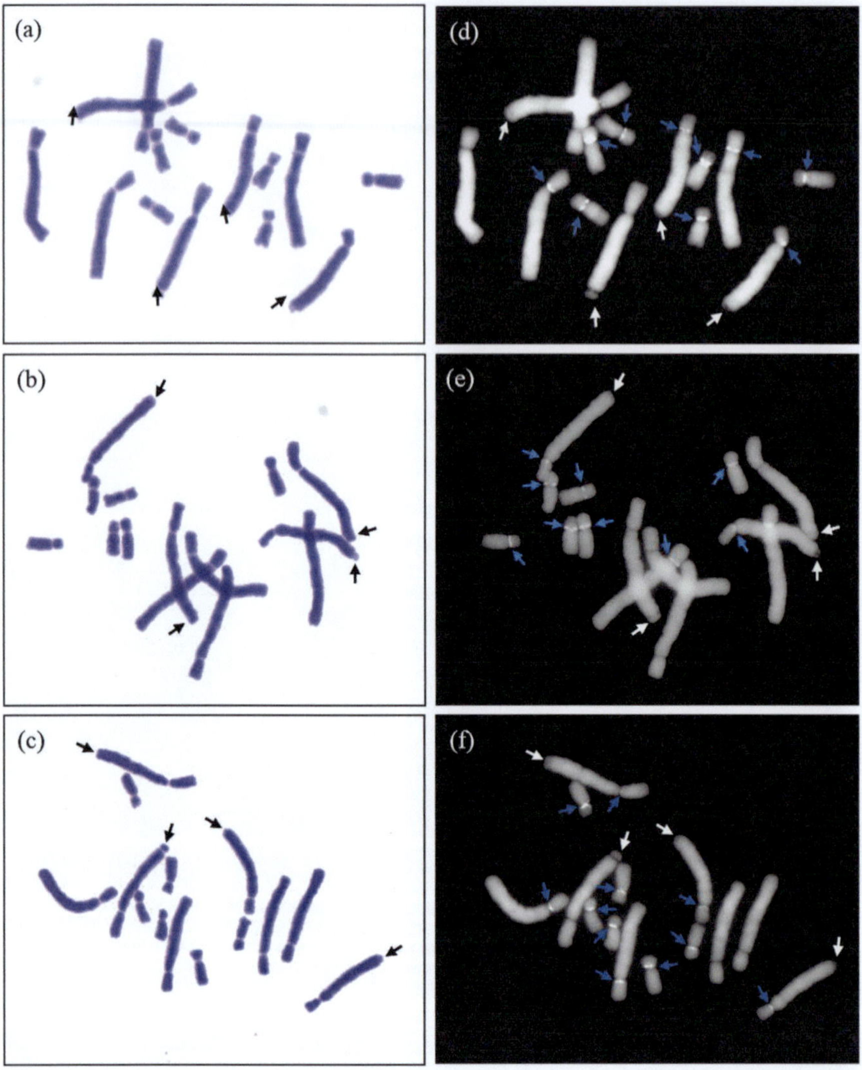

Fig. 4.13 EMA illustrates a somatic chromosome preparation based on Giemsa, DAPI, and CMA staining of metaphases in *Aloe vera* ($2n = 14$ chromosomes). The cytoplasm-free Giemsa-stained metaphases, which are well-scattered, display $2n = 14$ somatic chromosomes along with distinct chromosome morphology (**a**, **c**, **e**). Additionally, the DAPI-stained metaphases are shown (**b**, **d**, **f**). Black and white arrows indicate the terminal secondary constrictions on two pairs of chromosomes, while blue arrows highlight the positions of DAPI-positive bands on the chromosomes

4.5.5 Species: *Lens culinaris* Medik (Figs. 4.14 and 4.15)

Protocol parameters	Range of concentration and duration of treatment	Optimum concentration and duration of treatment for this germplasm
Time of enzymatic digestion at 37 °C	45–55 min	50 min
Concentration of Giemsa stain	1.5–2%	2%
Duration of Giemsa staining	10–14 min	12 min
Concentration of DAPI stain	0.1–0.2 μg mL^{-1}	0.1 μg mL^{-1}
Duration of DAPI staining	10–15 min	10 min
Concentration of Actinomycin D (AMD) stain	0.1–0.25 mg L^{-1}	0.25 mg L^{-1}
Duration of (AMD) staining	5–15 min	15 min
Concentration of CMA stain	0.1–0.3 mg mL^{-1}	0.1 mg mL^{-1}
Duration of CMA staining	45–50 min	50 min
[a]Concentration of distamycin A (DA)	0.1–0.25 mg L^{-1}	Not used
[a]Duration of DA staining	Not used	Not used

[a]Optional

Key Fluorescent Feature (Figs. 4.14 and 4.15)

Fluorescent karyotype analysis in more than 30 Indian cultivars of lentil unlocked diversity in the chromosome pairs carrying interstitial secondary constrictions and DAPI banding patterns. Analysis of more than 25 metaphase chromosome plates stained with contrasting DNA base-specific CMA and DAPI fluorochrome in this germplasm revealed the following fluorescent feature:

(i) A distinct and intense four CMA^{+ve} / $DAPI^{-ve}$ signal on the interstitial region of the fourth pair of chromosomes in this germplasm confirmed the position of secondary constrictions in this germplasm.
(ii) No other $DAPI^{+ve}$ or CMA^{+ve} signals were found on any pair of chromosomes.
(iii) The fluorochrome banding pattern of this particular *Lens culinaris* cultivar thus remains 2D + 12F.

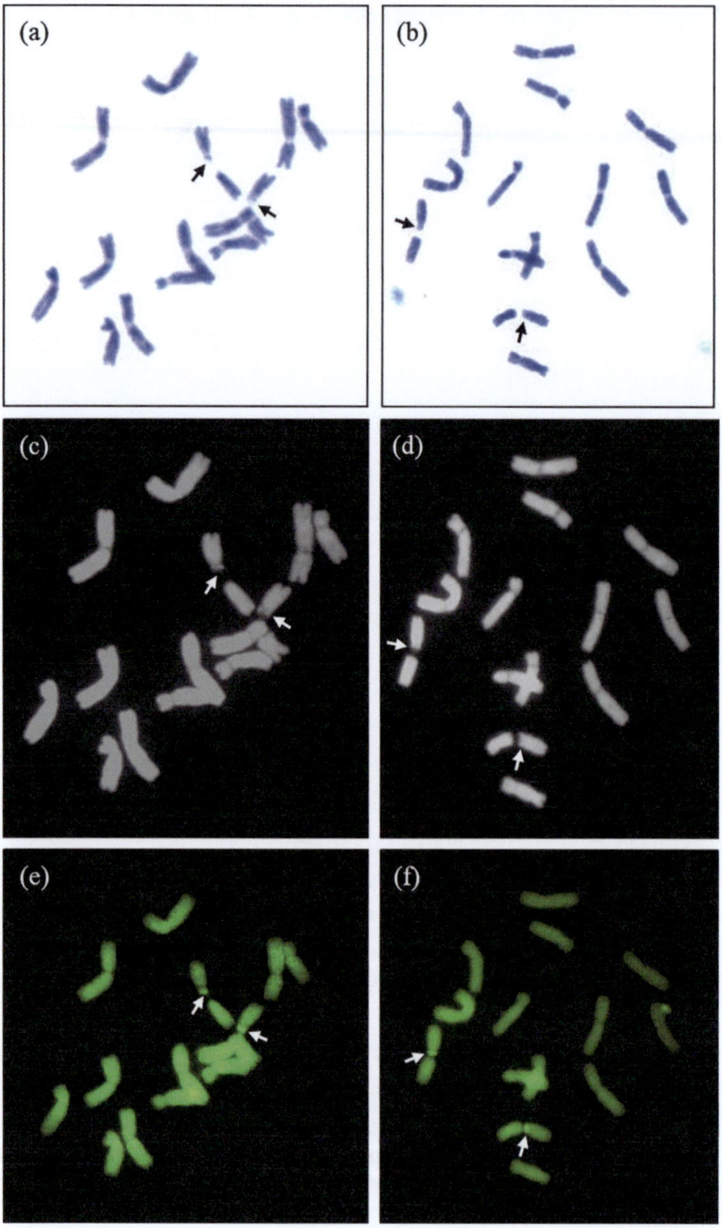

Fig. 4.14 EMA illustrates the preparation of somatic chromosomes from *Lens culinaris* ($2n = 14$ chromosomes) using Giemsa, DAPI, and CMA staining techniques. The cytoplasm-free, well-scattered Giemsa-stained metaphases reveal the $2n = 14$ somatic chromosomes, showcasing distinct chromosome morphology (**a**, **b**). Corresponding DAPI-stained metaphases are shown (**c**, **d**), along with those stained with CMA (**e**, **f**). Black and white arrows indicate the intercalary secondary constrictions present on one pair of chromosomes

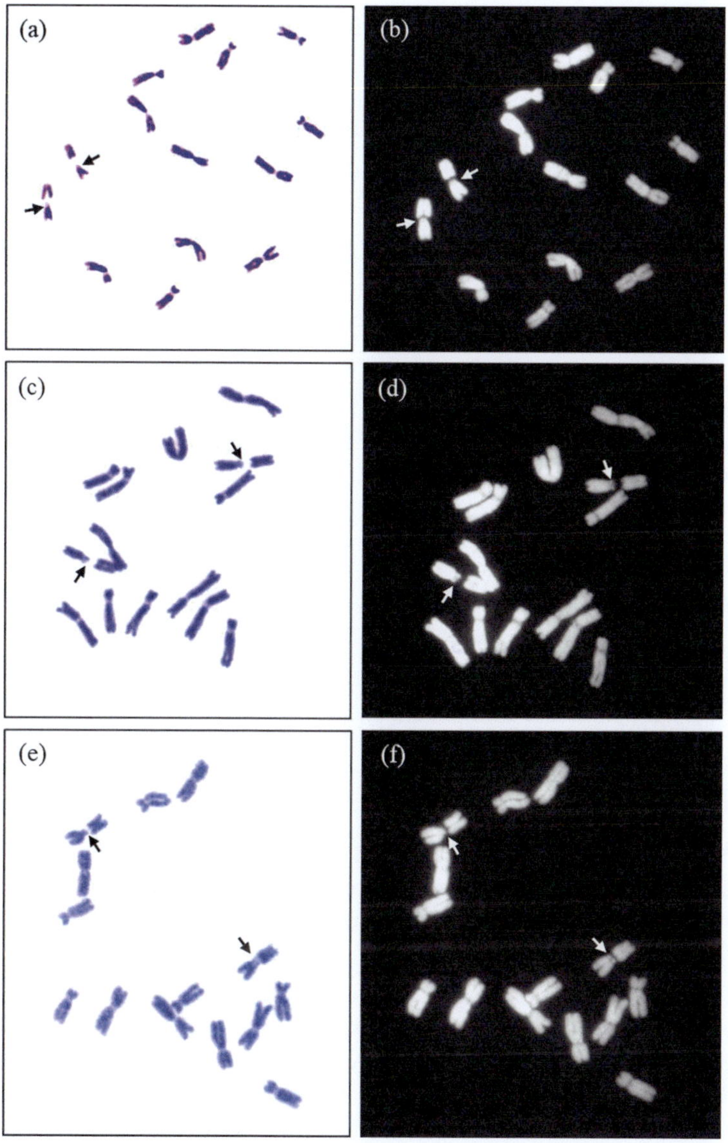

Fig. 4.15 EMA illustrates a somatic chromosome preparation of *Lens culinaris* ($2n = 14$ chromosomes), followed by Giemsa and DAPI staining. The cytoplasm-free Giemsa-stained metaphases reveal well-dispersed $2n = 14$ chromosomes with distinct morphological features (**a**, **c**, **e**), as seen in the DAPI-stained metaphases (**b**, **d**, **f**). The black and white arrows indicate intercalary secondary constrictions present on one pair of chromosomes

4.5.6 Species: *Nigella sativa* L. (Figs. 4.16 and 4.17)

Protocol parameters	Range of concentration and duration of treatment	Optimum concentration and duration of treatment for this germplasm
Time of enzymatic digestion at 37 °C	65–75 min	75 min
Concentration of Giemsa stain	1.5–2%	2%
Duration of Giemsa staining	10–15 min	10 min
Concentration of DAPI stain	0.1–0.2 μg mL^{-1}	0.1 μg mL^{-1}
Duration of DAPI staining	10–15 min	10 min
Concentration of Actinomycin D (AMD) stain	0.1–0.25 mg L^{-1}	0.25 mg L^{-1}
Duration of (AMD) staining	5–10 min	6 min
Concentration of CMA stain	0.1–0.5 mg mL^{-1}	0.5 mg mL^{-1}
Duration of CMA staining	70–100 min	80 min
[a]Concentration of distamycin A (DA)	0.1–0.25 mg L^{-1}	Not used
[a]Duration of DA staining	Not used	Not used

[a]Optional

Key Fluorescent Feature (Figs. 4.16 and 4.17)
Analysis of more than 25 metaphase chromosome plates stained with contrasting DNA base-specific CMA and DAPI fluorochrome in this germplasm revealed the following fluorescent feature:

(i) CMA-positive/DAPI-negative signals confirmed the presence of secondary constrictions on the three pairs of chromosomes (2nd, 3rd, and 6th).
(ii) No other $DAPI^{+ve}$ or CMA^{+ve} signals were found on any pair of chromosomes.
(iii) The fluorochrome banding pattern of *Nigella sativa* thus remains 6D + 6F.

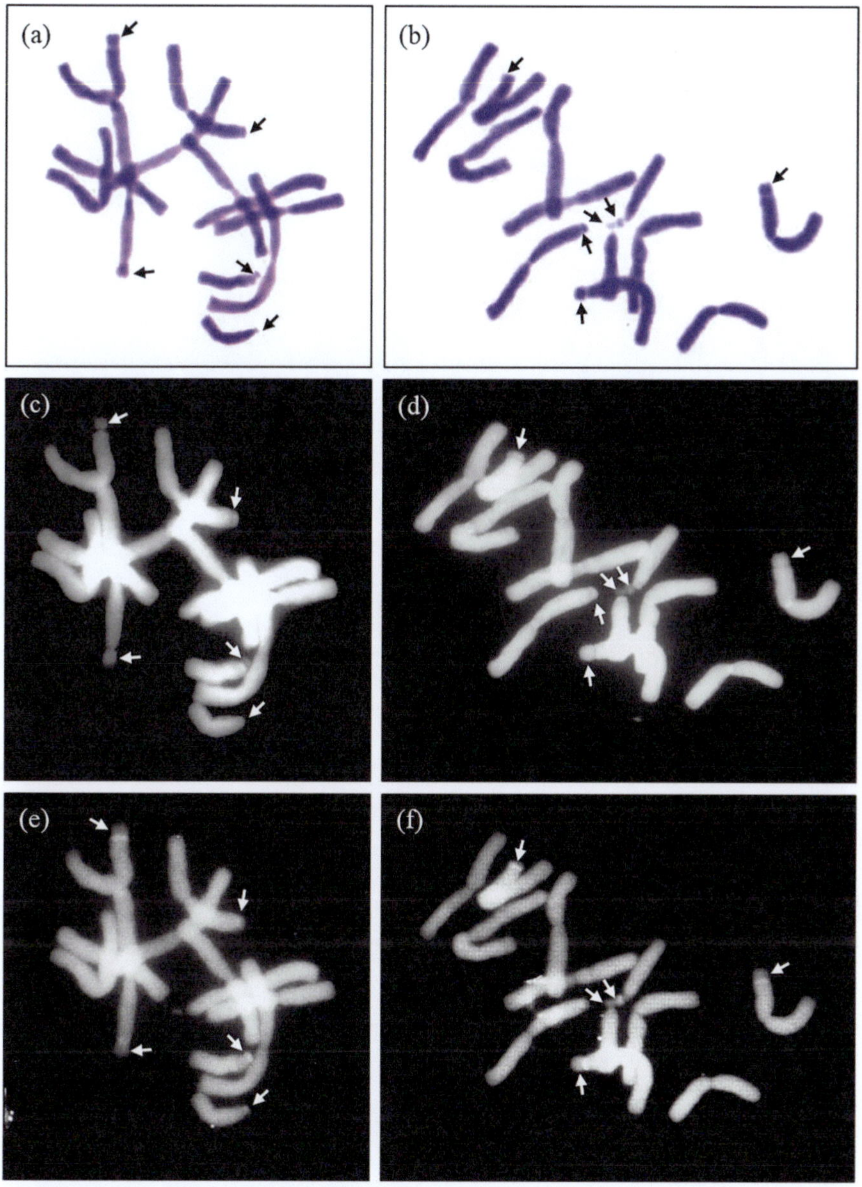

Fig. 4.16 EMA illustrates somatic chromosome preparation based on techniques followed by Giemsa, DAPI, and CMA staining of metaphases in *Nigella sativa*. The results show Giemsa-stained chromosomes (**a**, **b**), DAPI-stained chromosomes (**c**, **d**), and CMA-stained chromosomes (**e**, **f**), all displaying well-scattered $2n = 12$ somatic chromosomes characterized by distinct chromosomal morphology. Black and white arrows highlight terminal secondary constrictions present on three chromosome pairs

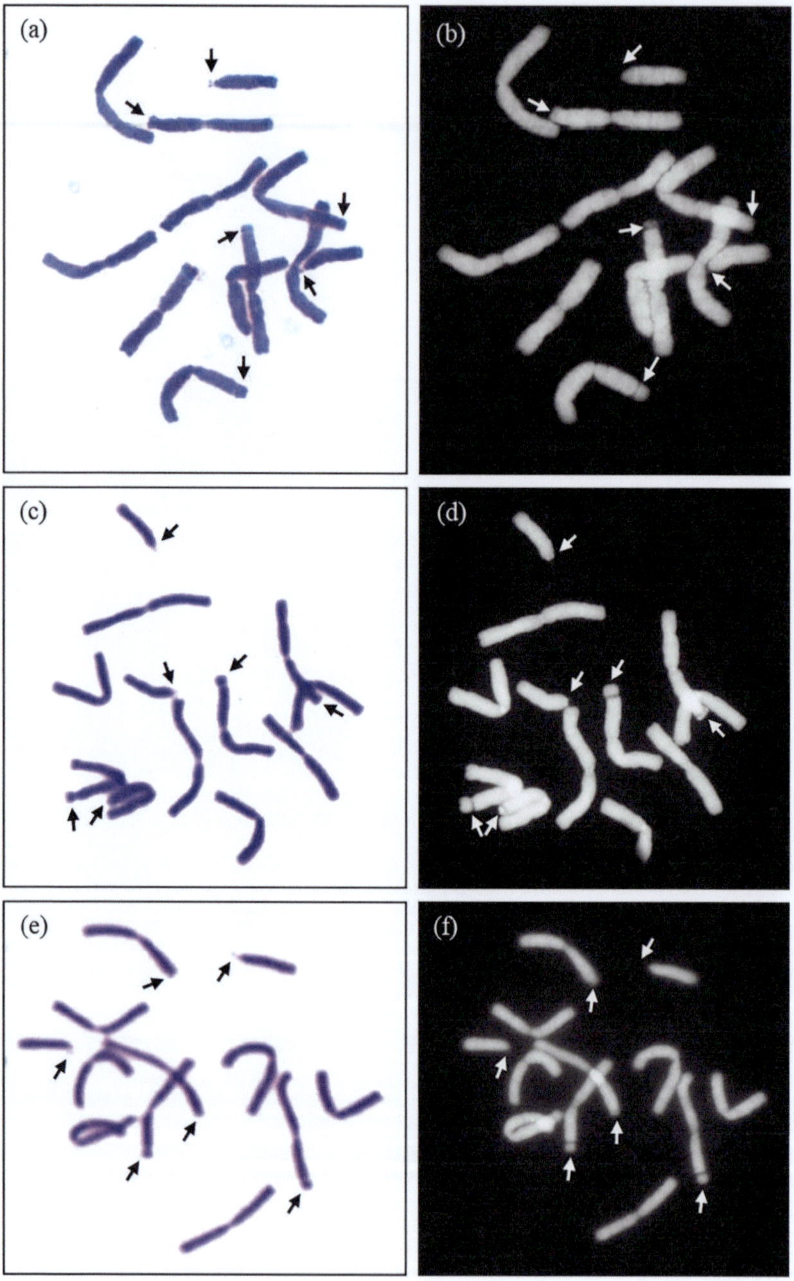

Fig. 4.17 EMA illustrates somatic chromosome preparations based on Giemsa and DAPI staining applied to the same metaphase plates of *Nigella sativa*. The chromosomes are shown as Giemsa-stained (**a**, **c**, **e**) and DAPI-stained (**b**, **d**, **f**), revealing well-dispersed 2n = 12 somatic chromosomes with distinct morphologies. Black and white arrows indicate the terminal secondary constrictions present on three pairs of chromosomes

4.5.7 Species: *Vicia faba* L. (Fig. 4.18)

Protocol parameters	Range of concentration and duration of treatment	Optimum concentration and duration of treatment for this germplasm
Time of enzymatic digestion at 37 °C	65–75 min	70 min
Concentration of Giemsa stain	1.5–2%	2%
Duration of Giemsa staining	10–15 min	10 min
Concentration of DAPI stain	0.1–0.2 μg mL^{-1}	0.2 μg mL^{-1}
Duration of DAPI staining	10–15 min	10 min
Concentration of Actinomycin D (AMD) stain	0.1–0.25 mg L^{-1}	0.25 mg L^{-1}
Duration of (AMD) staining	5–10 min	6 min
Concentration of CMA stain	0.1–0.5 mg mL^{-1}	0.5 mg mL^{-1}
Duration of CMA staining	60–80 min	80 min
[a]Concentration of distamycin A (DA)	0.1–0.25 mg L^{-1}	Not used
[a]Duration of DA staining	Not used	Not used

[a]Optional

Key Fluorescent Feature (Figs. 4.18 and 4.19)

Analysis of more than 25 metaphase chromosome plates stained with contrasting DNA base-specific CMA and DAPI fluorochrome in this germplasm revealed the following fluorescent feature:

(i) A distinct CMA^{+ve} / $DAPI^{-ve}$ signal on the interstitial region of the longest chromosome pair confirmed the position of secondary constrictions in this germplasm.

(ii) Two distinct and intense $DAPI^{+ve}/CMA^{-ve}$ bands were located in the longest chromosome pair.

(iii) Two distinct $DAPI^{+ve}/CMA^{-ve}$ bands were located in one pair of comparatively short chromosomes.

(iv) Based on the CMA and DAPI banding pattern, the model fluorescent karyotype remains 2B + 6F+ 2H.

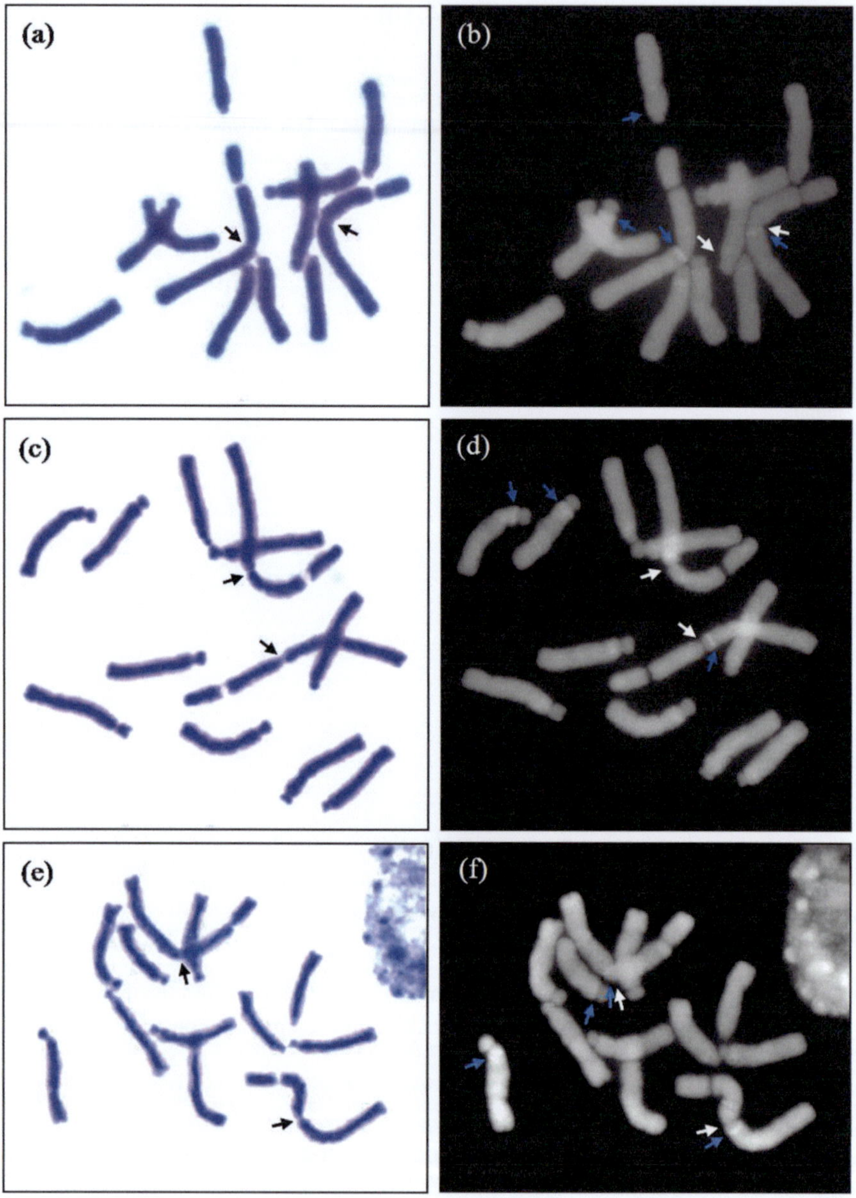

Fig. 4.18 EMA illustrates somatic chromosome preparation of *Vicia faba* based on Giemsa and DAPI staining applied to the same metaphases. The results show well-scattered $2n = 12$ somatic chromosomes with distinct morphology, as represented by Giemsa stains (**a**, **c**, **e**) and DAPI stains (**b**, **d**, **f**). Black and white arrows denote intercalary secondary constrictions on one chromosome pair, while blue arrows indicate the locations of DAPI-positive bands on the chromosomes

4.5.8 Species: *Ornithogalum virens/Albuca virens* (Fig. 4.19)

Protocol parameters	Range of concentration and duration of treatment	Optimum concentration and duration of treatment for this germplasm
Time of enzymatic digestion at 37 °C	70–80 min	7 min
Concentration of Giemsa stain	1.5–2%	2%
Duration of Giemsa staining	10–15 min	15 min
Concentration of DAPI stain	0.1–0.2 μg mL^{-1}	0.1 μg mL^{-1}
Duration of DAPI staining	15–20 min	20 min
Concentration of Actinomycin D (AMD) stain	0.1–0.25 mg L^{-1}	0.25 mg L^{-1}
Duration of (AMD) staining	5–10 min	10 min
Concentration of CMA stain	0.1— 0.5 mg mL^{-1}	0.5 mg mL^{-1}
Duration of CMA staining	60–80 min	70 min
[*]Concentration of distamycin A (DA)	0.1–0.25 mg L^{-1}	Not used
[*]Duration of DA staining	Not used	Not used

[a]Optional

Key Fluorescent Feature (Fig. 4.19)

Analysis of more than 25 metaphase chromosome plates stained with contrasting DNA base-specific CMA and DAPI fluorochrome in this germplasm revealed the following fluorescent feature:

(i) A distinct CMA^{+ve} / $DAPI^{-ve}$ signal on the interstitial region of the one pair of chromosomes confirmed the position of secondary constrictions in this germplasm.
(ii) No other CMA^{+ve} or $DAPI^{+ve}$ signals were observed on any other chromosomes.
(iii) Based on the CMA and DAPI banding pattern, the model fluorescent karyotype remains 2D + 4F.

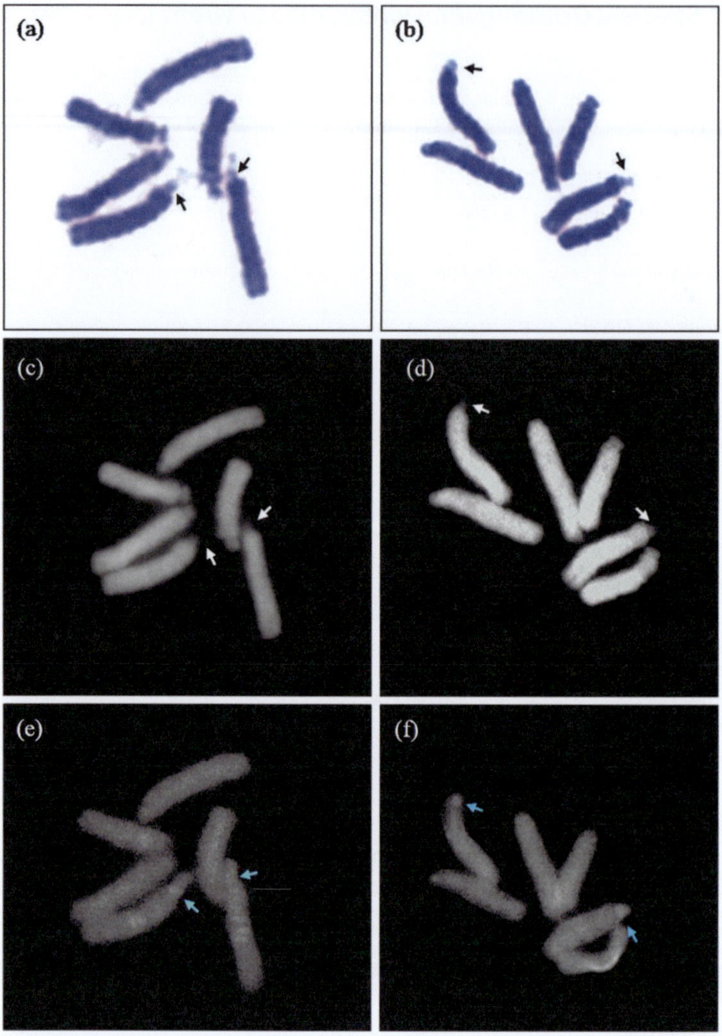

Fig. 4.19 EMA illustrates somatic chromosome preparation based on techniques followed by Giemsa, DAPI, and CMA staining of metaphases in *Ornithogalum virens/ Albuca virens*. The results show Giemsa-stained chromosomes (**a**, **b**), DAPI-stained chromosomes (**c**, **d**), and CMA-stained chromosomes (**e**, **f**), all displaying well-scattered $2n = 6$ somatic chromosomes characterized by distinct chromosomal morphology. Black and white arrows highlight intestinal secondary constrictions present on one chromosome pair, while blue arrows highlight the positions of CMA-positive bands on the chromosomes

4.6 Conclusions

In examining the repeated DNA sequence patterns in plant species, both CMA and DAPI can induce three distinct banding patterns along the length of chromosomes, particularly observed in the primary and secondary constrictions, as well as in the interstitial regions. These bands are categorized as positive, negative, and neutral, corresponding to $CMA^{+}DAPI^{-}$, $DAPI^{+}$ CMA^{-}, and CMA^{0} $DAPI^{0}$, respectively (Guerra et al. 2000). Our fluorescent banding studies involving various significant plant species—often utilized as practical teaching materials for understanding chromosome numbers and karyotype characteristics at the undergraduate and postgraduate levels—have uncovered many previously unknown molecular features (Jha et al. 2024b). Students have faced challenges in observing these molecular features on chromosomes due to limitations of conventional methods.

EMA-based chromosome preparation, followed by Giemsa and fluorescent banding techniques, has significantly expanded the horizons of traditional cytogenetics. For our students, molecular cytogenetics serves as an essential lens through which they can explore the intricate and captivating world of genomes. Our recent studies involving *Allium cepa* L., *Allium sativum* L., *Nigella sativa* L., *Trigonella foenum-graecum* L., *Aloe vera* L. Burm., and *Lens culinaris* Medik. have zeroed in on the technical aspects of chromosome preparation utilizing the EMA-based method, along with the staining and analysis of metaphases through non-fluorescent Giemsa and contrasting DNA base-specific fluorescent dyes, DAPI and CMA.

These investigations not only reaffirm the three banding patterns documented by Guerra (2000) but also reveal numerous novel molecular features that may pique interest in plant chromosome analysis and broader fields such as molecular biology and genomics. The documented evidence presented in the accompanying figures illustrates the number and locations of secondary constrictions in each Indian germplasm. We have noted both terminal and interstitial secondary constrictions across the studied species. The model karyotype formula mentioned in the figures for each species will enable students and researchers to compare their region-specific germplasms.

Centromeres are an indispensable part of chromosome morphology and are involved in the evolutionarily conserved function of maintaining chromosome integrity during cell division and directing chromosome segregation. In most species, the region usually contains highly repetitive DNA sequences. Based on its evolutionarily conserved function, it was considered to be a conserved heterochromatic site.

The organization and underlying DNA sequence of centromeres vary greatly; however, advances in molecular genetics and genomics reveal surprising diversity in the DNA sequence organization among different species, and no readily apparent conserved characteristics were detected (Ma et al. 2007; Melters et al. 2013; Barbosa et al. 2022; Zhou et al. 2022). The analysis of centromeric heterochromatin diversity at the chromosomal level through fluorescent banding in plant species is limited. Fluorescent banding with DNA base-specific DAPI and CMA in *Aloe vera* and *Trigonella foenum-graecum* revealed contrasting heterochromatic DNA sequence patterns. All centromeric regions of Trigonella chromosomes were CMA^{+ve}, i.e., GC-enriched. On the contrary, *Aloe vera*, a representative of bimodal karyotype with six small and eight long chromosomes, showed ten (6 small and 4 long) DAPI $^{+ve}$, i.e., AT-enriched centromeric regions with four CMA-DAPI neutral long chromosomes. The interesting observation under the microscope warrants further fluorescent banding analysis in other plant species to understand the heterochromatin diversity of DNA organization in centromeric regions. It will inspire our researchers and scientists to explore molecular interventions in this region, allowing them to verify the actual DNA sequence pattern and generate new ideas. Our findings indicate the presence of chromosome-specific DNA sequence diversity in the plant centromeric region, highlighting the importance of basic molecular cytogenetics in the era of genomics. A detailed protocol for EMA-based chromosome analysis and fluorescent banding is available for the above-noted plant species (Jha et al. 2024b). Hands-on training in preparing morphologically distinct chromosomes using the EMA method requires minimal time and can yield valuable complementary information for modern genomics.

References

Andras SC, Hartman TPV, Alexander J, Mcbride R, Marshall JA, Power JB, Cocking EC, Davey MR (2000) Combined PI-DAPI staining (CPD) reveals NOR asymmetry and facilitates karyotyping of plant chromosomes. Chromosom Res 8:387–391

Barbosa AC, Xu Z, Karari K, Williams W, Hauf S, Brown WR (2022) Mutation and selection explain why many eukaryotic centromeric DNA sequences are often A+ T rich. Nu Acid Res 50(1):579–596

Barcia JJ (2007) The Giemsa stain: its history and applications. Int J Surg Pathol 15(3):292–296

Biscotti MA, Olmo E, Heslop-Harrison JS (2015) Repetitive DNA in eukaryotic genomes. Chromosom Res 23(3):415–420

Cuadrado A, Cardoso M, Jouve N (2008) Physical organisation of simple sequence repeats (SSRs) in Triticeae: structural, functional and evolutionary implications. Cytogenet Genome Res 120:210–219

Fedak G, Kim NS (2008) Tools and methodologies for cytogenetic studies of plant chromosomes. Cytol Genet 42(3):189–203

Giemsa G (1904) Eine Vereinfachung und Vervollkommnung meiner Methylenblau-Eosin-Fa¨rbemethode zur Erzielung der Romanowsky-Nocht'schen Chromatinfa¨rbung. Centralblatt fu¨r Bakteriologie I Abteilung 32:307–313

Guerra M (2000) Patterns of heterochromatin distribution in plant chromosomes. Genet Mol Biol 23:1029–1041

Guerra M, Santos KGBD, Silva AEB, Ehrendorfer F (2000) Heterochromatin banding patterns in Rutaceae-Aurantioideae– a case of parallel chromosomal evolution. Am J Bot 87(5):735–747

Han YH, Zhang ZH, Liu JH, Lu JY, Huang SW, Jin WW (2008) Distribution of the tandem repeat sequencesand karyotyping in cucumber (*Cucumis sativus* L.) by fluorescence in situ hybridization. Cytogenet Genome Res 122(1):80–88

Heitz E (1928) Das Heterochromatin der Moose. I. Jahrb Wiss Bot 69:762–818

Heitz E (1929) Heterochromatin, Chromocentren, Chromomeren. Ber Botan Ges 47:274–284

Heslop-Harrison JS, Schwarzacher T (2013) Nucleosomes and centromeric DNA packaging. Proc Natl Acad Sci 110(50):19974–19975

Jha TB, Chakroborty P, Halder M (2024a) Karyotype analysis on *Gloriosa superba* using enzymatic maceration and air-drying-based Giemsa, DAPI, and CMA staining techniques. Cytologia 89(3):203–209

Jha TB, Halder M, Bhowmick BK (2024b) Giemsa-based chromosome staining and comparative fluorescent banding pattern in five valuable Indian plant species. Caryologia 77(3):37–45

Jorgenson KF, Van de Sande JH, Lin CC (1978) The use of base pair specific DNA binding agents as affinity labels for the study of mammalian chromosomes. Chromosoma 68(4):287–302

Kalendar R, Schulman A (2006) IRAP and REMAP for retrotransposon-based genotyping and fingerprinting. Nat Protoc 1:2478–2484

Kubis S, Schmidt T, Heslop-Harrison JS (1998) Repetitive DNA elements as a major component of plant genomes. Ann Bot 82S:45–55

Ma J, Wing RA, Bennetzen JL, Jackson SA (2007) Plant centromere organization: a dynamic structure with conserved functions. Trends Genet 23(3):134–139

Melters DP, Bradnam KR, Young HA, Telis N, May MR, Ruby JG, Sebra R, Peluso P, Eid J, Rank D, Garcia JF, DeRisi JL, Smith T, Tobias C, Ross-Ibarra J, Korf I, Chan SW (2013) Comparative analysis of tandem repeats from hundreds of species reveals unique insights into centromere evolution. Genome Biol 14(1):R10

Pardue ML, Gall JG (1970) Chromosomal localization of mouse satellite DNA. Science 168(3937):1356–1358

Ruiyang C, Song WQ, Ping'an XZ (1994) Chromosome G-banding in plants by inducing with trypsin and urea. Cell Res 4:79–87

Sahar E, Latt SA (1978) Enhancement of banding patterns in human metaphase chromosomes by energy transfer. Proc Natl Acad Sci USA 75(11):5650–5654

Schweizer D (1976) Reverse fluorescent chromosome banding with chromomycin and DAPI. Chromosoma 58(4):307–324

Schweizer D (1981) Counterstain-enhanced chromosome banding. Hum Genet 57:1–14

Schweizer D, Ambros P, Andrle M (1978) Modification of DAPI banding on human chromosomes by prestaining with a DNA-binding oligopeptide antibiotic, distamycin A. Exp Cell Res 111:327–332

Seabright M (1971) A rapid banding technique for human chromosomes. Lancet 2(7731):971–972

Setiawan AB, Wibowo A, Chee How Teo CH, Kikuchi S, Takato Koba T (2019) Repetitive DNA sequences accelerate molecular cytogenetic research in plants with small chromosomes. Indones J Biotechnol 24(2):82–87

She CW, Liu JY, Song YC (2006) CPD staining: an effective technique for detection of NORs and other GC-rich chromosomal regions in plants. Biotech Histochem 81(1):13–21

Tong P, Pidoux AL, Toda NR, Ard R, Berger H, Shukla M, Torres-Garcia J, Müller CA, Nieduszynski CA, Allshire RC (2019) Interspecies conservation of organisation and function between nonhomologous regional centromeres. Nat Commun 10(1):2343

Vosa CG, Marchi P (1972) Quinacrine fluorescence and Giemsa staining in plants. Nature New Biology 237(75):191–192

Wang C, Ma H, Baserga SJ, Pederson T, Huang S (2023) Nucleolar structure connects with global nuclear organization. Mol boil Cell 34(12):ar114

Wienert V, Lentner A, Böhler U (1993) Methodology and application of fluorescence techniques in dermatology. In: Frosch PJ, Kligman AM (eds) Noninvasive methods for the quantification of skin functions. Springer, Berlin, Heidelberg

Yamamoto M, Mukai Y (1989) Application of fluorescence in situ hybridization to molecular cytogenetics of wheat. Wheat Information Service 69:30–32

Zhang ZT, Yang SQ, Li ZA, Zhang YX, Wang YZ, Cheng CY, Li J, Chen JF, Lou QF (2016) Comparative chromosomal localization of 45S and 5S rDNAs and implications for genome evolution in *Cucumis*. Genome 59(7):449–457

Zhou J, Liu Y, Guo X, Birchler JA, Han F, Su H (2022) Centromeres: from chromosome biology to biotechnology applications and synthetic genomes in plants. Plant Biotechnol J 20(11):2051–2063

5 Karyotype and Karyotype Analysis

Abstract

This chapter explores the concept of karyotype and its analysis, a fundamental tool in understanding the genomic architecture of various species. Introduced by Russian scientist Levitsky in 1931, the term "karyotype" refers to the unique chromosome complement of each species, represented visually through cytogenetic methods. The relevance of karyotype analysis in the genomic era lies in its efficiency for visualizing critical genomic information such as chromosome number, morphology, and banding patterns. Karyotyping remains unaffected by environmental or developmental variables, thus it offers consistent insights into chromosomal characteristics. Plant karyotyping, however, presents distinct challenges due to factors like cell wall presence and polyploidy. This chapter addresses the objectives of karyotype analysis, including assessing chromosome features crucial for cytotaxonomy and plant breeding, and identifying chromosomal abnormalities. It also discusses karyotype symmetry and asymmetry, emphasizing size variability and its evolutionary implications, as proposed by Stebbins in his qualitative-quantitative framework. The classification into symmetric and asymmetric karyotypes further elucidates chromosomal organization, with symmetric karyotypes signifying a more primitive genomic state. Modern advancements in cytogenetic techniques, such as fluorescence in situ hybridization (FISH), have enhanced the understanding of plant genetic complexity, paving the way for more sophisticated research in karyotype analysis.

5.1 Introduction

In 1931, Russian scientist Levitsky introduced the term "karyotype" to describe the unique chromosome complement characteristic of each species. The term is derived from the Greek word "karyon," which means nucleus or seed, and specifically

T. B. Jha, M. Halder, *Classical and Molecular Plant Cytogenetics*,
https://doi.org/10.1007/978-981-95-8265-5_5

relates to chromosomes. A karyotype is a visual representation of the complete set of chromosomes for a particular species, which can be prepared using either conventional or molecular cytogenetic methods. Since chromosomes are considered discrete genomic units, the karyotype represents the highest level of structural and functional organization of genomes (Weiss-Schneeweiss and Schneeweiss 2013). Chromosomes serve as a complex instruction manual that guides the development and maintenance of species. Karyotyping involves meticulously examining each part of this manual, allowing scientists to understand how the instructions are organized and to identify any unusual or altered sections.

A fundamental question is why karyotype analysis remains relevant in the current genomic era. The answer is that karyotype provides a quick, cost-effective, and straightforward method to visualize essential information about a species' genome. This includes (i) the total number of chromosomes and (ii) chromosome morphology, which encompasses chromosome length, centromere position, the number of chromosomes with secondary constrictions, and the specific locations of these secondary constrictions on the chromosomes. Additionally, karyotype analysis can reveal banding patterns on chromosomes through specific staining techniques that aid in identifying individual chromosomes and detecting any chromosomal abnormalities. Notably, chromosomal features are not influenced by environmental factors or developmental stages. For example, the classic cytological studies on onions, conducted by undergraduate and postgraduate students in various countries over the past century, have consistently shown a chromosome number of $2n = 16$ (Guerra 2008).

In contrast to animal cells, chromosome analysis and karyotyping in plants present specific challenges, including the presence of cell walls, higher rates of polyploidy, greater genetic variability across plant species, and complex reproductive strategies that influence genetic makeup. Different plant species display variations in the size, shape, and number of chromosomes within their karyotypes. Similar to the evolution of morphological traits, it is believed that karyotypes in different species have evolved through natural selection.

Plant karyotyping is becoming increasingly sophisticated, allowing for a more comprehensive understanding of genetic complexity by employing modern advanced techniques, such as molecular cytogenetic methods, fluorescence in situ hybridization (FISH), chromosome painting, and high-resolution digital imaging.

5.2 Objective of Karyotype

The purpose of conducting a karyotype analysis is to evaluate various characteristics of chromosomes, including their number, size, and symmetry, as well as the position of centromeres, secondary constrictions, and repetitive DNA sequences (Heslop-Harrison and Schwarzacher 2011). These characteristics are useful for cytotaxonomy to understand evolutionary relationships, plant breeding programs for the development of new varieties, and the detection of chromosomal abnormalities.

5.3 Type of Karyotypes

Chromosome analysis reveals various species- or cultivar-specific information, with karyotype asymmetry being one of the key aspects. This concept arises from the significant variability in chromosome sizes found within the karyotype of any given plant species. Karyotype symmetry often involves changes in chromosome size and morphology, typically resulting from expansions or deletions in DNA sequences or through centric fusions or fissions, which are sometimes accompanied by disploidy (Weiss-Schneeweiss and Schneeweiss 2013). Over the years, numerous scientists have studied karyotype asymmetry. Notably, Stebbins (1971) in his influential work "Chromosomal Evolution in Higher Plants," proposed a qualitative-quantitative method for estimating karyotype asymmetry across twelve categories (Table 5.1). This approach considers four classes (ranging from 1 to 4) based on the increasing proportion of chromosomes with an arm ratio of less than 2:1, combined with three classes (from A to C) that reflect the increasing ratio of the largest to the smallest chromosome in a complement (Peruzzi and Eroğlu 2013). In contemporary studies, the coefficient of variation is now commonly employed as a statistically appropriate parameter to assess even minor variations in chromosome sizes within a karyotype.

Table 5.1 The classification of karyotypes in relation to their degree of asymmetry according to Stebbins (1971)

	Proportion of chromosomes with arm ratio			
Ratio Largest/smallest	1.00 (1)	0.99–0.51 (2)	0.50–0.01 (3)	0.00 (4)
<2:1 (A)	1A	2A	3A	4A
2:1–4:1 (B)	1B	2B	3B	4B
>4:1 (C)	1C	2C	3C	4C

5.3.1 Symmetric Karyotype

A karyotype is classified as symmetric when it has a higher number of metacentric and submetacentric chromosomes with relatively minimal variations in length between the smallest and largest chromosomes (Fig. 5.1a). This type of karyotype is considered to be a more primitive state. Symmetric karyotype is observed in *Allium cepa*.

5.3.2 Asymmetric Karyotype

Instead of uniformity among chromosomes, significant variability in chromosome size and a diverse array of centromere positions, often with a high proportion of acrocentric or telocentric chromosomes compared to metacentric and submetacentric chromosomes, are the remarkable features of an asymmetric karyotype (Fig. 5.1b). This type of karyotype represents greater karyotype heterogeneity

(Paszko 2006). According to Stebbins (1971), asymmetric karyotypes are a relatively advanced and derived trait. They arise from symmetrical karyotypes through the cumulative result of numerous chromosomal rearrangement processes that alter centromere position, the relative size of chromosome arms, or the chromosome number. Pericentric inversion, translocation, and differential amplification of heterochromatic regions are the main evolutionary drivers of asymmetric karyotype development. Pericentric inversion is considered the most potent contributor to intrachromosomal asymmetry. In plants, differential amplification of heterochromatic regions across different chromosomes can dramatically alter the relative sizes of chromosomes and increase overall asymmetry without significantly changing coding gene content. Additional structural changes, such as paracentric inversions, deletions, duplications, and reciprocal translocations, also contribute to karyotype diversification.

- *Bimodal Karyotype*: In certain instances, inter-chromosomal asymmetry can manifest significantly, resulting in noticeable differences in chromosome lengths. This leads to a karyotype that can be categorized into two distinct groups: one comprising larger chromosomes and the other consisting of smaller ones. The term "bimodal" was first introduced by Avdulov in 1931 to describe karyotypes that display two highly discontinuous chromosomal subsets: one with large chromosomes and the other with small chromosomes. Although asymmetry and bimodality are related concepts, they are not identical. An asymmetrical karyotype is not necessarily bimodal, but a bimodal karyotype always exhibits some degree of asymmetry (Araújo et al. 2024).

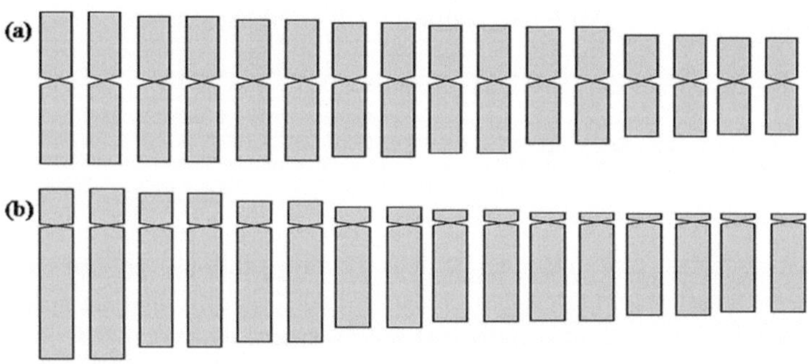

Fig. 5.1 Ideograms of the hypothetically possible (**a**) symmetric and (**b**) asymmetric karyotypes with $2n = 16$ chromosomes

Bimodal karyotypes have been observed in several plant genera, including *Aloe vera* ($2n = 14$ chromosomes), which consists of eight long and six small chromosomes (Fig. 5.2a, b), and *Agave cupreata* ($2n = 60$ chromosomes), which contains five pairs of large homologous chromosomes and 25 pairs of small homologous chromosomes. Additionally, *Hypochaeris catharinensis*, a member of the Asteraceae family, exhibits a bimodal karyotype with $2n = 8$ chromosomes, comprising two pairs of long chromosomes and two pairs of small chromosomes. Bimodal karyotypes have also been reported in various animal groups, including birds, reptiles, and amphibians.

The evolutionary significance of the origin and evolution of bimodal karyotypes remains a topic of ongoing study. Presence of bimodal karyotype in some evolutionary lineages has evolutionary significance as it may be related to various evolutionary processes, such as chromosomal reorganization or rearrangements, allopolyploidy, and differential accumulation of repetitive DNA in a specific chromosome subset (Vosa 2005; Vimala et al. 2021; Ramos et al. 2023; Araújo et al. 2024).

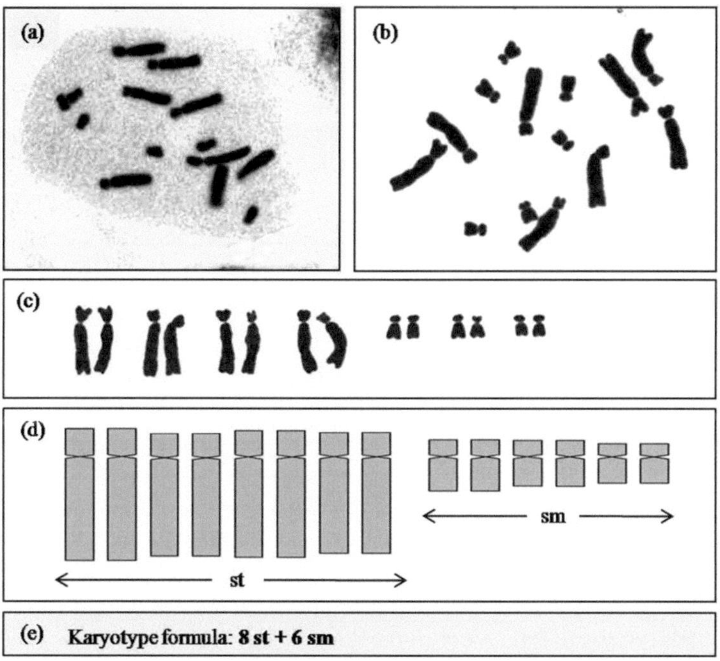

Fig. 5.2 Orcine-stained (**a**) and Giemsa-stained (**b**) metaphases display a bimodal karyotype of *Aloe vera* with eight long and six short chromosomes. The karyogram (**c**), ideogram (**d**), and karyotype formula (**e**) are derived based on these observations

- *Alpha-Level Karyology*: Alpha-level karyology focuses on the fundamental morphological analysis of chromosomes through relatively straightforward, cost-effective, and accessible techniques. This foundational approach involves the characterization of key chromosomal parameters, such as counting the chromosome numbers in an organism, determining basic ploidy levels, assessing chromosome size and shape, identifying centromere positions, and classifying chromosomes based on absolute length, relative length, centromere index, and arm ratios. It is particularly useful for initial species characterization, basic studies of genetic diversity, and assessments of reproductive compatibility.
- *Beta-Level Karyology*: Beta-level karyology represents a more advanced approach that delves deeper into the molecular understanding of chromosomal structure and genetic organization through sophisticated analytical methods. This method goes beyond basic morphological observations, offering the potential to uncover specific chromosomal regions, pinpoint gene locations, analyze genetic variations, and map molecular markers. It employs techniques such as molecular-level chromosome analysis, detailed genetic mapping, and chromosomal banding methods (including G-banding, C-banding, and R-banding), as well as molecular cytogenetic investigations using fluorescence techniques. Beta-level investigations are essential for diagnosing genetic disorders, researching evolutionary biology, developing precision breeding programs, and comprehending the molecular mechanisms underlying diseases.

5.4 Methods for Karyotype Analysis

The primary objective of karyotype preparation is to obtain high-quality chromosome plates that exhibit detailed karyomorphological features. These features are essential for measuring, analyzing, and interpreting chromosomal characteristics. The preparation process involves several systematic steps, including pretreatment, fixation, staining, slide preparation (through squashing or maceration), observation, documentation, analysis, and interpretation (Fig. 5.3). Each plant species requires a standardized protocol for chromosome preparation and staining to yield high-quality chromosome plates. Detailed methodologies for pretreatment, fixation, staining, and slide preparation have been discussed in Chap. 3.

Once the somatic chromosomes have been prepared, the slides are examined under a microscope. Cells that exhibit well-scattered metaphase chromosomes with distinct karyomorphological features are digitally documented by capturing

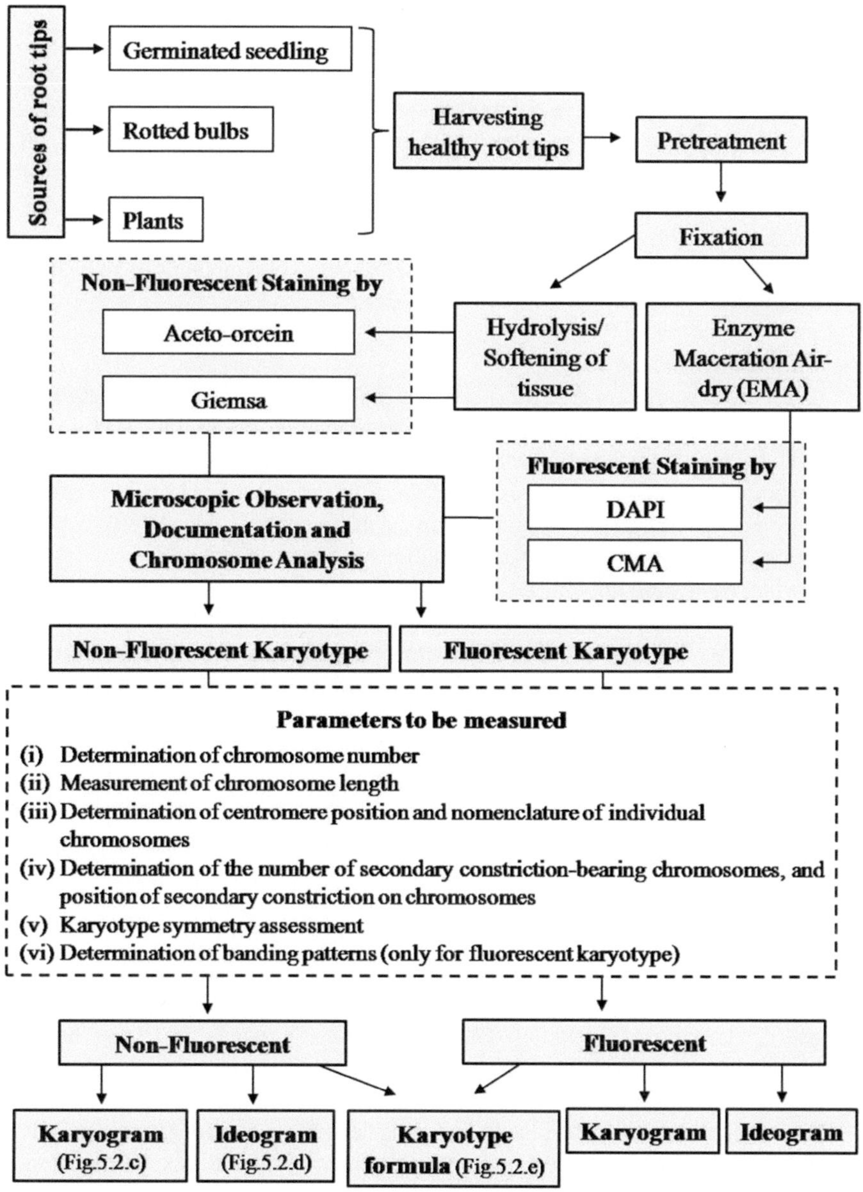

Fig. 5.3 Steps of karyotype analysis by classical and molecular cytogenetics approaches

photographs using a camera attached to the microscope. Alternatively, these metaphase plates can be rendered through manual drawing techniques, such as using a prism or a camera lucida, which aids in understanding the karyotype. Subsequently, chromosomal characterization can be performed by analyzing the photographs and manually creating camera lucida drawings, or by utilizing software tools for further analysis.

5.4.1 Karyomorphometric Parameters

Karyomorphometric analysis enhances our understanding of the karyotype. According to Levin (2002), a comprehensive description of the karyotype should include chromosome number, total length of haploid chromosome complements, absolute and relative sizes of the chromosomes, the number and positions of secondary constrictions or satellites, and the karyotype formula. Table 5.3 provides a sample datasheet of *Allium cepa*, highlighting typical parameters used for karyotype analysis. Common karyomorphometric parameters and methods of analysis are discussed further below.

(i) *Chromosome Number*

The chromosome number for a specific plant species, variety, germplasm, or race can be determined by counting the total number of chromosomes in each cell. It is recommended to count at least 25 metaphase plates to confirm the chromosome number of the plant sample. Any variability in chromosome numbers among the cells of single root tips or different root tips should be noted, along with the frequency of such variations.

(ii) *Chromosome Length*

The parts of chromosomes on either side of the centromere are called chromosome arms. The total number of chromosome arms in a karyotype is referred to as the fundamental arm number (FN). The lengths of the two arms of a chromosome may differ. The longer arm is designated as "q," while the shorter arm is referred to as "p." The lengths of the long arm (L) and the short arm (S) of each chromosome are measured. By convention, the p arm is always positioned at the top in a karyotype. For chromosomes with secondary constrictions, the length of the additional arm formed by this feature is also measured. The total length of individual chromosomes is calculated by adding the lengths of all the arms together. It is important to analyze numerous metaphase plates, as the level of chromosome condensation may vary from cell to cell.

From these measurements, other karyological parameters can be derived, such as absolute chromosome length (CL), relative chromosome length (RL), average chromosome length (ACL), total chromatin length (TCL), and haploid chromosome length (HCL). When conducting karyomorphometric analyses, all measurements should be expressed in microns. While software used for chromosome measurement typically provides lengths in microns, measurements obtained with a camera Lucida are in millimeters. Students should convert millimeter measurements to microns by dividing the arm length measured in millimeters by the magnification of the image drawn.

Based on chromosome length across various species, Lima de Faria (1980) categorized chromosomes into four classes: very small ($\leq$ 1 μm), small ($>$ 1 μm and $\leq$4 μm), intermediate ($>$ 4 μm and $<$12 μm), and large ($>$ 12 μm and $\leq$60 μm). In contrast, Fukui (1996) considered chromosomes to generally be either large (8–10 μm or more) or small (1–3 μm), although he acknowledged the potential existence of intermediate chromosome types in many plant species.

- *Individual chromosome length (CL)*: The absolute length of an individual chromosome can be calculated by adding the lengths of all arms of the chromosome.

$$\text{Total length of each chromosome} = \text{Long arm length}(\text{L}) + \text{Short arm length}(\text{S}) + \text{Additional arm length}(\text{SC if present})$$

- *Relative chromosome length (RL)*: The relative chromosome length is calculated as the length of a specific chromosome multiplied by 100 and divided by the total length of all chromosomes in the haploid set. This is expressed as a percentage and indicates the proportion of the entire genome represented by that particular chromosome. However, it is essential to note that RL is a linear measurement and does not take into account mass or volume.
- *Total chromatin length (2n) of a karyotype (TCL)*: Total chromatin length is determined by adding the absolute total chromosome length of individual chromosomes of the karyotype.

$$\text{Total chromatin length }(\text{TCL}) = \Sigma(\text{CL}_1 + \text{CL}_2 + \text{CL}_3 + \text{CL}_4 + \ldots \text{CL}_N \text{Chromosomes})$$

N = Number of chromosomes

- *Haploid chromosome length (HCL)*: Haploid chromosome length (HCL) of a karyotype is determined by the total chromosome length divided by the haploid number of chromosomes of the species.

$$\text{Haploid chromosome length }(\text{HCL}) = \text{Total chromatin length }(\text{TCL}) / n$$

n = Haploid number of chromosomes

- *Average chromosome length (ACL)*: The average chromosome length (ACL) is determined by total chromatin length (TCL) divided by total number of diploid chromosomes (N = 2n).

$$\text{Average chromosome length }(\text{ACL}) = \text{Total chromatin length }(\text{TCL}) / \text{total number of diploid chromosomes }(N)$$

- *Disparity index (DI)* of chromosomes in a karyotype is calculated following the formula of Mohanty et al. (1991).

$$\mathrm{DI} = (\text{Longest chromosome} - \text{Shortest chromosome}) / (\text{Longest chromosome} + \text{Shortest chromosome}) \times 100$$

- *Variation coefficient (VC)* among chromosome complements is determined by the following formula:

$$\mathrm{VC} = (\text{Standard deviation} / \text{The mean length of chromosome}) \times 100$$

(iii) *Centro*mere *Position and Nomenclature of Individual Chromosomes*
Determination of centromere position is foundational information for a karyotype. The centromere divides a chromosome into two arms and appears as a cytologically visible primary constriction and site of the kinetochore. The position of the centromere can be calculated based on the arm ratio (r) or the centromeric index (CI). These parameters are used for chromosome classification and nomenclature, such as metacentric (m), sub-metacentric (sm), subtelocentric (st), and telocentric (t) (Levan et al. 1964; Stebbins 1971; Guerra 1986).

- *Arm ratio (r)*: Arm ratio is defined as the length of the longer arm of the chromosome divided by the length of the shorter arm.

$$\text{Arm ratio}(r) = \text{Length of the long arm } (L) / \text{Length of the short arm } (S)$$

It's calculated to determine the centromeric position (Levan et al. 1964, subsequently modified by Stebbins (1971) and Guerra (1986)). The arm ratio is always greater than or equal to one and serves as an alternative method to the centromeric index for describing the positions of the centromere and the relative lengths of the two arms.

- *Centromeric index (CI)*: The centromeric index is an important measurement in cytogenetics that measures the position of the centromere within a chromosome. It represents the ratio of the length of the short arm of a chromosome to the total length of that chromosome, multiplied by 100. Based on the centromeric index, the chromosomes can be classified as metacentric, sub-metacentric, and telocentric. It indicates the relative lengths of the two arms of a chromosome. Mathematically, it can be expressed as

$$\text{Centromeric index}\left(\text{CI}\right) = \left\{\begin{array}{l}\text{Length of the shorter arm of a chromosome}\left(\text{S}\right) \\ /\text{Total length of the chromosome}\left(\text{L}+\text{S}\right)\end{array}\right\} \times 100$$

Its value can range from 0.5 (if S = L) to 0 (if S = 0) and is expressed as a percentage. It is fundamentally used in calculating *the total form percentage or the mean centromeric index value (TF%)* with the help of the following formula (Huziwara 1962).

$$\text{TF\%} = \left(\text{Total sum of the short arm length}\right) / \left(\text{Total sum of chromosome length}\right) \times 100$$

Different researchers proposed different methods of chromosome classification and nomenclature based on the position of the centromere. Table 5.2 illustrates chromosome classification and nomenclature of Levan et al. (1964), which was later modified by Stebbins (1971), and Guerra (1986).

The position of the centromere may change within and between related species members due to natural structural rearrangements, such as pericentric inversion and translocation. With enormous diversity in plants, cytogeneticists have observed that in a few genera and species, the centromere may not be located in a single position on the chromosomes. They may be dispersed over the entire length of the chromosome. Such chromosomes are called holocentric or holokinetic found in *Carex*, *Rhynchospora* (Cyperaceae), *Cuscuta* (Convolvulaceae), etc.

Table 5.2 Chromosome classifications and nomenclature of Levan et al. (1964), Stebbins (1971), and Guerra (1986)

Levan et al. (1964) Arm ratio			Stebbins (1971), and Guerra (1986) CI value	
Metacentric	Median point	1	M	50
	Median region	≥ 1–1.7	m	50–39.5
Submetacentric	Submedian region	≥ 1.7–3	sm	39.5–25
Acrocentric	Subterminal region/acrocentric region	≥ 3.0–7.0	st / Ac	25–12.5
Telocentric	Terminal region	≥ 7–∞	t	12.5–0

(iv) *Number of Secondary Constriction-Bearing Chromosomes and Their Position on Chromosomes*
The identification of chromosomes with secondary constrictions can be achieved through the simple visualization of numerous metaphase plates. The relative position of these constrictions can be determined using either the "r value" or the "CI value" calculations. However, classical chromosomal preparation and staining techniques may have limitations in clearly identifying secondary constriction regions under the microscope, especially when they are located at intercalary positions. Staining methods, such as CMA and DAPI, can significantly enhance the detection and confirmation of secondary constriction presence and its position.

(v) *Karyotype Symmetry Assessment*
According to the Paszko (2006) and Stebbins's classification (Stebbins 1971) the symmetry of a karyotype can be analyzed by the following methods:

- *Mean centromeric asymmetry (MCA)*: This is calculated from the formula, Mean $(L - S) / (L + S) \times 100$, where L = total length of long arms and S = total length of short arms.
- *Coefficient of variation of chromosome length (CV_{CL})*: This measures inter-chromosomal asymmetry and is calculated as follows:

$$\mathrm{CV}_{\mathrm{CL}} = \left(S_{\mathrm{CL}} / X_{\mathrm{CL}}\right) \times 100,$$

  where S_{CL} = standard deviation of chromosome length in a chromosome set and X_{CL} = mean chromosome length.
- *Coefficient of variation of the centromeric index (CV_{CI})*: This is calculated by the formula $(S_{CI} / X_{CI}) \times 100$. Where S_{CI} = standard deviation; X_{CI} = mean centromeric index.

(vi) *Karyogram Preparation*
A karyogram is a photographic representation of an organism's chromosomes, with the individual homologous pairs sorted, grouped, and arranged in descending order of length. If sex chromosomes are present, they are typically placed at the end of the autosomes. This arrangement offers a comprehensive overview of an individual organism's chromosomal makeup, including the total chromosome count, sex chromosomes (if present), and chromosomal abnormalities, such as deletions, duplications, or translocations.

(vii) *Preparation of Ideogram*
A chromosomal ideogram is an effective way to visualize genomic data through a diagrammatic representation of the karyotype of a species, population, or individual. It provides a visual summary of individual chromosome features, including chromosome number, length, position of centromere, and banding patterns. Ideograms are essential tools in cytogenetics, helping researchers to identify chromosomal abnormalities, such as translocations,

deletions, duplications, and inversions, as well as the evolution of the organism.

Karyograms and ideograms can be created from high-resolution digital microphotographs of stained chromosomes, captured with a digital camera, using specialized software or Camera Lucida drawing techniques. Chromosomes are paired and arranged in descending order based on their total length, considering centromeric positions, the location of secondary constrictions, and significant chromosomal landmarks. Generally, chromosomes with secondary constrictions are placed after those without, regardless of their length.

(viii) *Karyotype Formula*

Every organism possesses a unique chromosomal signature that serves as a blueprint for the organism. The karyotype formula is a standardized method used to represent an individual's complete set of chromosomes in a concise yet comprehensive manner. The karyotype formula includes the total number of chromosomes within a species, their length, and specifies the types of chromosomes (M, m, sm, st, and T) present in a karyotype. Additionally, it may indicate the presence of sex chromosomes and any chromosomal numerical aberrations (such as trisomy and monosomy) if applicable. Variability in the karyotype formula is not random. It often reflects evolutionary history and footprints.

5.5 Fluorescent Karyotype

Advancements in molecular biology have significantly enhanced our understanding of plant genomes. A substantial portion of plant genomes consists of repetitive DNA sequences (Biscotti et al. 2015). These repetitive sequences are now recognized as valuable molecular markers for analyzing genetic diversity and exploring genome evolution (Adhikari et al. 2017). Their specific chromosomal locations vary by species and can be identified in centromeric (Koo et al. 2010), pericentromeric, or subtelomeric regions (Han et al. 2008), as well as being dispersed throughout various chromosomal regions (Kubis et al. 1998). As a result, these sequences have been utilized as cytogenetic probes to examine chromosomal organization in plants (Cuadrado et al. 2008), serve as karyotyping markers (Han et al. 2008), and enable cytogenetic comparative analysis among closely related species (Zhang et al. 2016). The accumulation of molecular data regarding chromosomes has expanded the field of molecular cytogenetics and rejuvenated research on plant chromosomes (Setiawan et al. 2019).

In addition to classical karyotyping, fluorescent karyotyping is an advanced cytogenetic technique that uses various non-fluorescent and fluorescent staining techniques, such as G-banding, R-banding, and C-banding, to reveal distinct banding patterns along the chromosomes. Traditional chromosome staining techniques employed basic dyes that produced uniform single-colored light and dark bands on chromosomes. In contrast, fluorescent karyotyping offers a more detailed and vibrant colorful visualization and analysis of chromosomal structure, composition, and potential abnormalities under a fluorescence microscope equipped with the appropriate filters. Two commonly used contrasting DNA base-specific fluorescent dyes are 4′-6-diamidino-2-phenylindole (DAPI) and Chromomycin A3 (CMA). These dyes are highly effective for visualizing the position and nature of diverse heterochromatin blocks that consist of repetitive DNA sequences found directly on chromosomes. Notably, the secondary constriction of nucleolar organizer regions (NORs) serves as chromosomal landmarks composed of tandem repeats of ribosomal genes (rRNA). In eukaryotes, each unit includes three genes coding for 18S, 5.8S, and 28S ribosomal RNA, which are separated by two intergenic spacers and an external transcribed spacer. CMA can confirm the location of secondary constrictions, while fluorescent in situ hybridization (FISH) is capable of identifying rRNA genes in NOR regions. As a result, the term "fluorescent karyotype" has been introduced to represent the molecular features of chromosomes in ideograms. Based on the preferential binding of CMA and DAPI to GC- and AT-rich sequences, Guerra (2000) and Barros e Silva and Guerra (2010) have identified three possible types of heterochromatic signals/bands: GC-rich (CMA^+ / $DAPI^-$), AT-rich ($DAPI^+$ / CMA^-), or AT/GC-neutral ($DAPI^0$ / CMA^0) in different species.

High-resolution images can be captured through digital imaging systems, and computer software assists in analyzing the unique pattern of bands or signals on chromosomes from the documented images. The number, position, intensity, and size of these bands are relevant in karyotype analysis as they serve as unique, chromosome-specific identifiers that can be determined for karyotyping. In addition to the TCL and r and CI values, specific banding patterns can help identify and pair homologous chromosomes (Tables 5.3 and 5.4).

Fluorescent karyotyping using fluorescent dyes or fluorescence in situ hybridization (FISH) technology enhances chromosome visualization by introducing color and precision. This technique improves the detection of subtle chromosomal abnormalities associated with genetic disorders that traditional staining methods may overlook. Additionally, it helps in understanding chromosomal evolution by comparing the structures of chromosomes across different species. Various fluorescent-based methods can be employed in molecular cytogenetic research. Some of these methods are discussed in the next sections.

5.6 Software of Karyotype Analysis

In modern cytogenetic research, specialized software programs, such as ISCN (International System for Human Cytogenetic Nomenclature) software, CytoVision, and Ideo-Kar (Mirzaghaderi and Marzangi 2015) are used to facilitate the precision of chromosomal measurements and analysis, interpretation, and the generation of high-quality ideograms from digitally captured or scanned images.

Table 5.3 Model karyo-morphometric parameters used for karyotype analysis [example: *Allium cepa*]

Sl. No.	Species name	Chromosome no. (2n)	Length of longest chromosome (μm)		Length of shortest chromosome (μm)		Total chromatin length (2n) (mean ± S.D)	Average chromosome length (mean ± S.D)	[a]Number and position of SAT chromosome	Ordering no. of the SAT bearing pair	Symmetry	Diploid Karyotype formula
			Absolute (mean ± S.D)	Relative (mean ± S.D)	Absolute (mean ± S.D)	Relative (mean ± S.D)						
1	*Allium cepa*	16	3.21 ± 0.09	5.56 ± 0.08	1.86 ± 0.02	3.22 ± 0.10	57.7 ± 0.96	2.40 ± 0.04	2, T erminal	10th	2A	10 m + 10sm + 2St + 2st.sat

Table 5.4 Karyo-morphometric parameters generally used for the preparation of fluorescent karyotype [example: *Allium cepa*]

Sl. No.	Species name	Chromosome no. (2n)	Fluorescent banding patterns				Total number of fluorescent signals (CMA^{+ve}& $DAPI^{-ve}$)	Fluorescent karyotype (2n)[a]	Remarks if any
			Types and number of signals		Position of the signals	Chromosome number			
			$DAPI^+$	CMA^+					
1	*Allium cepa*	16	0	32	Terminal region	1-8th	34	14A + 2B	
			0	2	Satellite region	6th			

[a]Fluorescent karyotype – A: CMA-positive terminal signal, B: CMA-positive signal at the terminal part and the secondary constriction region

5.7 Statistical Analysis

Statistical analysis is an essential component of karyomorphological studies, as variations in the absolute length of chromosome arms can arise due to differences in chromosomal condensation between cells. To ensure a comprehensive analysis, all karyomorphological features should be reported as mean ± standard deviation. Additionally, a one-way analysis of variance (ANOVA) should be conducted to assess whether the observed differences are statistically significant at a threshold of $p \leq 0.05$. The SPSS statistical package (SPSS Inc. (IBM), Chicago) is a suitable tool for performing ANOVA and subsequent post hoc analyses. The concept and procedure of ANOVA analysis are given in Chap. 11.

5.8 Applications of Plant Karyotyping

The plant karyotyping represents an intriguing area of botanical research and genetics. Karyotype constancy ensures the accurate transfer of genetic material to progeny, while variation facilitates evolutionary changes (Weiss-Schneeweiss and Schneeweiss 2013). The advent of molecular cytogenetics and advancements in microscopic and statistical software has significantly expanded the scope of karyotype analysis beyond what was possible just a few decades ago. It now encompasses not only numerical variations but also intricate karyograms and ideograms, detailing the total length of haploid and diploid chromosome complements, absolute and relative sizes of chromosomes, centromere positioning, and the number and location (terminal or intercalary) of secondary constrictions or satellites. Additionally, it allows for the identification of heterochromatic blocks rich in AT or GC repetitive sequences and the direct localization of rDNA genes on chromosomes. By comparing the chromosomes of different taxa, researchers can gain insights into the patterns and mechanisms of karyotype evolution and understand the significance of these processes for diversification and speciation (Weiss-Schneeweiss et al. 2008). This evolving understanding suggests that karyotypes may not be exclusively species-specific but also specific to cultivars or populations. Karyotyping in plants serves several essential functions that enhance our understanding of plant genetics, evolution, and reproductive characteristics.

The implications of plant karyotyping are profound, aiding in developing climate-resistant crops, enhancing our understanding of biodiversity, preserving endangered plant species, and improving agricultural productivity.

(a) *Chromosome Characterization and Identification*: Plant karyotyping allows researchers to accurately count and determine chromosome numbers, basic structure, and morphology of chromosome in any plant species. This process helps distinguish different types of chromosomes within a karyotype and identify unique chromosomal characteristics of particular plant species. By counting the chromosome numbers, researchers can assess the polyploidy level of any plant. Additionally, in some species, karyotyping can detect the presence of B chromosomes.

(b) *Genetic Diversity and Evolutionary Studies*: Karyotype analysis offers valuable insights into the evolutionary relationships among various plant species by examining the structural and numerical variations in their chromosomes. This analysis can reveal the mechanisms behind chromosomal evolution and speciation, including structural and numerical variability in chromosome. Additionally, it helps identify polyploid species and trace their evolutionary origins.

(c) *Applications in Plant Breeding and Agriculture*: In agricultural and horticultural contexts, karyotype information can be used to select plants with specific chromosomes or chromosomal regions associated with desirable genetic traits (Marker-Assisted Selection), identify potential plant breeding lines with optimal chromosomal configurations, and develop new hybrid crop varieties with improved agronomic characteristics by understanding chromosome compatibility between different species for hybridization.

(d) *Detection of Chromosomal Abnormalities*: Plant karyotyping can reveal structural abnormalities (such as deletions, duplications, and inversions) and numerical abnormalities (such as aneuploidy), which can have significant genetic consequences.

(e) *Species Identification and Taxonomic Classification*: Subtle differences in karyotypes can distinguish closely related species that may appear morphologically similar. This is particularly valuable in identifying cryptic species. By analyzing chromosome number, size, and morphology, karyotypes contribute to plant taxonomy, known as cytotaxonomy, by.
   (i) Providing additional criteria for species classification.
   (ii) Helping to distinguish between closely related plant species.
   (iii) Supporting phylogenetic research.
   (iv) Complementing morphological and molecular classification methods when morphology and other characteristics are insufficient for differentiation.

(f) *Conservation Genetics*: Karyotype analysis can be employed to evaluate genetic diversity within and between plant populations. This information is crucial for conservation efforts, as it allows for the identification of populations exhibiting high genetic diversity, which is vital for their long-term survival.

References

Adhikari S, Saha S, Biswas A, Rana TS, Bandyopadhyay TK, Ghosh P (2017) Application of molecular markers in plant genome analysis: a review. Nucleus 60(3):283–297

Araújo LF, Dornelas CS, Felix LP, Nollet F (2024) What defines a bimodal karyotype? Bimodality revisited. Caryologia 77(4):43–60

Barros e Silva AE, Guerra M (2010) The meaning of DAPI bands observed after C-banding and FISH procedures. Biotech Histochem 85(2):115–125

Biscotti MA, Olmo E, Heslop-Harrison JS (2015) Repetitive DNA in eukaryotic genomes. Chromosom Res 23:415–420

Cuadrado A, Cardoso M, Jouve N (2008) Increasing the physical markers of wheat chromosomes using SSRs as FISH probes. Genome 51(10):809–815

Fukui K (1996) Plant chromosomes at mitosis. In: Fukui K, Nakayama S (eds) Plant chromosomes: laboratory methods. CRC Press, Boca Raton, pp 1–17

Guerra MD (1986) Reviewing the chromosome nomenclature of Levan et al. Braz J Genet 9:741–743

Guerra M (2000) Patterns of heterochromatin distribution in plant chromosomes. Genet Mol Biol 23:1029–1041

Guerra M (2008) Chromosome numbers in plant cytotaxonomy: concepts and implications. Cytogenet Genome Res 120(3–4):339–350

Han YH, Zhang ZH, Liu JH, Lu JY, Huang SW, Jin WW (2008) Distribution of the tandem repeat sequences and karyotyping in cucumber (*Cucumis sativus* L.) by fluorescence in situ hybridization. Cytogenet Genome Res 122(1):80–88

Heslop-Harrison JS, Schwarzacher T (2011) Organisation of the plant genome in chromosomes. Plant J 66(1):18–33

Huziwara Y (1962) Karyotype analysis in some genera of Compositae. VIII. Further studies on the chromosomes of Aster. Am J Bot 49(2):116–119

Koo DH, Nam YW, Choi D, Bang JW, De Jong H, Hur Y (2010) Molecular cytogenetic mapping of *Cucumis sativus* and *C. Melo* using highly repetitive DNA sequences. Chromosom Res 18:325–336

Kubis S, Schmidt T, Heslop-Harrison JS (1998) Repetitive DNA elements as a major component of plant genomes. Ann Bot 82:45–55

Levan A, Fredga K, Sandberg AA (1964) Nomenclature for centromeric position on chromosomes. Hereditas 52:201–220

Levin DA (2002) The role of chromosomal change in plant evolution. Oxford University Press, USA

Lima-De-Faria A (1980) Classification of genes, rearrangements and chromosomes according to the chromosome field. Hereditas 93(1):1–46

Mirzaghaderi G, Marzangi K (2015) IdeoKar: an ideogram constructing and karyotype analyzing software. Caryologia 68(1):31–35

Mohanty BD, Ghosh PD, Maity S (1991) Chromosomal analysis in cultured cells of barley (*Hordeum vulgare* L.) structural alterations in chromosomes. Cytologia 56(2):191–197

Paszko B (2006) A critical review and a new proposal of karyotype asymmetry indices. Pl Syst Evol 258(1):39–48

Peruzzi L, Eroğlu HE (2013) Karyotype asymmetry: again, how to measure and what to measure? Comp Cytogenet 7(1):1–9

Ramos LC, Báez M, Fuchs J, Houben A, Carvalho R, Pedrosa-Harand A (2023) Differential repeat accumulation in the bimodal karyotype of *Agave* L. Genes (Basel) 14(2):491

Setiawan AB, Wibowo A, Teo CH, Kikuchi S, Koba T (2019) Repetitive DNA sequences accelerate molecular cytogenetic research in plants with small chromosomes. Indones J Biotechnol 24(2):82–87

Stebbins GL (1971) Chromosomal evolution in higher plants. Edward Arnold Ltd., London

Vimala Y, Lavania S, Lavania UC (2021) Chromosome change and karyotype differentiation–implications in speciation and plant systematics. Nucleus 64(1):33–54

Vosa CG (2005) On chromosome uniformity, bimodality and evolution in the tribe Aloineae (Asphodelaceae). Caryologia 58(1):83–85

Weiss-Schneeweiss H, Schneeweiss G (2013) Karyotype diversity and evolutionary trends in Angiosperms. In: Greilhuber J, Dolezel J, Wendel JF (eds) Plant genome diversity, vol 2. Springer, Vienna, pp 209–230

Weiss-Schneeweiss H, Tremetsberger K, Schneeweiss GM, Parker JS, Stuessy TF (2008) Karyotype diversification and evolution in diploid and polyploid South American *Hypochaeris* (Asteraceae) inferred from rDNA localization and genetic fingerprint data. Ann Bot 101(7):909–918

Zhang L, Yang X, Tian L, Chen L, Yu W (2016) Identification of peanut (*Arachis hypogaea*) chromosomes using a fluorescence in situ hybridization system reveals multiple hybridization events during tetraploid peanut formation. New Phytol 211(4):1424–1439

Gametic Chromosome Analysis

6

Abstract

This chapter delves into the process of gametic chromosome analysis in diploid higher plants, focusing on the role of flowers as the primary source of haploid gametes required for sexual reproduction. It outlines the meiotic division occurring in the anthers, where pollen mother cells (PMCs) generate haploid daughter nuclei. The chapter emphasizes the importance of using fresh flower buds to prepare meiotic chromosomes, highlighting Barbara McClintock's method as a notable advancement. The preparation techniques, notably the conventional carmine smear method, are detailed, providing a step-by-step approach for creating cytological slides. The process requires specific materials, such as flower buds, slides, and aceto-carmine solution, as well as instruments such as needles and scalpels. The meticulous collection and preparation of flower buds, removal of anthers, and the application of staining solutions are outlined to achieve clear visualization of the meiotic chromosomes. It also touches upon the variance in bivalent numbers and the physical characteristics of haploid cells. Through documented case studies of various plant materials, the chapter serves as a practical guide for undergraduate and postgraduate students in understanding meiotic chromosome preparations. The methodologies enabled an efficient observation of different meiotic stages, enhancing comprehension of gametic chromosome analysis. Ultimately, the chapter highlights the crucial elements for successful gametic chromosome studies, contributing to the broader understanding of plant reproductive biology and genetics.

T. B. Jha, M. Halder, *Classical and Molecular Plant Cytogenetics*,
https://doi.org/10.1007/978-981-95-8265-5_6

6.1 Introduction

In diploid (2n) higher plants, flowers serve as the exclusive source of haploid gametes (n) required for sexual reproduction and fertilization. This process entails meiotic cell division occurring within the anthers and gynoecium of the flowers. Each meiotic cell contains a haploid number of chromosomes. Male meiocytes, also known as pollen mother cells (PMCs), develop within the anthers and undergo meiotic division. A significant number of PMCs are produced in the anther, and each PMC gives rise to four haploid daughter cells, each retaining the haploid chromosome number. Consequently, gametic chromosome analysis is predominantly performed using materials obtained from the anthers. It is important to note that gametic cell divisions within and between anthers may not occur synchronously. In contrast, the gynoecium produces a limited number of haploid gametes, which are more difficult to access and are rarely utilized in routine meiotic analysis. Details regarding meiotic cell divisions can be found in Chap. 2.

Barbara McClintock (1929) introduced a notable enhancement to the classical method for preparing meiotic (gametic) chromosomes in maize (Indian corn). Compared to preparations for mitotic chromosomes, meiotic chromosome preparations are less time-consuming because they do not require a pretreatment step and can be performed with fresh, unopened flower buds. By ensuring optimal sizes and timing (preferably in the morning), the various stages of division can be fixed by directly placing the tissues in a fixative solution.

In any species, the gametic, or haploid, chromosome number is represented by several bivalents, which may vary in number and be either even or odd. Haploid cells typically exhibit oval or round shapes rather than rectangular shapes. Moreover, the configuration of meiotic chromosomes significantly differs from that of mitotic chromosomes. The preparation technique is called "smearing," and the chromosomes are commonly stained with a 1% or 2% solution of acetocarmine. Along with the classical meiotic chromosome analysis, modern meiotic analysis using fluorescence dye also been reported in different plant species (Ross et al. 1996; Zhong et al. 1996; Hizume et al. 2014).

6.2 Methods of Meiotic Chromosome Preparation

6.2.1 Cytological Slide Preparation by Conventional Carmine Smear Method

6.2.1.1 Requisition

(i) Flower buds, (ii) grease-free slides, (iii) preferably square cover slips (22 mm), (iv) blotting paper, (v) 2% aceto-carmine solution, (vi) 45% acetic acid, (vii) spirit lamp, (viii) paraffin block or wax, (ix) marker, (x) arrow-headed needle, (xi) scalpel, (xii) forceps, and (xiii) needle.

6.2.1.2 Steps of Conventional Carmine Smear Preparation (Fig. 6.1)

(i) *Collection of Flower Buds:* Collect various sizes of unopened flower buds in the morning, preferably between 6 am and 11 am. These buds can be fixed in a solution of acetic alcohol (1:2 or 1:3) overnight at room temperature and preserved at −20 °C for long-term use. For short-term use, change them to 70% alcohol. Although gametic chromosome preparation can be performed

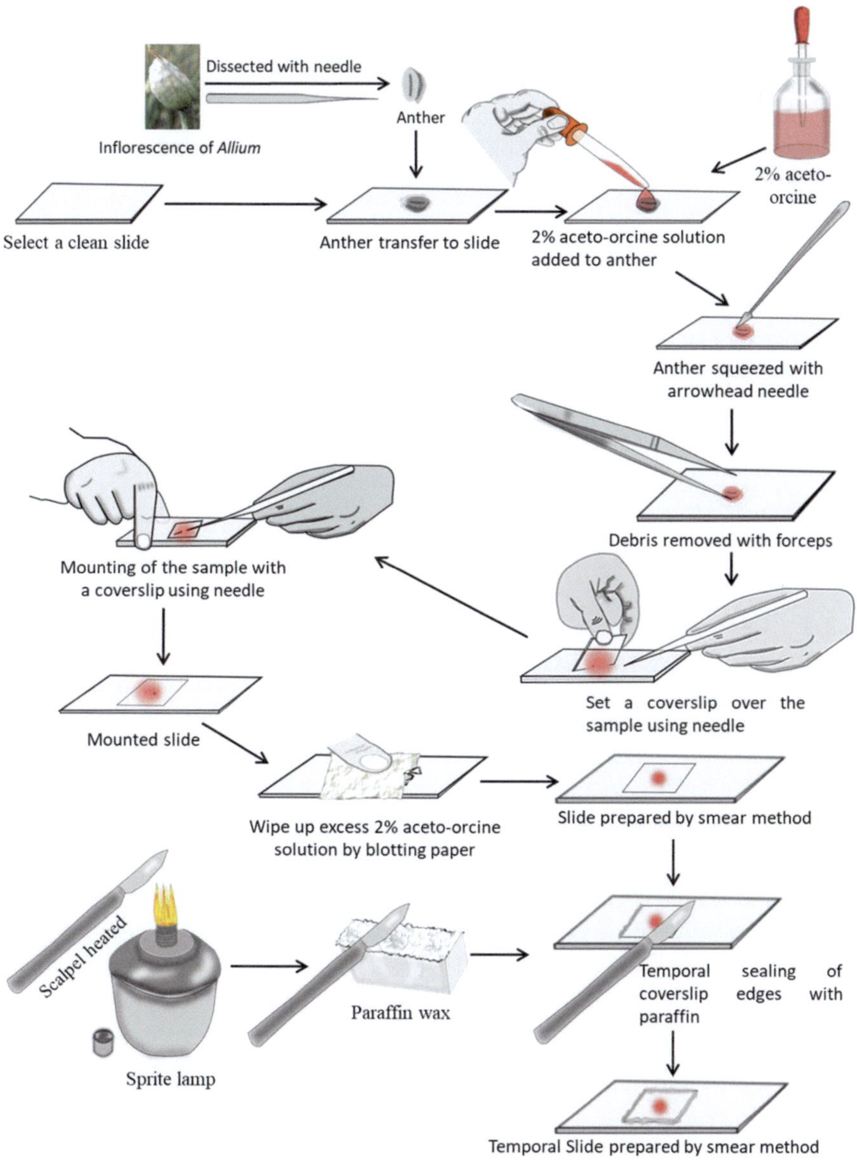

Fig. 6.1 Steps of meiotic slide preparation by the carmine smear preparation method

using either fresh or fixed anthers from unopened flower buds, fresh flower buds are recommended for hands-on training in undergraduate and postgraduate classes.

(ii) *Preparation for Meiotic Analysis:* Arrange the flower buds of different sizes on a clean glass slide, ensuring that neither very small nor very large buds are used as starting material.

(iii) *Anther Removal:* Using a fine needle, carefully remove two to three anthers and place them on a clean, grease-free slide. The remaining flower bud with its anthers should be placed in water.

(iv) *Application of Aceto-Carmine Solution:* Add one drop of 2% aceto-carmine solution to the anthers placed on a slide. To eliminate excess stain (if required), introduce a drop of 45% acetic acid at the edge of the cover glass.

(v) *Smearing Process:* Use a small iron needle, arrowhead needle, or scalpel to gently squeeze or smear the anthers in the aceto-carmine solution to release the microspores.

(vi) *Covering the Slide:* Use forceps to remove the anther walls and place a square cover glass (preferably 22 mm) over the preparation.

(vii) *Heating the Preparation:* Mildly heat the slide by passing it quickly over a flame of an alcohol lamp. This process effectively increases staining intensity and chromosome scattering. Gently press the cover glass to flatten the cells and blot away any excess fluid.

(viii) *Seal the Cover Glass:* Carefully seal the cover glass with paraffin or nail polish to avoid air bubble formation.

(ix) *Microscopic Examination:* View the preparation under a bright-field microscope.

6.2.2 Case Studies

Following the above-noted protocol, we have documented each essential step for gametic chromosome preparations in the following plant materials. These materials are familiar with gametic chromosome analysis at the undergraduate and postgraduate levels. We have documented and included different meiotic stages in each material for observation and to enhance their understanding of gametic chromosome preparations.

(i) *Ornithogalum virens/Albuca virens,* $n = 3$ bivalents (Fig. 6.2),
(ii) *Nigella sativa,* $n = 6$ bivalents (Fig. 6.3),
(iii) *Vicia faba,* $n = 6$ bivalents (Fig. 6.4),
(iv) *Aloe vera,* $n = 7$ bivalents (Fig. 6.5),
(v) *Allium cepa* (onion), $n = 8$ bivalents (Fig. 6.6),
(vi) *Setcreasea pallida* ("Purple Heart"), $n = 12$ bivalents (Fig. 6.7),
(vii) *Rhoeo discolor,* $n = 12$ chromosomes (Fig. 6.8).

Fig. 6.2 Meiosis study of *Albuca virens* ($n = 3$ bivalents). (**a**) Potted-grown *Albuca virens* plant, (**b**) inflorescence with opened and unopened flower buds, (**c**) opened flower with six anthers, and (**d–g**) PMC showing three bivalents at metaphase I stage of meiosis division

Fig. 6.3 Meiosis study of *Nigella sativa* (n = 6 bivalents). (**a**) Potted-grown *Nigella sativa* plants with unopened and opened flowers, (**b**) opened flower with many anthers, (**c**) isolated anthers, (**d–f**) PMC showing six bivalents at diakinesis substage of prophase I, (**g**) PMC showing six bivalents at metaphase I stage of meiosis division, (**h**) PMC showing anaphase I stage of meiosis division, and (**i**) PMC showing anaphase II stage of meiosis division

Fig. 6.4 Meiosis study of *Vicia faba* (n = 6 bivalents). (**a**) Field-grown plants, (**b**) enlarged view of flowers, (**c**) PMC showing diakinesis substage of prophase I, (**d**–**g**) PMC showing six bivalents in metaphase I stage, and (**h**) PMC showing telophase I stage of meiosis division

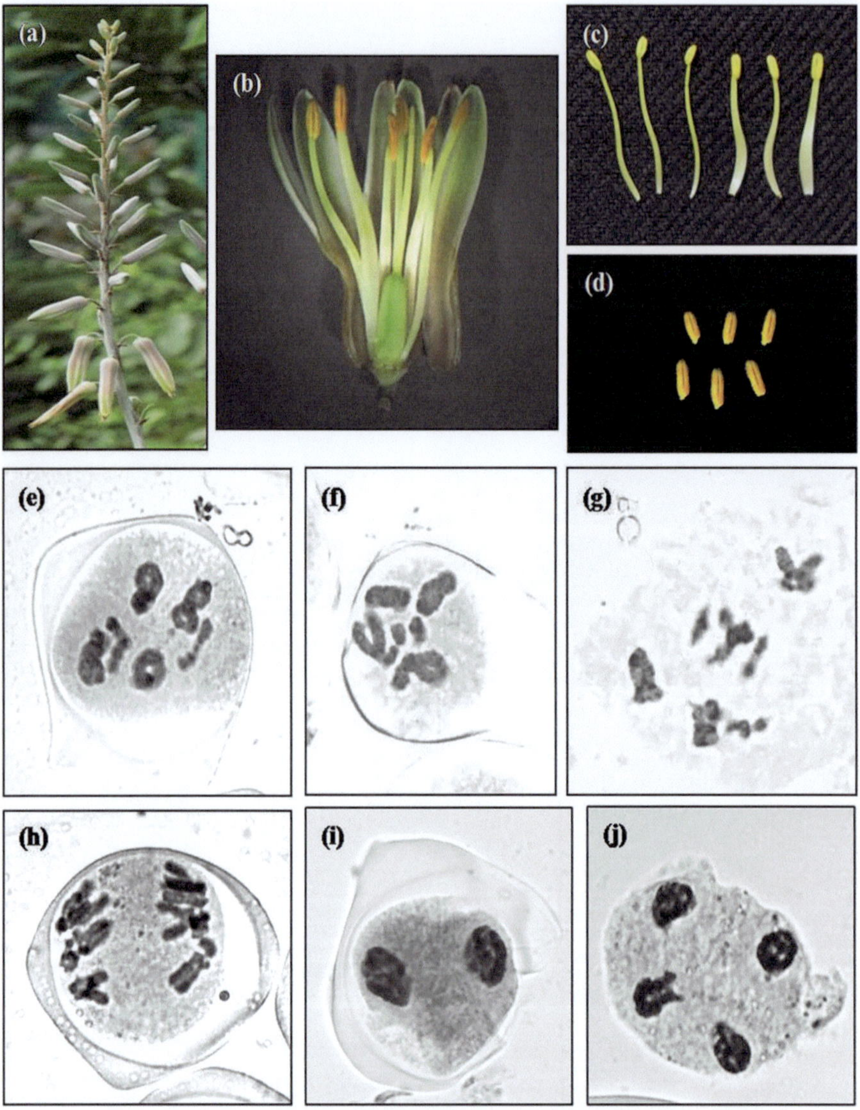

Fig. 6.5 Meiosis study of *Aloe vera* ($n = 7$ bivalents). (**a**) inflorescence of *Aloe vera* with opened and unopened flower buds, (**b**) dissected flower bud showing stamens and carpel, (**c**) isolated stamens (filament plus anther), (**d**) isolated anthers, (**e**) PMC showing diakinesis substage of prophase I, (**f**, **g**) PMC showing seven bivalents in metaphase I stage, (**h**) PMC showing anaphase I stage of meiosis division, (**i**) PMC showing telophase I stage of meiosis division, and (**j**) PMC showing telophase II stage of meiosis division

Fig. 6.6 Meiosis study of *Allium cepa* ($n = 8$ bivalents). (**a**) Field-grown *Allium cepa* plants, (**b–d**) different developmental stages of inflorescence, (**e**) sorting of unopened flower buds based on size, (**f**) anthers, (**g**) PMC showing eight bivalents at diplotene substage of prophase I, (**h–f**) PMC showing eight bivalents in metaphase I stage of meiosis division, (**j**) PMC showing anaphase I stage of meiosis division, (**k**) PMC showing telophase I stage of meiosis division, and (**l**) PMC showing telophase II stage of meiosis division

Fig. 6.7 Meiosis study of *Setcreasea pallida* (n = 12 bivalents). (**a**) Garden-grown *Setcreasea* plants with flowers, (**b**) inflorescence, (**c**) opened flower, (**d**) size-based sorted flower buds, (**e**) individual anthers, (**f**) PMC showing diakinesis substage of prophase I with 12 bivalents, (**g**) PMC showing metaphase I stage of meiosis division, (**h**) PMC showing metaphase II stage of meiosis division, (**i**) PMC showing anaphase I stage of meiosis division, (**j**) PMC showing telophase I stage of meiosis division (diad stage), and (**k**) PMC showing telophase II stage of meiosis division (tetrad stage)

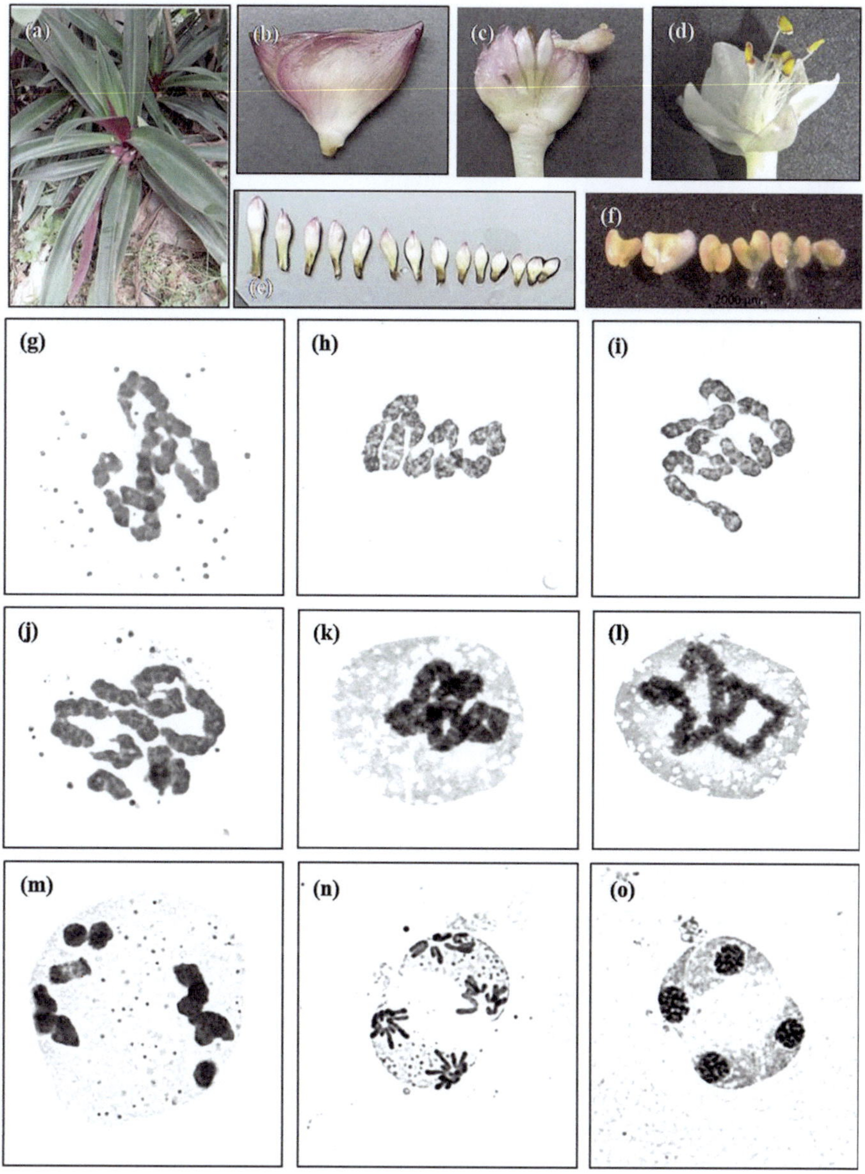

Fig. 6.8 Meiosis study of *Rhoeo discolor* (n = 12 chromosomes). (**a**) Potted-grown *Rhoeo discolor* plants with flowers, (**b**) side view of inflorescence with bracts, (**c**) top view of inflorescence with bracts, (**d**) inflorescence without bracts, (**e**) size-based sorted flower buds, (**f**) isolated anthers, (**g–l**) different configuration of ring structure formed by 12 univalent chromosomes, (**m**) anaphase I stage of meiosis division, (**n**) anaphase II stage of meiosis division, and (**o**) telophase II stage of meiosis division

6.2.3 Cytological Slide Preparation by Non-conventional Methods of Hizume et al. (2014)

(i) Collect appropriate sizes of fresh flower buds.
(ii) Transfer the flower buds to a 5–10 mL glass vial based on their size.
(iii) Add 2 mM KCl as a hypotonic solution.
(iv) Squeeze the flower buds with tweezers to release PMC in the hypotonic solution.
(v) Filter the PMC suspension through a coarse filter to remove debris.
(vi) Keep the filtered suspension in a 2 mM KCl solution for 30 min and then centrifuge at 1000 rpm for 3–5 min.
(vii) Decant half of the supernatant volume and keep the PMCs in the residual solution.
(viii) Add ice-cold fixative (1:3 acetic alcohol) and gently mix. Keep the mixture in crushed ice for 10 min.
(ix) Decant the supernatant and add fresh ice-cooled fixative to maintain the required density of PMCs.
(x) Use a pipette to drop an aliquot of the suspension on a clean, grease-free slide.
(xi) Allow the suspension to spread concentrically on the glass slide by itself.
(xii) Keep the slide for evaporation of the fixative.
(xiii) The slide is now ready to be stained with conventional carmine solution or with fluorescent dye DAPI and CMA.
(xiv) Add a drop of carmine and keep it covered with a glass cover for 5–10 min. Subsequently, observe it under a microscope. Blot off excess fluid through soft pressing with filter paper.
(xv) For fluorescent staining, add one drop of 45% acetic acid to the spread, cover it with a cover glass, gently heat it on a flame, and softly squash it with filter paper for a meiotic spread.
(xvi) Keep the slide at a frozen temperature or with dry ice to remove the cover glass and let it air dry overnight.
(xvii) Stain the slide with 0.1 μg mL^{-1} DAPI or 0.1 mL^{-1} CMA for the required time.

6.2.4 Cytological Slide Preparation by Improved Modified Method of Dang et al. (2015)

(i) Flower buds were collected in the morning.
(ii) Fixed in Acetic methanol (1:3) and kept overnight at 4 °C.
(iii) Fixed flower buds were washed in distilled water, cut into two sections, and transferred to a microcentrifuge tube containing 200 μL of distilled water.

(iv) With the help of fine tweezers, PMCs were squeezed into the distilled water, and the another wall was removed using forceps.
(v) The PMC suspension was centrifuged at 2000 × g for 3 min, and the supernatant was removed using a micropipette.
(vi) Enzyme mixture containing 3% cellulose, 0.3% pectinase, and 1% snailase (W/V) was prepared, and a minimum of 300–500 μL was added.
(vii) The microtubes were kept inside a BOD or immersed in a 37 °C water bath for the species-specific time (1–3 h).
(viii) After enzyme digestion, the suspension was centrifuged at 2000 × g for 3 min, the supernatant was removed, distilled water was added, and the suspension was centrifuged again.
(ix) Finally, the precipitate was resuspended in 100 μL acetic methanol (1:3) solution and kept for 10 min.
(x) 10 μL of suspension was dropped from above on every grease-free slide. The fixative will disperse and dry rapidly using an alcohol flame.
(xi) Slides were stained with 5% Giemsa stain.

6.3 Pollen Mitosis Through Smear Preparation

Pollens or pollen grains are haploid (n), and they develop from diploid (2n) pollen mother cells via meiotic division. They represent the first cell of the male gametophytic generation. Further development of pollen or microspores is a post-meiotic process. Pollen grains are typically shell-shaped and feature distinct cell walls. A unicellular microspore undergoes mitotic division, a process known as pollen mitosis, which occurs within the male gametophyte. The process of chromosome processing to study pollen mitosis is almost similar to meiosis chromosome preparation using flower buds, except for the size of the flower buds. More mature floral buds are required for the pollen mitosis study. Figure 6.9 illustrates pollen mitotic studies in four commonly used materials: *Allium cepa* ($n = 8$), *Rhoeo discolor* ($n = 6$), *Setcreasea pallida* ($n = 12$), and *Aloe vera* ($n = 7$).

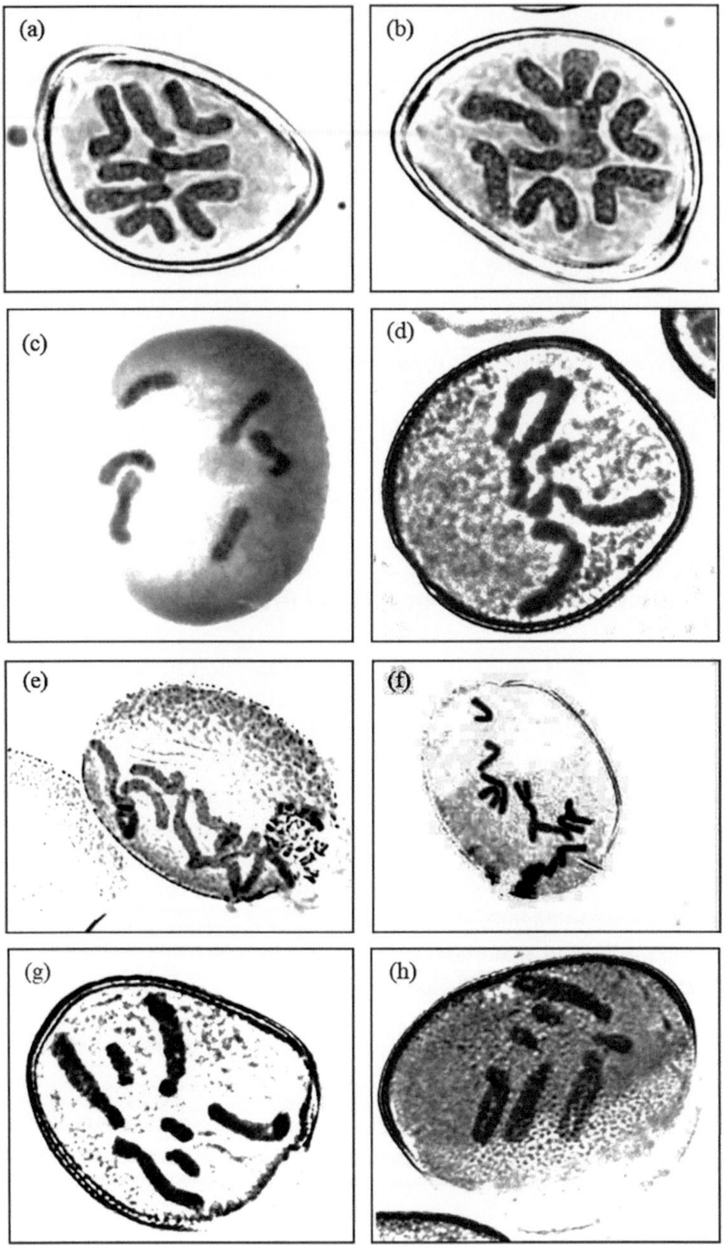

Fig. 6.9 Pollen mitosis in various plant species. Pollen mitotic metaphases displaying $n = 8$ chromosomes in the pollen cells of *Allium cepa* (**a**, **b**); pollen mitotic metaphases exhibiting $n = 6$ chromosomes in the pollen cells of *Rhoeo discolor* (**c**, **d**); pollen mitotic metaphases revealing $n = 12$ chromosomes in the pollen cells of *Setcreasea pallida* and (**e**, **f**); and pollen mitotic metaphases showing $n = 7$ (three small and four long) chromosomes in the pollen cells of *Aloe vera* (**g**, **h**). (Image **g** is taken from Das et al. 2010. Nucleus 53:89–94. With permission)

Salient Features of Pollen Mitosis

(i) Cells are boat-shaped and loosely arranged, with prominent exine and intine cell wall layers.
(ii) Nucleolus and nuclear membrane are absent.
(iii) Two haploid sets of daughter chromosomes are visible towards the two opposite poles and are away from the equatorial plane.
(iv) Chromosomes are comparatively less condensed, longer, and countable.
(v) Centromeres are toward the poles with arms projecting towards the equatorial plane.
(vi) Haploid number of chromosomes visible in pollen cell.

References

Dang J, Zhao Q, Yang X, Chen Z, Xiang S, Liang G (2015) A modified method for preparing meiotic chromosomes based on digesting pollen mother cells in suspension. Mol Cytogenet 8(1):80–88

Das A, Mukherjee P, Ghorai A, Jha TB (2010) Comparative karyomorphological analyses of in vitro and in vivo grown plants of *Aloe vera* L. BURM. f. Nucleus 53(3):89–94. https://doi.org/10.1007/s13237-010-0001-1

Hizume M, Kaneko K, Miyake T (2014) A method for preparation of meiotic chromosomes of conifers and its applications. Chromosome Bot 9(3):83–88

McClintock B (1929) A cytological and genetical study of triploid maize. Genetics 14(2):180–222

Ross KJ, Fransz P, Jones GH (1996) A light microscopic atlas of meiosis in *Arabidopsis thaliana*. Chromosome Res 4(7):507–516

Zhong XB, Hans de Jong J, Zabel P (1996) Preparation of tomato meiotic pachytene and mitotic metaphase chromosomes suitable for fluorescence in situ hybridization (FISH). Chromosome Res 4(1):24–28

Chromosome Studies from Permanent Slides with Salient Features

7

Abstract

This chapter explores the study of plant chromosomes through the analysis of permanent slides, providing essential insights for undergraduate and postgraduate students in plant science. Utilizing conventional orcein/carmine methods, students engage in hands-on training to identify and differentiate between normal and abnormal mitotic and meiotic stages. By examining chromosomal features in somatic and gametic cells, the chapter emphasizes the importance of chromosome morphology in understanding genetic principles. Key points of identification include examining the type of cells—somatic, gametic, or microspores—and recognizing the chromosomal configurations associated with each cell type. The characterization of somatic chromosomes, typically rod-shaped and with distinct constrictions, contrasts with the condensed, bivalent forms of gametic chromosomes. The provided table summarizes the morphological differences and chromosome numbers across cell types, enhancing clarity for students. Additionally, the chapter outlines slide preparation, focusing on the salient features of various divisional stages, including normal and anomalous cases arising from natural occurrences. The chapter also incorporates the "Allium test" to assess the genotoxic effects of various non-biodegradable compounds, including effluents, despite treatment to meet discharge standards, on *Allium cepa* chromosomes. The chapter will further enrich their practical knowledge on how the non-biodegradable compounds are impacting our ecosystems. Overall, this comprehensive approach fosters a deeper comprehension of chromosomal anatomy and its implications in plant genetics, preparing students for future research and applications in the field.

7.1 Introduction

In plant science practical classes at the undergraduate and postgraduate levels, students are required to gain hands-on training in preparing and analyzing plant chromosomes using conventional orcein/carmine methods and making them permanent

T. B. Jha, M. Halder, *Classical and Molecular Plant Cytogenetics*,
https://doi.org/10.1007/978-981-95-8265-5_7

for future studies. Identifying different normal and abnormal mitotic and meiotic stages from permanent slides or printed photographs provides an opportunity for students to demonstrate their knowledge of plant chromosomes. To improve understanding of normal and abnormal chromosomal features during mitotic and meiotic cell division, the following pages will present notable features of different divisional stages along with clear microphotographs taken from commonly used class teaching materials.

Chromosome studies provide foundational insights into genetics. While chromosomes from somatic and gametic cells are examined in different tissues, their composition remains consistent. However, they exhibit distinct morphological differences in somatic cells, gametic cells, and microspores. Additionally, the chromosomal characteristics of microspores align with the Mendelian concept that, although expressions may vary, the original traits re-emerge at the earliest opportunity. Ultimately, the study of somatic and gametic chromosomes represents a genetic cycle of alternation and generation. Table 7.1 provides valuable information in this regard.

Points to Note Before Identification

- Examine the microscopic field to identify the chromosomal stage. At the beginning, the process involves identifying the type of cells: somatic, gametic, or microspores (pollens).
- If the cells are somatic, try to identify the divisional stages by observing chromosomal configurations (normal prophase, metaphase, anaphase, telophase, or with anomalous features).
- If the cells are gametic, try to identify the divisional stages by observing chromosomal configurations (normal diplotene, diakinesis, metaphase I, anaphase I, diad, metaphase II, anaphase II, or with abnormal features).
- Pay attention to the morphological configuration of the chromosomes. Somatic chromosomes are rod-shaped, long, medium, or small in size, with clear primary constriction regions. If the plate shows well-scattered condensed chromosomes or chromosomes with few overlaps, try to count them to confirm the diploid chromosome ($2n$) number. Normal diploid ($2n$) somatic cells will always contain an even number of chromosomes. On the other hand, gametic or haploid (n) cells may contain an odd or even number of chromosomes.
- Look at the chromosomes for their distinct size differences (bimodal karyotype such as *Aloe vera* ($2n = 14$; eight long and six small)). If possible, count the number of chromosomes that appear to be metacentric (M/m), submetacentric (sm), and acrocentric (ac). Note down if you can find a secondary constricted chromosome. Generally, they are terminal in *Allium cepa* but interstitial or intercalary in *Allium sativum* and *Lens culinaris*.

Table 7.1 Comparative overview of identifying features of metaphase division of mitotic, meiotic, and pollen mitosis

Characters	Mitotic metaphase	Meiotic metaphase	Pollen mitosis
Cell shape	 Somatic cells are usually rectangular or hexagonal in shape with a thin cell wall	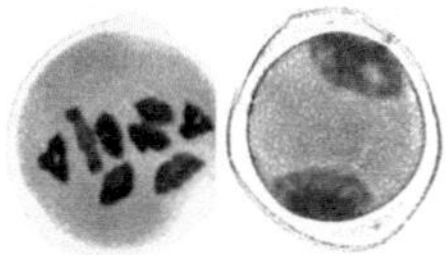 Gametic cells or PMCs are usually slightly spherical, oval, or round in shape with a thick cell wall	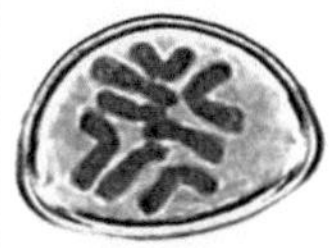 Microspores or pollen cells are generally boat-shaped with a very thick cell wall
Morphological configuration of a chromosome	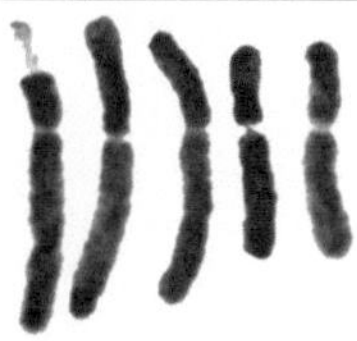 Chromosomes may be long, medium/ small sized, condensed, rod shaped, with distinct constrictions. Represents 2*n* numbers	 Chromosomes are short, thick, and highly condensed without any distinct visible constrictions. They are in a bivalent form (*n*) with some exceptions and represent n numbers	Reappearance of the somatic mitotic nature of the chromosomes. Highly condensed rod-shaped with distinct constrictions, but represent n numbers
Chromosome number	Normal diploid (2*n*) cells contain an even number of homologous chromosomes	Normal gametic cells (*n*) may contain an even or an odd number of bivalents	A pollen cell (*n*) may contain an odd or even number of univalent chromosomes

The various stages of somatic and gametic chromosomal configurations are described below, in accordance with the conceptual framework outlined previously.

7.2 Preparation of Somatic and Gametic Chromosomal Stages and Their Salient Features

The preparation of slides for somatic and gametic chromosomes from various plant materials was conducted using classical cytogenetic techniques. Specific stages were carefully selected, photographed, and described, emphasizing their key features. The anomalous stages presented from the gametic chromosome preparations are natural occurrences and not induced. In addition, we performed the "*Allium* test" using chemicals to evaluate their genotoxic effects on the chromosomes of *Allium cepa*. The protocol for the "*Allium* test" is outlined under Sect. 7.3.

7.2.1 Normal Divisional Stages from Somatic Chromosomes

7.2.1.1 Specimen-1: Mitotic Prophase (Fig. 7.1)

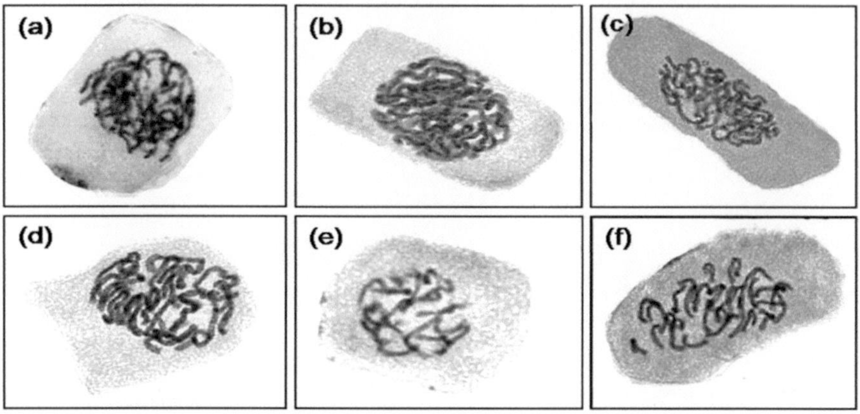

Fig. 7.1 (**a**–**f**) Mitotic prophase stage in somatic cells

The provided slide or photograph (Fig. 7.1) illustrates the following characteristics:

(i) The nucleus appears significantly larger and occupies more area within either rectangular or hexagonal cells compared to the interphase nucleus.
(ii) The widely spread chromatins resemble round, woolly balls, with a nucleolus that is typically non-detectable, though it may be visible on rare occasions, along with the nuclear membrane.
(iii) The chromatins are more distinct, thicker, and less coiled than those seen during interphase due to a gradual process of condensation. The level of this condensation may vary based on their position during prophase.

Therefore, the supplied slide or photograph depicts the mitotic prophase stage in somatic cells (Fig. 7.1).

7.2.1.2 Specimen-2: Mitotic Metaphase (Fig. 7.2)

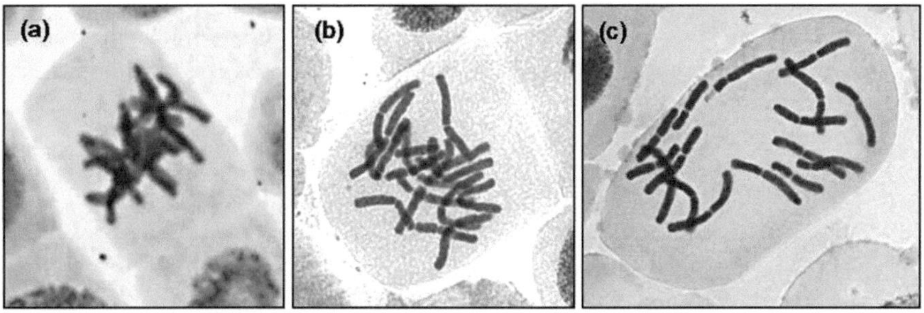

Fig. 7.2 (**a**–**c**) Mitotic metaphase stage in somatic cells: (**a**, **b**) side view and (**c**) polar view

The provided slide or photograph (Fig. 7.2) reveals the following characteristics:

(i) There are no round or circular structures visible within the rectangular cells, indicating the complete absence of the nucleolus and nuclear membrane.
(ii) The chromatins have undergone maximum condensation, transformed into long, highly condensed, rod-shaped structures characterized by a distinct primary constriction, known as chromosomes.
(iii) The chromosomes, whether countable or uncountable, are located at the equatorial plate (as seen from a side view) or dispersed throughout the cytoplasm (in the polar view).

Therefore, the supplied slide or photograph depicts the mitotic metaphase stage (either from the polar view or side view) in somatic cells (Fig. 7.2).

7.2.1.3 Specimen-3: Mitotic Anaphase (Fig. 7.3)

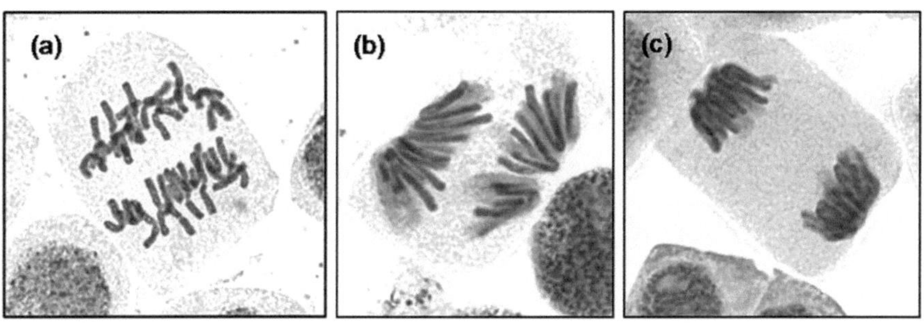

Fig. 7.3 (**a**–**c**) Mitotic anaphase stage in somatic cells

The provided slide or photograph (Fig. 7.3) illustrates the following characteristics:

(i) There are no round or circular structures visible within the rectangular cells, indicating the complete disappearance of both the nucleolus and the nuclear membrane.
(ii) Two sets of condensed, rod-shaped daughter chromosomes are evident, resulting from the separation of sister chromatids from each metaphase chromosome.
(iii) These two sets of non-countable or countable chromosomes can be found distributed anywhere between the equatorial plane and the two opposing poles, as each set is migrating toward the opposite poles.
(iv) During the movement of the chromosomes, the centromeres are oriented toward the poles while the chromatid arms extend toward the equatorial plane. This positioning can cause the chromosomes to appear in "V," "L," "j," or "i" shapes, depending on the location of the centromeres.

Therefore, the slide or photograph depicts the mitotic anaphase stage in somatic cells (Fig. 7.3).

7.2.1.4 Specimen-4: Mitotic Telophase (Fig. 7.4)

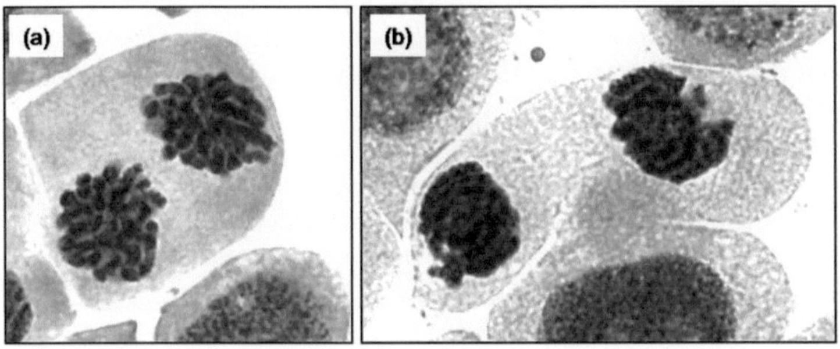

Fig. 7.4 (a–b) Mitotic telophase stage in somatic cells

The provided slide or photograph (Fig. 7.4) illustrates the following features:

(i) Two sets of non-countable chromosomes are positioned at opposite poles of the rectangular cell.
(ii) The outline of each chromosome set appears smooth, indicating the reformation of the nuclear membrane around them.
(iii) The chromosomes have lost their distinct rod-like shape and become more coiled.

Therefore, the slide or photograph clearly depicts the mitotic telophase stage in somatic cells (Fig. 7.4).

7.2.2 Normal Divisional Stages from Gametic Chromosomes

7.2.2.1 Specimen-5: Diplotene Substage of Meiotic Prophase I (Fig. 7.5)

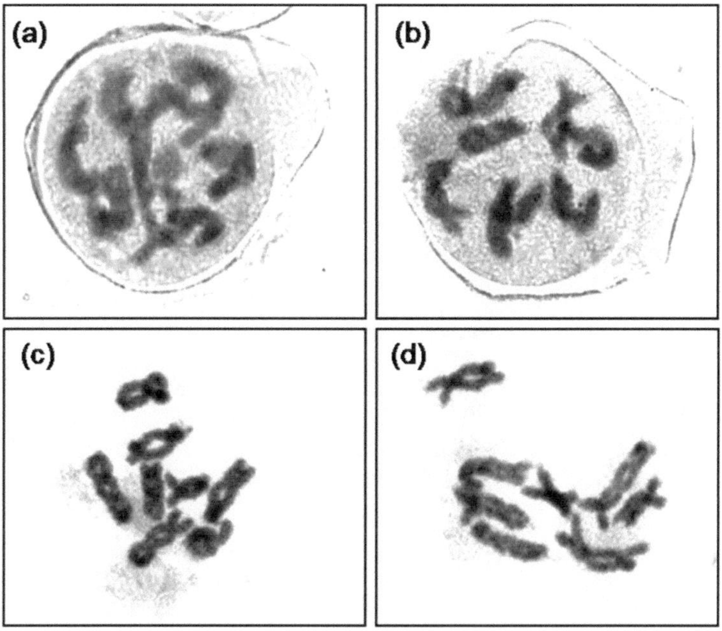

Fig. 7.5 (**a**–**d**) Diplotene substage of meiotic prophase I in PMC

The provided slide or photograph (Fig. 7.5) displays the following characteristics:

(i) The cell plate or partition wall is absent in the loosely arranged, thick-walled, spherical, round, or oval-shaped cells.
(ii) Within these cells, chromosomes are thick, short, and condensed, without distinct constrictions, forming bivalents.
(iii) The bivalents are comparatively less condensed and exhibit several intercalary X-shaped chiasmata along the lengths of homologous chromosome pairs.
(iv) Occasionally, the presence of a nucleolus can be observed.

Therefore, the supplied slide or photograph represents the diplotine substage of meiotic prophase I in PMC (Fig. 7.5).

7.2.2.2 Specimen-6: Diakinesis Substage of Meiotic Prophase I (Fig. 7.6)

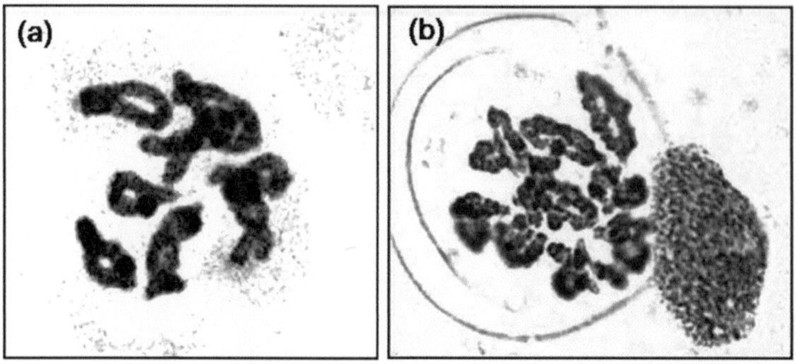

Fig. 7.6 (**a**–**b**) Diakinesis substage of meiotic prophase I in pollen mother cells (PMC)

The provided slide or photograph (Fig. 7.6) depicts the following characteristics:

(i) The presence of a cell plate or partition wall is absent in the loosely arranged, thick-walled, spherical, round, or oval-shaped cells.
(ii) Within these cells, chromosomes appear as thick, short bivalents that are relatively less condensed and lack distinct constrictions.
(iii) Bivalents are dispersed throughout the cytoplasm, exhibiting varying degrees of terminalized chiasmata, which manifest as X-, Y-, or eye-shaped configurations.
(iv) The nucleolus is not visible, and the bivalents may be difficult to count.

Therefore, the supplied slide or photograph represents the diakinesis substage of meiotic prophase I in pollen mother cells (Fig. 7.6).

7.2.2.3 Specimen-7: Metaphase I of Meiosis (Fig. 7.7)

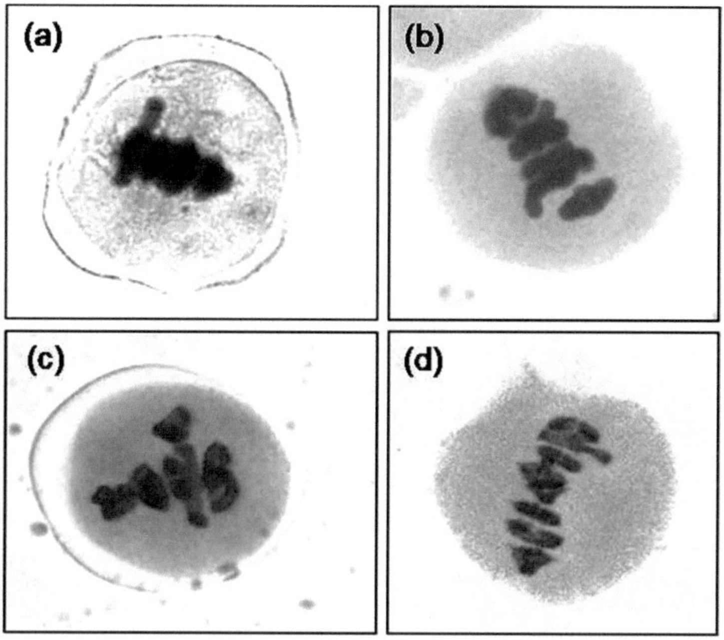

Fig. 7.7 (**a–d**) Metaphase I of meiosis cell division in pollen mother cells

The provided slide or photograph (Fig. 7.7) depicts the following characteristics:

(i) The cell plate or partition wall is absent in loosely arranged, thick-walled, spherical, round, or oval-shaped cells.
(ii) The nuclear membrane and nucleolus are not present in these cells.
(iii) Chromosomes appear as very thick, short, and highly condensed bivalents, lacking distinct constrictions.
(iv) These highly condensed bivalents are arranged linearly along the equatorial plane in the lateral view (Fig. 7.7a, b, d) or scattered within the cytoplasm of the cell in the polar view (Fig. 7.7c).
(v) The bivalents may or may not be distinct and countable, which indicates the "n" number.

Therefore, the provided slide or photograph displays the metaphase I stage of meiosis cell division in a pollen mother cell from either a lateral or a polar perspective (Fig. 7.7).

7.2.2.4 Specimen-8: Anaphase I of Meiosis (Fig. 7.8)

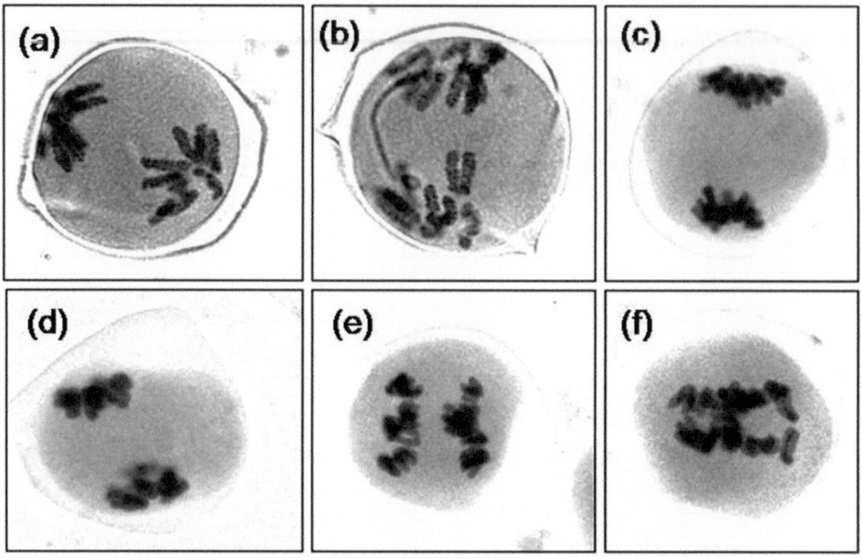

Fig. 7.8 (**a–f**) Anaphase I stage of meiosis cell division in pollen mother cells

The provided slide or photograph (Fig. 7.8) displays the following characteristics:

(i) The cell plate or partition wall is absent in the loosely arranged, thick-walled, spherical, round, or oval-shaped cells.
(ii) Neither the nuclear membrane nor the nucleolus is present in the cell.
(iii) The chromosomes appear very thick, short, and highly condensed, lacking distinct constrictions.
(iv) Two sets of chromosomes are situated between the equatorial plane and the opposite poles of the cell, with the arms of the chromosomes oriented toward the equator during their polar movement.

Thus, the slide or photograph depicts the anaphase I stage of meiosis cell division in pollen mother cells (Fig. 7.8).

7.2.2.5 Specimen-9: Telophase I of Meiosis (Fig. 7.9)

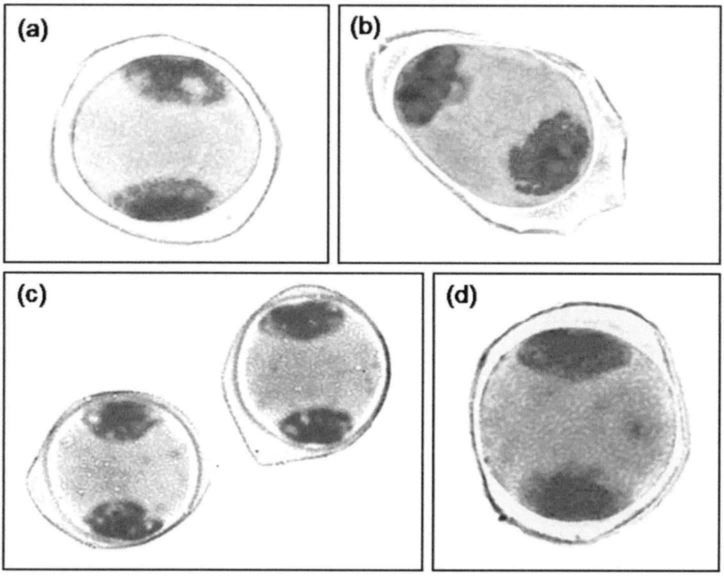

Fig. 7.9 (**a–d**) Telophase I of meiosis cell division in pollen mother cells

The provided slide or photograph (Fig. 7.9) displays the following characteristics:

(i) A cell plate or partition wall is absent in the loosely arranged, thick-walled cells that are spherical, round, or oval in shape.
(ii) Distinct individual chromosomes are not visible; instead, two sets of chromosomes are positioned at opposite poles of the cell, appearing as clumps.
(iii) The smooth outline of the two sets of chromosomal mass suggests the reappearance of the nuclear membrane surrounding them.

Therefore, the slide or photograph illustrates the telophase I stage of meiosis in pollen mother cells (Fig. 7.9).

7.2.2.6 Specimen-10: Metaphase II of Meiosis (Fig. 7.10)

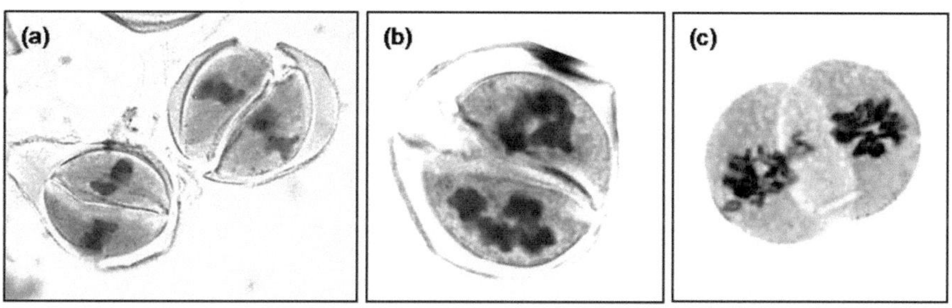

Fig. 7.10 (**a–c**) Metaphase II stage of meiosis cell division in pollen mother cells

The provided slide or photograph (Fig. 7.10) displays the following characteristics:

(i) A cell plate or partition wall is evident in loosely arranged, thick-walled cells that exhibit a distinct two-celled diad condition.
(ii) The nuclear membrane and nucleolus are absent in these cells.
(iii) Two haploid sets of very thick, short, and highly condensed chromosomes, lacking distinct constrictions, are linearly arranged (in lateral view; Fig. 7.10a) along the equatorial planes or scattered (in polar view; Fig. 7.10b) within the cytoplasm of each diad.
(iv) The chromosomes may or may not be distinctly countable.

Thus, the supplied slide or photograph illustrates the metaphase II stage (in either lateral or polar view) of meiosis cell division in pollen mother cells (Fig. 7.10).

7.2.2.7 Specimen-11: Anaphase II of Meiosis (Fig. 7.11)

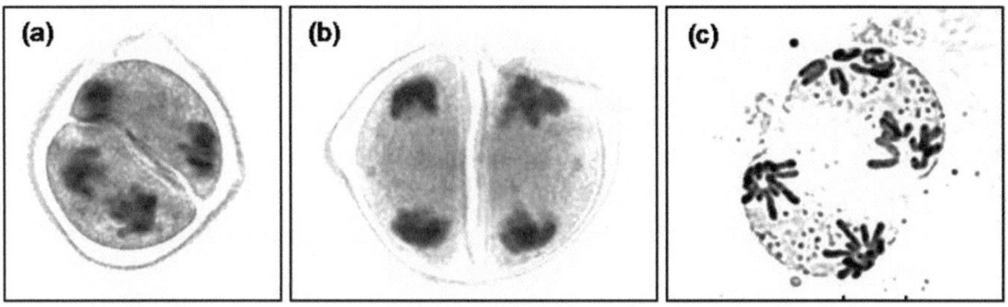

Fig. 7.11 (a–c) Anaphase-II stage of meiosis cell division in pollen mother cells

The provided slide or photograph (Fig. 7.11) illustrates the following characteristics:

(i) A cell plate or partition wall is evident in the loosely arranged, thick-walled, spherical or oval-shaped cells, indicating a distinct two-celled diad condition.
(ii) The nuclear membrane and nucleolus are absent in these cells.
(iii) Each cell of the diad contains two sets of indistinguishable separating chromatids (daughter chromosomes).
(iv) Each set of daughter chromosomes is positioned between the equatorial plane and the two poles, with the arms extending toward the equatorial plane.

Therefore, the slide or photograph represents the anaphase II stage of meiosis cell division in pollen mother cells (Fig. 7.11).

7.2.2.8 Specimen-12: Telophase II of Meiosis (Fig. 7.12)

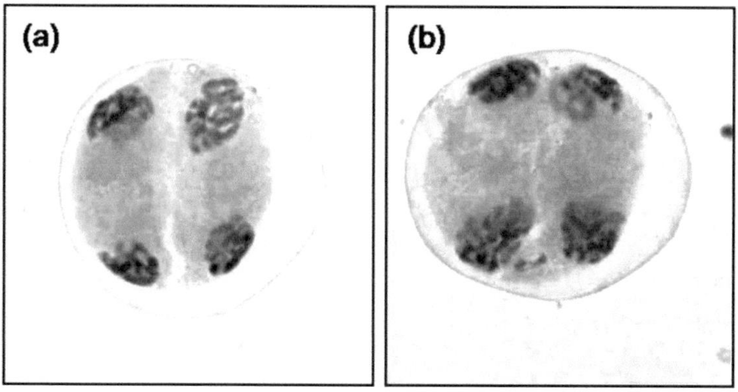

Fig. 7.12 (**a**–**b**) Telophase II stage of meiosis cell division in pollen mother cells

The provided slide or photograph (Fig. 7.12) depicts the following features:

(i) A cell plate or partition wall is evident in loosely arranged, thick-walled, spherical or oval-shaped cells, indicating a distinct two-celled diad condition.
(ii) The nuclear membrane and nucleolus are absent in these cells.
(iii) Four compact daughter chromosomal masses with smooth outlines are visible at the opposite poles of each cell within the diad.
(iv) The formation of the second cell plate is not yet complete within the diad, meaning the pollen mother cells have not yet evolved into tetrads, nor have they been released to form individual haploid microspores or pollen grains.

Thus, the supplied slide or photograph illustrates the telophase II stage of meiosis in the pollen mother cells (Fig. 7.12).

7.2.3 Abnormal Divisional Stages from Gametic Chromosomes

7.2.3.1 Specimen-13: Lagging Chromosome (Fig. 7.13)

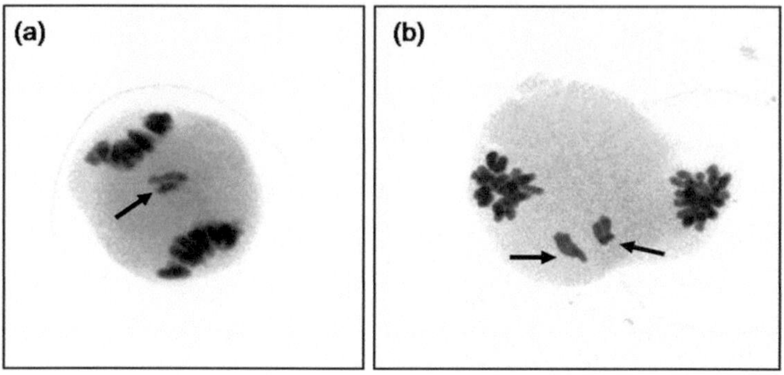

Fig. 7.13 Abnormal meiotic cell division in PMC showing (**a**) one lagging bivalent and (**b**) two lagging bivalents in anaphase I

The provided slide or photograph (Fig. 7.13) displays the following characteristics:

(i) A cell plate or partition wall is absent in the loosely arranged, thick-walled, spherical or oval-shaped cells.
(ii) The nuclear membrane and nucleolus are not visible within this cell.
(iii) Two sets of short, thick chromosomal masses are present at or near the two opposite poles, lacking distinct constriction. The outlines of these chromosomal masses appear irregular, not smooth and circular, suggesting that the nuclear membrane has not reappeared yet.
(iv) One bivalent chromosome (Fig. 7.13a) or two univalent chromosomes (Fig. 7.13b) are located far away from the other chromosomal masses, possibly due to delayed movement relative to the rest.

Therefore, the supplied slide or photograph illustrates abnormal meiotic cell division in PMC, showing lagging chromosomes (bivalent) in anaphase I (Fig. 7.13).

7.2.3.2 Specimen-14: Chromosomal Bridge in Anaphase I (Fig. 7.14)

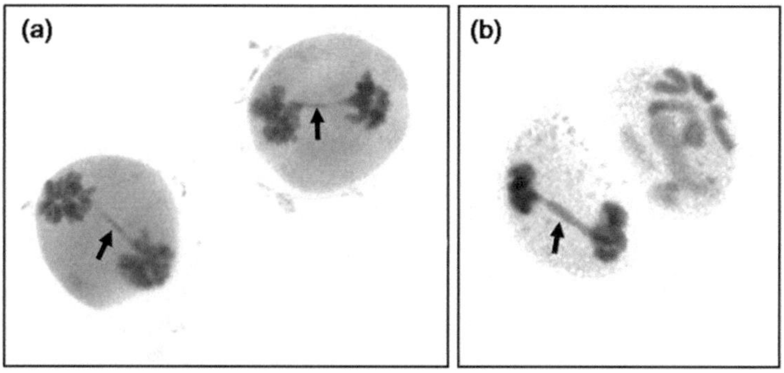

Fig. 7.14 Abnormal meiotic cell division in PMC showing chromosomal bridge formation in anaphase I

The provided slide or photograph (Fig. 7.14) displays the following characteristics:

(i) There is an absence of a cell plate or partition wall in the loosely arranged, thick-walled, spherical, or oval-shaped cells.
(ii) The nuclear membrane and nucleolus are not discernible in these cells.
(iii) Two sets of short, thick chromosomal masses with indistinct constrictions are located at or near opposite poles. The outlines of these chromosomal masses may appear irregular, not smooth, and circular, as the nuclear membrane has not reappeared yet.
(iv) A thread-like chromosomal connection (bridge) is visible, traversing the equatorial plane, linking the two masses of separating or separated chromosomes.

Therefore, the supplied slide or photograph illustrates abnormal meiotic cell division in PMC showing chromosomal bridge formation in anaphase I (Fig. 7.14).

7.2.3.3 Specimen-15: Late Separation of Chromosome (Fig. 7.15)

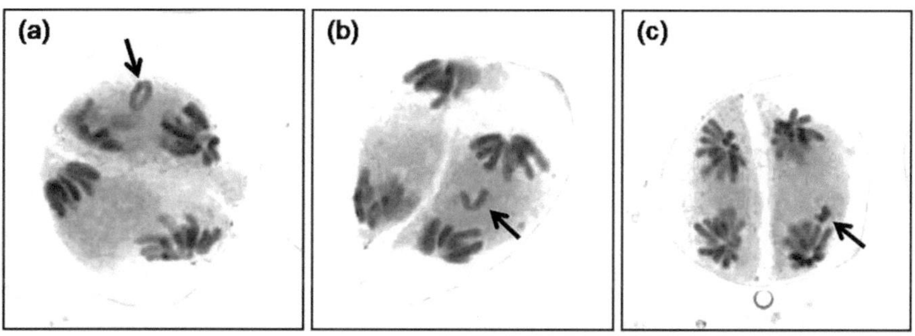

Fig. 7.15 (**a–c**) Abnormal meiotic cell division in PMC showing late separation of chromosomes in anaphase II

The provided slide or photograph (Fig. 7.15) illustrates the following features:

(i) There is a presence of a cell plate or partition wall in the loosely arranged, thick-walled, spherical, or oval-shaped cells.
(ii) The nuclear membrane and nucleolus are not discernible in these cells.
(iii) Most chromosomes are positioned at two poles with centromeres oriented toward the pole and arms projecting toward the equatorial plane (anaphase).
(iv) A chromosome can be observed lagging with respect to other daughter chromosomes, situated apart from the pole.

Thus, the provided slide or photograph depicts the meiotic anaphase II stage, characterized by the late separation of chromosomes (Fig. 7.15).

7.2.3.4 Specimen-16: Ring Chromosomes (Fig. 7.16)

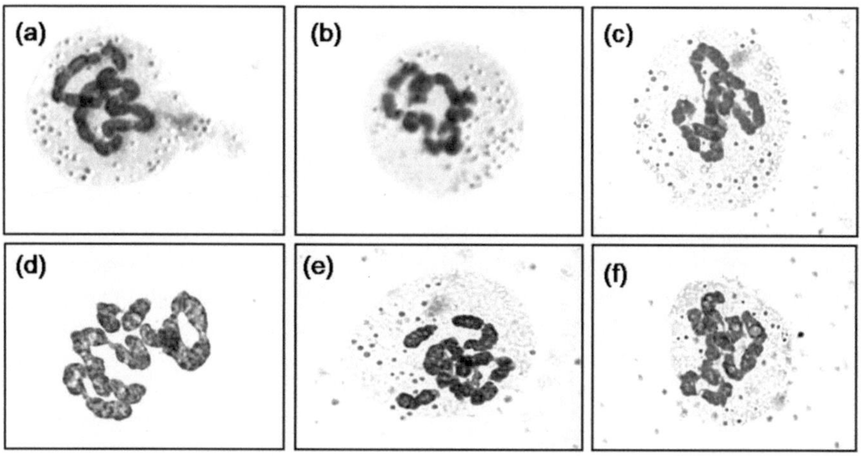

Fig. 7.16 (a–f) Meiotic metaphase I stage showing different configuration of multivalent ring chromosomes

The provided slide or photograph (Fig. 7.16) displays the following characteristics:

(i) The absence of a cell plate or partition wall in the loosely arranged, thick-walled, spherical, or oval-shaped cells.
(ii) The nuclear membrane and nucleolus are absent in these cells.
(iii) The individual chromosomes are short, thick, and highly condensed, exhibiting indistinct constrictions. They form a distinct, large multivalent ring configuration through the end-to-end pairing or attachment of 12 chromosomes (as seen in *Rhoeo*). This configuration indicates a translocation heterozygote resulting from a reciprocal translocation.

Therefore, the supplied slide or photograph illustrates the meiotic metaphase I stage, characterized by multivalent ring chromosomes (Fig. 7.16).

7.2.3.5 Specimen-17: Micronuclei in Meiotic Prophase I (Fig. 7.17)

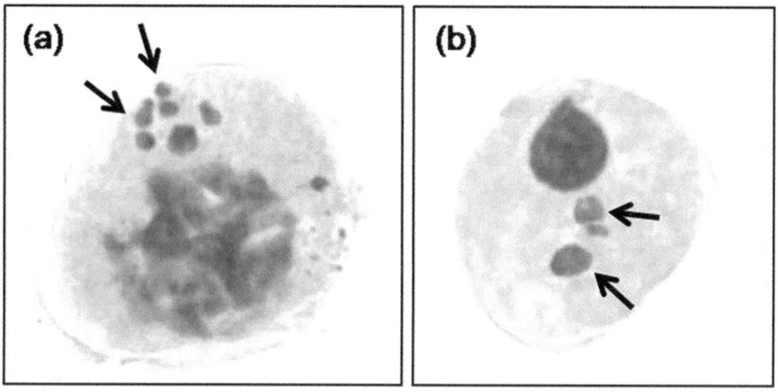

Fig. 7.17 (**a**–**b**) Micronuclei formation in meiotic prophase I

The provided slide or photograph (Fig. 7.17) displays the following characteristics:

(i) The absence of a cell plate or partition wall in the loosely arranged, thick-walled, spherical, or oval-shaped cells.
(ii) The nuclear membrane is present in these cells.
(iii) Several micronuclei, which are significantly smaller than the primary nucleus, are observed in the same cytoplasm.

Therefore, the supplied slide or photograph illustrates micronuclei (Fig. 7.17).

7.3 Effect of Chemicals on Somatic Chromosomes of *Allium cepa*

The use of chemicals in soil and various industries, both directly and indirectly, is not a recent phenomenon. Many of these chemicals are not environmentally friendly and can lead to both short-term and long-term effects on plants, animals, and human health. Herbicides and insecticides used on plants are often non-biodegradable or degrade very slowly. Conversely, industrial effluents are expected to be discharged into the environment in accordance with discharge standards. However, once these treated effluents enter the natural ecosystem, their impact on different organisms is rarely monitored or recorded. Certain compounds, including organic chemicals and heavy metals, can have both chronic and acute effects on organisms in natural environments, even at low concentrations (Wijeyaratne and Wickramasinghe 2020).

Therefore, it is crucial to monitor the impact of this diverse array of chemicals. Various known qualitative and quantitative bioassays are available in the literature. However, plant bioassays, in general, have proven more sensitive than most other systems in detecting the genotoxic effects of environmental pollutants (Ma 1999). There are several reasons for utilizing plant systems: they are easy to store and handle, economically viable, and do not require complex laboratory setups (Fiskesjo 1985). Additionally, the effect of these chemicals can be directly monitored on the behavior of the chromosomes, which are highly condensed, distinct, and species-specific structures of genomic DNA. The plant-based environmental monitoring method was introduced by Levan in 1938 using the plant *Allium cepa*, and since then, the test has been known as the "*Allium* test (AT)." Root meristem cells of *A. cepa* are highly stable plant material that contains $2n = 16$ long chromosomes and are sensitive to genetic damage caused by chemicals. They are easily observed under a light microscope and thus treated as valuable hands-on training resources for monitoring or testing environmental pollutants in educational institutions. The test is also known as an in vivo cytotoxicological test, and its result can be used to assess genotoxicity in plants and eukaryotes in general, including humans (Bonciu et al. 2018). The US Environmental Protection Agency (EPA) recognized *Allium* root tip assays as one important plant bioassay for monitoring or testing environmental pollutants (Ma 1999; Ma et al. 2005). *Allium* test is a simple bioassay and does not require any sophisticated instruments. To conduct the *Allium* test in any laboratory, the following simple materials are needed.

(i) *Selection of Testing Chemical:* Liquid or solid.
(ii) *Concentrations of Testing Chemicals:* The printed label on the container of the testing chemical provides detailed information about its use. For example, it may specify a concentration of 2% pesticide in water, or indicate that 50 mL or 50 gm of pesticide should be mixed with 100 L of water, or that 4 L or 4 kg of pesticide is necessary for 100 L of water. However, to test the genotoxic effect of any chemical, the concentration is very important. The concentration range may vary from 10% to 100% of the prescribed dose. The amount of chemicals can be mixed or dissolved in water. PPM (parts per million) is a standard unit often used interchangeably with milligrams per liter (mg L^{-1}), i.e., 1 ppm = 1 mg L^{-1}, since one liter of water has a mass of one million milligrams. One milligram is one thousandth of a gram, whereas one microgram (μg) is one millionth of a gram. Similarly, 1 mL is one thousandth of a liter and one microliter (μL) is one millionth of a liter.
(iii) *Plant Material:* Small- to medium-sized *Allium cepa* bulbs of about 1.5–4 cm in diameter.
(iv) Bottomless glass tubes, 2.5–4 cm in diameter and with a length of 5–7 cm.
(v) Temperature-controlled incubator (optional).
(vi) Slides and cover glasses.
(vii) Fixative, 45% acetic acid, or 1 N HCl and 2% aceto-orcein.
(viii) Light microscope equipped with a camera and a computer for photomicrography.

7.3.1 Experimental Steps

Method 1

(i) Select good-quality *Allium* bulbs. Remove its outer scales, and clean the root growing from the bottom disc, keeping the ring of root primordia intact.
(ii) Place the bulbs in fresh water so that the root-inducing disc part remains moist.
(iii) Bulbs are directly placed in glass tubes filled with normal water and kept until roots are induced.
(iv) *Allium* bulbs with 0.5–1 cm roots are transferred to glass tubes containing the various concentrations of testing chemicals below the EC_{50} for various time periods (1–48 h). Roots are harvested after completion of 1, 2, 3, 4, 24, and 48 h.
(v) Wash the roots in water and directly fix them in a 1:2 or 1:3 acetic acid: alcohol solution for at least one hour at a low temperature.
(vi) Roots are incubated in 45% acetic acid for 20–25 min, and then stained with a mixture of 2% aceto-orcein and 1 N HCl (9:1). Gentle heat is applied for a few seconds, and roots are incubated for more than one hour.
(vii) Squash the root tip on a clean, grease-free slide.
(viii) Repeat the same procedure after 2, 3, 4, 24, and 48 h if needed.
(ix) Observe the slides under the microscope; count the number of cells along with dividing cells within one field. The minimum number of cells should be at least 1000. Categorize the dividing cells as metaphase, anaphase, telophase, and types of abnormalities.

Calculate mitotic index (MI), a measurement to determine the percentage of cells undergoing mitosis, following the formula:

$$\text{Mitotic index}(\text{MI}) = \text{No. of dividing cells}(\text{X}) \times 100 \div \text{Total number of cells}(\text{Y})$$

Testing chemicals are expected to disturb the normal cell division pattern. However, recovery patterns of the cells can further be studied with the following steps-

(i) The bulbs with remaining root tips are transferred to the glass tubes containing freshly prepared Knop solution.
(ii) Recovery pattern may be studied after 24, 48, and 72 h following the stage "v".
(iii) The experiments should be performed in a relatively constant room temperature of approximately 20–25 °C and protected from direct sunlight.

Method: 2

(i) Select good-quality *Allium* bulbs. Remove the outer scales, and clean the root growing bottom disc, keeping the ring of root primordia intact.
(ii) Place the bulbs in fresh water so that the root-inducing disc part remains moist.

(iii) Bulbs are directly placed in test tubes containing the various concentrations of test chemicals below the EC_{50} for various time periods (4–48 h). Roots are harvested after completion of experimental period.
(iv) Clean the roots in water after 24 and 48 h of treatment and place them in fresh water or knop solution for another 24, 48, and 72 h.
(v) Rooted bulbs (0.5–1 cm in length) are transferred to test tubes containing the various concentrations of testing chemicals below the EC_{50} for various time periods (1–48 h). Roots are harvested after completion of 1, 2, 3, 4, 24, and 48 h.
(vi) Wash the roots in water and directly fix them in a 1:2 or 1:3 acetic acid: alcohol solution for a minimum of one hour at low temperature.
(vii) Change the roots in 45% acetic acid for 20–25 min, stained with a mixture of 2% aceto-orcein and 1 N HCl mixture (9:1) with gentle heat and incubated for more than one hour.
(viii) Squash the root tip on a clean, grease-free slide.

For root growth inhibition studies, you have to monitor the growth of the *Allium cepa* roots in different concentrations of testing agents for 96 h.

Points to Note

Preparation of Knop's solution: It is a simple nutrient solution containing the following ingredients. To prepare a 100 mL Knop solution stock:

(i) Measure 100 mL of distilled or deionized water into a 500 mL beaker or flask.
(ii) Add 0.3 gm of calcium nitrate [$Ca(NO_3)_2$] to the water.
(iii) Add 0.1 gm of magnesium sulfate ($MgSO_4$) to the water.
(iv) Add 0.1 gm of potassium nitrate (KNO_3) to the water.
(v) Add 0.1 gm of potassium phosphate (KH_2PO_4) to the water.
(vi) For immediate use, add 500 mL of distilled or deionized water to the original 100 mL stock solution. [Note: This becomes a ready-to-use 1% knop solution.]
(vii) Autoclave the solution before use.

$*EC_{50}$ is an essential pharmacological parameter that assesses the potency of a compound. It represents the concentration required to elicit a 50% response, which corresponds to "half of the maximum effective concentration." A lower EC_{50} value indicates greater potency, signifying that a smaller amount of the compound is needed to achieve the desired effect. Traditionally, EC_{50} is expressed as a molar concentration.

Cytotoxicity: It refers to the harmful effects of any chemical agent on living cells.

Genotoxic: It refers to the fatal or harmful genetic changes at the chromosomal or gene level within a cell.

7.3.2 *Effects of the Allium Test on Chromosomal and Morphological Features*

The *Allium* test, when performed according to the above-noted protocols, typically induces both macroscopic and microscopic changes in root morphology and plant chromosomes.

The macroscopic features observed in the "*Allium* test" primarily pertain to root length and morphology (Rank 2003). These characteristics are crucial for assessing the effects of various substances on plant growth and can provide insights into the overall health and viability of the plant system being studied. Root length can serve as an indicator of growth inhibition or enhancement.

While chromosomal behavior is a microscopic feature. Various cellular phenomena may be observed in addition to the typical stages of metaphase and anaphase during cell division. These may include early anaphase events, the presence of sticky chromosomes, chromosomal bridges, fragmentation, and c-mitosis. The occurrence and severity of these phenomena can be influenced by several factors, including the dosage of the chemical treatment, the duration of exposure, and the specific recovery phases following treatment. Understanding these variables is crucial for elucidating the impact of chemical agents on cellular division and chromosomal integrity.

7.3.2.1 Specimen-18: Early Separation of Chromosomes (Fig. 7.18)

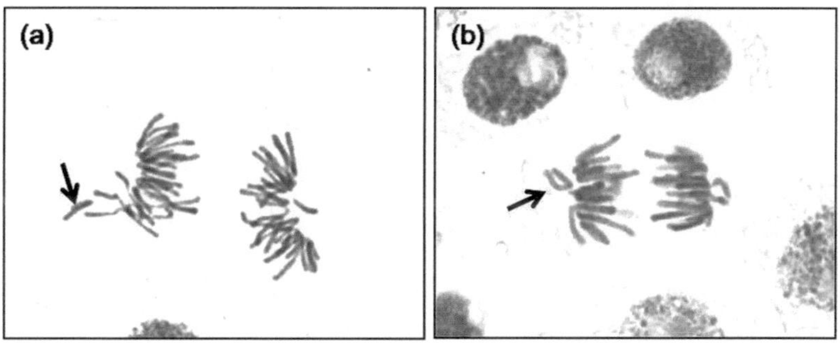

Fig. 7.18 (**a**–**b**) Abnormal mitotic anaphase stage showing early separation of chromosomes

The provided slide or photograph (Fig. 7.18) illustrates the following features:

(i) The cells exhibit a rectangular shape and are arranged compactly.
(ii) The chromosomes appear long, condensed, and rod-shaped, featuring distinct primary constrictions.
(iii) Both the nucleolus and nuclear membrane are absent.
(iv) Most chromosomes are positioned between the equatorial plane and the poles, with centromeres oriented toward the pole and arms projecting toward the equatorial plane (anaphase).

(v) A small number of chromosomes can be observed slightly ahead of other daughter chromosomes, situated very close to the pole.

Thus, the provided slide or photograph depicts the mitotic anaphase stage, characterized by the early separation of chromosomes (Fig. 7.18).

7.3.2.2 Specimen-19: Sticky Bridge (Fig. 7.19)

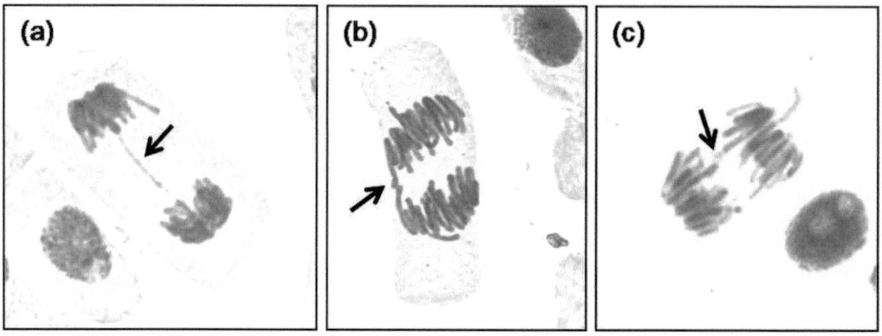

Fig. 7.19 (a–c) Abnormal mitotic anaphase stage showing a sticky bridge

The provided slide or photograph (Fig. 7.19) illustrates the following features:

(i) Cells are rectangular and arranged compactly.
(ii) Chromosomes are long, condensed, and rod-shaped, featuring distinct primary constrictions.
(iii) The nucleolus and nuclear membrane are absent.
(iv) Two sets of daughter chromosomes are visible, positioned far from the equatorial plane, near the two opposite poles.
(v) Centromeres are located toward the poles, with the arms extending toward the equatorial plane.
(vi) A thread-like chromosome bridge connects the two sets of separating chromosomes as it traverses the equatorial plane.
(vii) Chromosomes exhibit a sticky nature.

Therefore, the provided slide or photograph depicts the mitotic anaphase stage, characterized by a sticky bridge, indicating an abnormal stage (Fig. 7.19).

7.3.2.3 Specimen-20: Chromosome Fragmentation (Fig. 7.20)

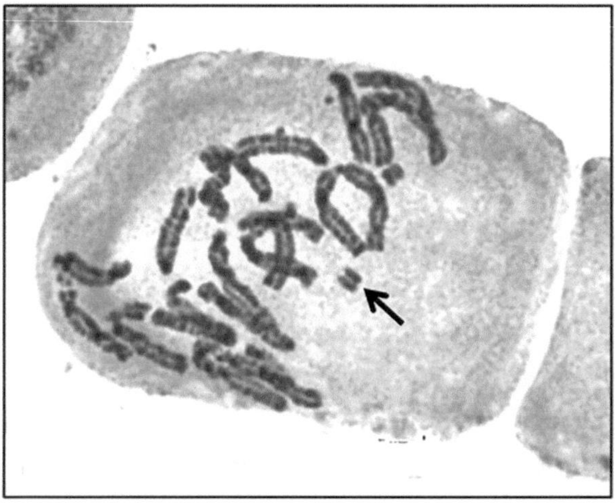

Fig. 7.20 Abnormal mitotic metaphase stage showing chromosome fragmentation

The provided slide or photograph (Fig. 7.20) illustrates the following features:

(i) Cells are rectangular in shape and are compactly arranged.
(ii) Long, condensed, and rod-shaped chromosomes with distinct primary constrictions are scattered throughout the cytoplasm or arranged in the equatorial plane.
(iii) The nucleolus and nuclear membrane are absent.
(iv) A small chromosomal fragments are visible. The fragmented chromosomes are smaller in size and appear to be acentric.

Thus, the provided slide or photograph depicts the mitotic metaphase stage with evidence of chromosome fragmentation, indicating an abnormal stage (Fig. 7.20).

7.3.2.4 Specimen-21: Multipolarity (Fig. 7.21)

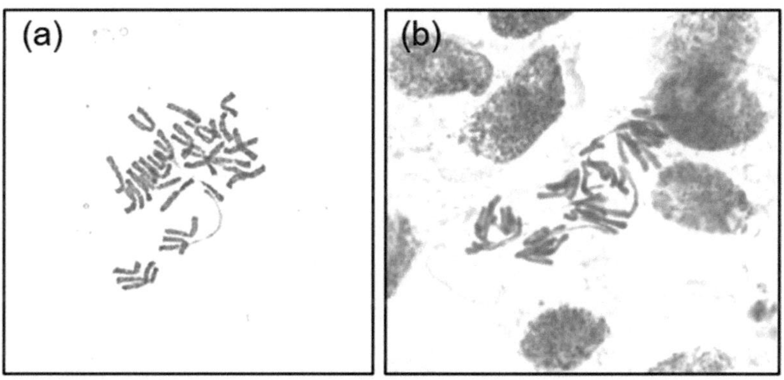

Fig. 7.21 (**a**–**b**) Abnormal mitotic anaphase stage showing multipolarity

The provided slide or photograph (Fig. 7.21) illustrates the following features:

(i) The cells are rectangular in shape and are closely packed together.
(ii) The chromosomes are rod-shaped and condensed, exhibiting constrictions. Both the nucleolus and nuclear membrane are absent.
(iii) Multiple sets of long, condensed chromosomes can be observed.
(iv) More than two poles are visible within the dividing cell.
(v) Each pole has a star-shaped configuration, comprised of daughter chromosomes with radiating arms.
(vi) An uneven distribution of chromosomes is noted at different poles.

Thus, the provided slide or photograph illustrates the mitotic anaphase stage with multipolarity, indicating an abnormal stage (Fig. 7.21).

7.4 Salient Features of Pollen Mitosis

7.4.1 Specimen-22: Pollen Mitosis (Fig. 7.22)

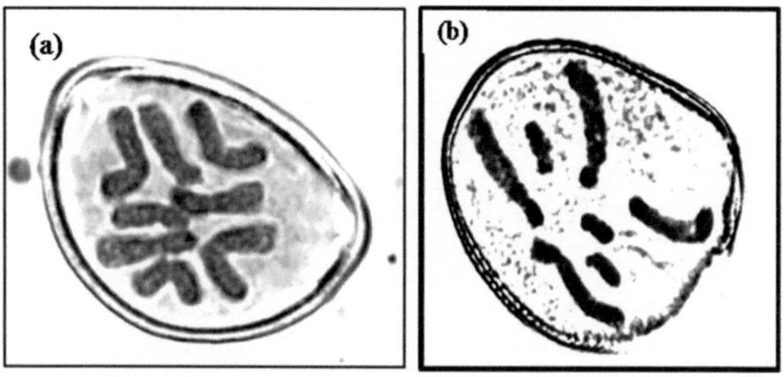

Fig. 7.22 Pollen mitosis showing (**a**) a haploid set of eight chromosomes from *Allium cepa*, and (**b**) a haploid set of seven chromosomes from *Aloe vera*. (Image **b** is taken from Das et al. 2010. Nucleus 53:89–94. With permission)

The provided slide or photograph (Fig. 7.22) displays the following characteristics:

(i) Loosely arranged, boat-shaped cells with prominent exine and intine walls are visible.
(ii) The nucleolus and nuclear membrane are absent.
(iii) A haploid (n) set of long, condensed, and countable chromosomes (eight chromosomes in Fig. 7.22a and seven chromosomes in Fig. 7.22b), featuring distinct constrictions, is observed.

Therefore, the supplied slide or photograph (Fig. 7.22) illustrates pollen mitosis.

References

Bonciu E, Firbas P, Fontanetti CS, Wusheng J, Karaismailoğlu MC, Liu D, Menicucci F, Pesnya DS, Popescu A, Romanovsky AV, Schiff S (2018) An evaluation for the standardization ofthe *Allium cepa* test as cytotoxicity and genotoxicity assay. Caryologia 71(3):191–209

Das A, Mukherjee P, Ghorai A, Jha TB (2010) Comparative karyomorphological analyses of in vitro and in vivo grown plants of *Aloe vera* L. BURM. f. Nucleus 53(3):89–94. https://doi.org/10.1007/s13237-010-0001-1

Fiskesjo G (1985) The *Allium* test as a standard in environmental monitoring. Hereditas 102(1):99–112

Ma TH (1999) The international program on plant bioassays and the report of the follow-up study after the hands-on workshop in China. Mutat Res 426(2):103–106

Ma TH, Cabrera GL, Owens E (2005) Genotoxic agents detected by plant bioassays. Rev Environ Health 20(1):1–14

Rank J (2003) The method of *Allium* anaphase telophase chromosome assay. Acta Biologica Iugoslavica Serija D Ekologija 1:38–42

Wijeyaratne WMDN, Wickramasinghe PGMU (2020) Chromosomal abnormalities in *Allium cepa* induced by treated textile effluents: spatial and temporal variations. J Toxicol 3:1–10

Genome Size and C-Value Paradox in Plants

8

Abstract

This chapter explores the concepts of genome size and the C-value paradox in plants, tracing the historical evolution of genetic understanding from Mendel's foundational work on heredity to the elucidation of chromosomes by Boveri and Sutton. The chapter defines the genome as the complete set of genetic information in an organism and highlights the structural differences between prokaryotic and eukaryotic genomes. It introduces the term "C-value," which quantifies DNA content and emphasizes its constancy across a species' cells. Advancements in knowledge and the availability of high-throughput instruments over the past few decades have significantly enhanced our ability to measure the actual C-value of DNA within cells. Advances in measurement techniques over recent decades have led to significant insights into plant genome sizes, which can vary remarkably—up to 3000-fold—across the plant kingdom. Diversity in eukaryotic genome size also revealed a range from 2.3 Mbp in the parasitic microsporidian *Encephalitozoon intestinalis* to approximately 150,000 Mbp in the plant *Paris japonica.* The C-value has emerged as a vital parameter in research on biodiversity, taxonomy, and phylogeny. The chapter also addresses the complexities surrounding C-value consistency, particularly in cases such as maize, where supernumerary chromosomes and heterochromatin raise questions about the relationship between DNA content and chromosome number. Simultaneously, DNA size estimation in prokaryotes revealed substantial variation, ranging from 140 kilobase pairs (Kbp) to nearly 15 megabase pairs (Mbp). Eukaryotic genome sizes showed even greater diversity, with a range more than 60,000-fold, from 2.3 Mbp in the parasitic microsporidian *Encephalitozoon intestinalis* to approximately 150,000 Mbp in the plant *Paris japonica.* Ultimately, the chapter underscores how variations in genome size and structure reflect evolutionary processes and continue to intrigue scientific inquiry, paving the way for ongoing exploration in plant genomics.

T. B. Jha, M. Halder, *Classical and Molecular Plant Cytogenetics*,
https://doi.org/10.1007/978-981-95-8265-5_8

8.1 Genome

In his landmark discovery, Gregor Mendel (1865) narrated the basic principles of heredity based on experiments with pea plants long before the concepts of chromosomes, DNA, and genes were understood. He hypothesized that parents contribute some substance or factors in their reproductive cells to their offspring, determining all heritable characteristics.

At the beginning of the twentieth century, Theodor Boveri and Walter Sutton's (1902) experiments unlocked the nature of Mendelian factors as chromosomes. They proposed the "Chromosomal Theory of Inheritance," which later gained more clarity with the discovery of DNA in 1953. This discovery confirmed the nature of chromosomes as discrete condensed genomic units. Since then, it has become clear that Mendelian factors are highly condensed gene-bearing DNA sequences that control and maintain an organism's hereditary characteristics.

The term "chromosome" was established more than a hundred years ago, whereas the term "genome" was coined by Hans Winkler to describe the study of genes and chromosomes (Deakin et al. 2019). A genome is now defined as an organism's complete set of genetic information. In eukaryotic organisms, the genome is stored in the nucleus and packaged into multiple linear chromosomes. It comprises protein-coding exons and noncoding introns, the latter of which are composed of bulky repetitive DNA sequences. In contrast, the absence of true nuclei causes the prokaryotic genome dispersed throughout the cytoplasm as a single circular DNA molecule, along with additional components, such as plasmids and mobile insertion elements. Irrespective of whether a nucleus is present, all living organisms on this planet contain genomic DNA, and every genome has a specific size.

Genome: The complete set of genetic information in an organism.

Eukaryotic genome is stored inside the nucleus and packaged into multiple long linear molecules of DNA called chromosomes.

Prokaryotic genome is dispersed in the cytoplasm as a single circular DNA.

8.2 Genome Size and C-Value Paradox

From the mid-twentieth century onward, scientists measured genome size using biochemical and microscopic techniques. Many scientists, such as Swift (1950) and Vendrely (1955), worked on plants, such as *Tradescantia* and *Zea*, and either proposed or supported the theory of DNA constancy. In 1965, Bachmann and Cowden proposed a similar theory based on their studies of animals, such as frogs, mice, and grasshoppers. They noted that the nuclear DNA content is constant within a given set of chromosomes across a species. Karyological studies initiated by Tischler also indicated that chromosome number, total chromosome length, and nucleus size are highly species-specific (Šmarda and Bureš 2010).

During this period, the term "C-value" emerged as the unit for estimating DNA size or amount. As the "C-value" has been popularized, it refers to the DNA content of the un-replicated reduced chromosome complement (the 1C-value). The letter "C" in "C-value" stands for constant or characteristic, signifying that genome size remains constant within the cells of a species. This is why the DNA content in haploid (gamete) and diploid (fertilized zygote/somatic) cells is unequivocally referred to as 1C and 2C, respectively. The C-value reinforces that DNA is the only stable hereditary molecule, distinct from the highly transient proteins.

Since the C-value represents a quantitative aspect of the genome, it has become a fundamental parameter for addressing practical and predictive questions related to biodiversity, taxonomy, and phylogeny. Advancements in knowledge and the availability of high-throughput instruments over the past few decades have significantly enhanced our ability to measure actual C-value, DNA content, or amount of the DNA within cells. Estimating genome size in plants has become a topic of research, with genome size expressed either as the mass of DNA in picograms (pg) or as base pairs (Table 8.1). The relationship between the mass of DNA in picograms and its length in base pairs is 1 pg of DNA $\approx$ 1 billion base pairs, or 1000 Mbp.

Research in this domain flourished during the twentieth century, led by renowned scientists such as Jim Price, Mike Bennett, and Johann Greilhuber, and gained further momentum in the twenty-first century, with influential contributions from researchers such as M. D. Bennett, I. J. Leitch, Jaroslav Doležel, and David W. Galbraith. Astonishingly, almost 3000-fold variation in DNA content has been reported across the plant kingdom (Flavel 2021). Moreover, remarkable variations in genome size within and between species members became apparent and were established as an evolutionary feature. This divergence in the accumulation of C-value data in plants has surprised and intrigued scientists.

The discovery of supernumerary chromosomes and heterochromatin knobs within the same genome of maize raised questions about C-value consistency (Kuwada 1925; Longley 1927, 1938). Furthermore, it became evident that the 1C and 2C-values do not correlate with the haploid (n) and diploid (2n) chromosome numbers. It is consistent with earlier observations of Kuwada (1925) and Longley (1927, 1938) that additional supernumerary chromosomes and heterochromatin knobs in the maize genome may change genomic size.

Simultaneously, DNA size estimation in prokaryotes revealed substantial variation, ranging from 140 kilobase pairs (Kbp) to nearly 15 megabase pairs (Mbp). Eukaryotic genome sizes showed even greater diversity with a range more than 60,000-fold, from 2.3 Mbp in the parasitic microsporidian *Encephalitozoon intestinalis* to approximately 150,000 Mbp in the plant *Paris japonica* (Table 8.1; Pellicer et al. 2010; Elliott and Gregory 2015). This enormous diversity raises a legitimate question about the constancy of the C-value in genomic DNA. It challenges the argument that organismal complexity and more complex organisms have progressively larger genomes because the amount of DNA in a haploid genome (1C-value) does not seem to correlate strongly with the complexity of an organism, exhibiting extreme variability. These perplexing differences in C-values across related microbes, animals, and plants have contributed to confusion. Thomas (1971) dubbed it the "C-value paradox." Thus, the "C-value paradox" or "enigma" becomes a reality in the highly diversified plant kingdom (Gregory 2005).

Table 8.1 Genome size of some common plants (data collected from Plant DNA C-values Database; Leitch et al. 2019 https://cvalues.science.kew.org/)

Plant species	Common name	Genome size	
		Mbp/1C	pg/1C
[a]*Ostreococcus tauri* (Bathycoccaceae)	(Unicellular green alga algae)	12.56	0.01
Chlorella vulgaris (Chlorellaceae)	(Unicellular green algae)	38.80	0.04
Chara fragilis (Characeae)	Stonewort (multicellular algae)	18,914	19.30
Riccia crystallina (Ricciaceae)	(Bryophyte)	509.60	0.52
Anthoceros punctatus (Anthocerotaceae)	Dotted hornwort (Bryophyte)	176.40	0.18
Funaria hygrometrica (Funariaceae)	Bonfire moss (Bryophyte)	395	0.40
Lycopodium clavatum (Lycopodiaceae)	Club moss (Pteridophyte)	2800.35	2.86
Selaginella rupestris (Selaginellaceae)	Rock spike-moss (Pteridophyte)	107.80	0.11
Marsilea quadrifolia (Marsileaceae)	Water clover (Pteridophyte)	3993.50	4.08
Pinus sylvestris (Pinaceae)	Pine tree (Gymnosperm)	22520.40	22.98
Pinus lambertiana (Pinaceae)	Sugar pine (Gymnosperm)	28,959	29.55
Genlisea tuberosa (Lentibulariaceae)	Carnivorous plant (Angiosperm)	65	0.07
Genlisea aurea (Lentibulariaceae)	Carnivorous plant (Angiosperm)	67.2	0.07
Glycine max (Fabaceae)	Soybean (Angiosperm)	1107.40	1.13
Pisum sativum (Fabaceae)	Garden Pea (Angiosperm)	4802	4.90
Arabidopsis thaliana (Brassicaceae)	Thale cress (Angiosperm)	156.8	0.16
[b]*Paris japonica* (Melanthiaceae)	Japanese canopy plant (Angiosperm)	149,185	152.23
Oryza sativa (Poaceae)	Rice (Angiosperm)	490	0.50
Zea mays (Poaceae)	Maize/corn (Angiosperm)	2646	2.70
Hordeum vulgare (Poaceae)	Barley (Angiosperm)	8971.90	9.16
Triticum aestivum (Poaceae)	Bread wheat (Angiosperm)	16,954	17.30
Allium cepa (Amaryllidaceae)	Onion (Angiosperm)	17,542	17.90

[a] One of the smallest known plant genomes, [b]one of the largest known plant genomes

Haploid DNA amounts within the genus *Oryza sativa* (rice), Sorghum, and *Allium cepa* (onion) revealed three- to eight-fold differences. Further, the genome size of cultivated *Zea mays* (Corn) has expanded by about 50% compared to its wild relative, *Zea luxurians*. DNA content concerning the polyploid taxa reveals that the common hexaploid wheat has $2n = 6x = 42$ chromosomes and presents a 12C DNA content in somatic cells at the G1 stage, while its triploid primary endosperm nucleus ($3n = 9x = 63$ chromosomes) has a 6C DNA content. Surprisingly, the microspores or pollen of this species, after meiosis ($n = 3x = 21$ chromosomes), contain an equivalent amount of 6C DNA. Similar patterns have been observed in animals: primitive amphibians and fish contain more than 20 times the DNA per nucleus as humans, even though mammals display greater developmental complexity than primitive fish. This indicates that they must possess more genes.

Understanding the vocabulary related to the genome and genome size is crucial for scientific data analysis. In most diploid animals and plants, "genome" and "genome size" are used interchangeably. However, recent observations indicate that while the term "genome" originally suggested that polyploids may have more than one genome (Winkler 1920), Bennett and Leitch (2005) used the term "genome size" for the DNA content of a monoploid chromosome set only. This poses challenges for explaining the large number of polyploid plants with multiple genomes. For example, the "wheat genome" exhibits this complexity. The genome and genome size of wheat are not confined to one of its monoploid genomes, A, B, or D, but rather to the entire set of $2n = 42$ or $n = 21$ chromosomes. This example highlights the need to use both "genome" and "genome size" in an inclusive or exclusive sense. A genome can be generatively polyploid or monoploid, reduced or non-reduced, and replicated or non-replicated, yet the term "genome" remains appropriate (Greilhuber et al. 2005). Similarly, confusion arises when genome size and C-value are calculated for triploid saffron (*Crocus sativus*, $2n = 3x = 24$) and diploid and triploid *Musa* species ($2n = 33$).

To address these confusions, scientists at the 2003 Plant Genome Size Workshop, held at the Royal Botanic Gardens, Kew, reviewed the modern usage of terminology related to nuclear DNA content. They discussed the practical and more relaxed approach to the use of the terms "genome" and "genome size" while explaining concepts such as "genome size variation," "genome size research," "genome size measurements," "genome size workshop," and "genome size database." Eventually, they proposed "holoploid genome size" (from the Greek word holos, meaning complete) and the abbreviation "C-value." The term "holoploid genome size" serves as a shortcut for the "DNA content of the whole complement of chromosomes characteristic for the organism, irrespective of the degree of generative polyploidy, aneuploidies, etc." The term "holoploid genome size" is logical, as an organism always has a holoploid genome, even if it may not have a polyploid or monoploid genome (Greilhuber et al. 2005). The comparison between monoploid and holoploid genomes is discussed in Table 8.2. Figure 8.1 illustrates the status of C-value and chromosome number during different stages of mitotic, meiotic, and pollen mitotic division in higher plant cells, which change during the life cycle of higher plants.

Points to Remember:

- *Holoploid genome:* The term was derived from the Greek word "holos," which means complete. It defines the whole chromosome complement with chromosome number n, irrespective of its ploidy level.
- *Monoploid genome:* The DNA content of one set of chromosomes (the base number x) of an organism.
- *C-value:* It was initially employed to mean the DNA content of the "unreplicated haploid chromosome complements".
- *1C-value:* The DNA content of one non-replicated holoploid genome (chromosome number n)

- *2C-value:* The DNA content of a non-replicated, non-reduced holoploid genome (chromosome number 2n)
- *Cx-value:* The DNA content of a monoploid genome (with base chromosome number x) in polyploids and non-polyploids.
- *1Cx-value:* The DNA content of one non-replicated monoploid genome with chromosome number *x*. [1Cx = 2C/level of ploidy]
- *Genome size:* an all-inclusive term for C-values and Cx-values. The C-value is equal to the Cx-value in diploids, while these are different in polyploids.

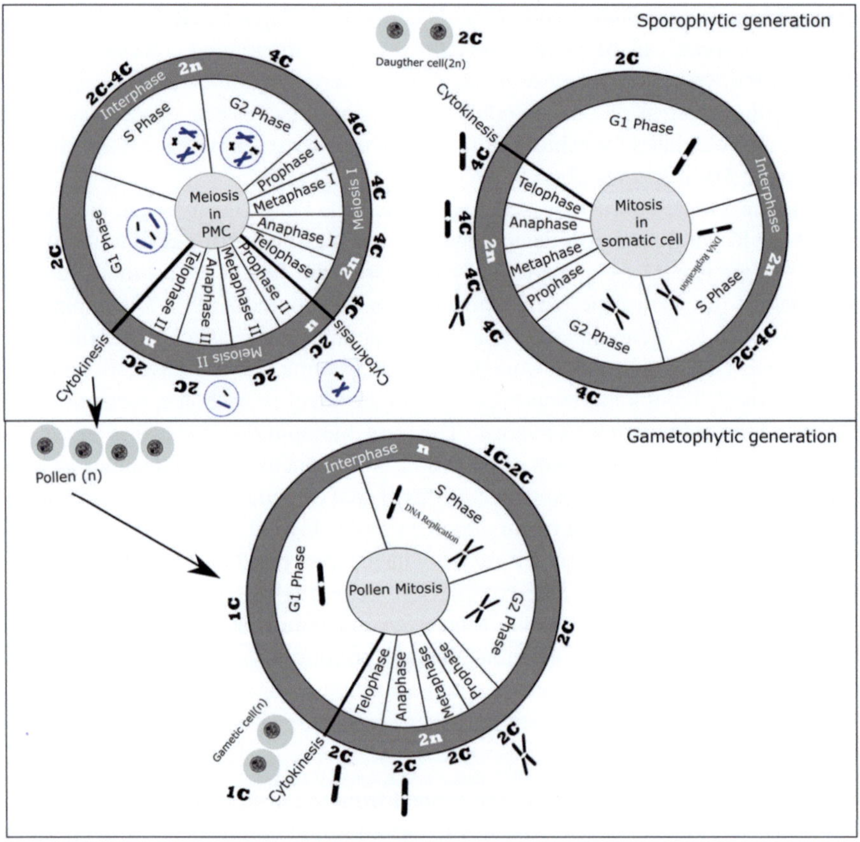

Fig. 8.1 Status of C-value and chromosome number during different stages of mitotic, meiotic, and pollen mitotic division in higher plant cells

Table 8.2 Comparison between monoploid and holoploid genome

Features	Monoploid genome	Holoploid genome
Genome status	Related to genome size	Related to genome size
Chromosome number	Basic set of haploid chromosome number (x) *Triticum aestivum* $2n = 6x = 42$, $x = 7$ Also referred to as Cx-value, and for the non-replicated state as 1Cx-value	Whole haploid set of chromosome numbers (n) *Triticum aestivum* $2n = 6x = 42$, n = 21 Also referred to as C-value, the non-replicated gametic cell is 1C, and the somatic cell is 2C
Genomic DNA content	Helps to understand the monoploid genome size	Helps to understand the overall gametic or somatic genome size

8.3 Present Status of C-Value Paradox

The term C-value paradox was coined by Thomas Jr. in 1970 but, he noted that the C-value paradox reflected three unexplainable issues: (i) why do many "lower" organisms have significantly higher C-values than the more evolved "higher" organisms; (ii) why do the C-values between some related species with comparable morphology and body organization differ by one or more orders of magnitude; and (iii) why is the proportion of DNA that does not code for proteins (noncoding or ncDNA) so high (up to ~98%) even in genomes of species that carry the "basal" c-value characteristic of the given group.

Over the years, scientists have been working to understand the reason behind the C-value paradox, or C-value enigma, in organisms. Through traditional chromosomal analysis, they have found that differences in genome size within a species may be due to various factors, like numerical variations in chromosome number caused by euploidy and aneuploidy, the presence of B-chromosomes and sex chromosomes, or different chromosomal rearrangements.

Advancements in molecular biology, molecular cytogenetics, and genomics have clarified that nuclear genomes, whether monoploid or holoploid, contain both coding and noncoding DNA sequences. The noncoding sequences include introns, promoters, and regulatory DNA sequences. A significant portion of various classes of noncoding DNA sequences is composed of repetitive DNA motifs, which may be present in hundreds or even thousands of copies across the genome (Fig. 8.2). These repetitive DNA motifs are located in specific regions along the chromosomes. Although the process and reasons behind the accumulation of abundant noncoding DNA in the nuclear genome remain unclear, this accumulation has occurred over many decades as an evolutionary consequence. These repetitive sequences evolve more rapidly than coding sequences and have contributed to variations in genome sizes within and between species (Biscotti et al. 2015).

It has been discovered that over half of the total nuclear genomic DNA in many plant species consists of repetitive DNA sequences, which do not code for proteins. These repetitive DNA sequences have been identified within and between plant species through chromosomal-based molecular cytogenetics.

While addressing the C-value enigma or paradox, scientists found that variation in C-value within and between species may have developed from the accumulation of noncoding DNA sequences. This raises a key question: if these variations have no biological significance, why are they accumulated throughout the evolutionary process? The exploration of noncoding DNA variation is still ongoing, as its biological relevance remains a critical area of research. The relationship between genome size and organismal complexity is complex and multifaceted. Initially, the C-value paradox drew attention to the disconnect between these two factors. However, recent research has unveiled the diverse roles of noncoding DNA, including its involvement in chromatin and the three-dimensional organization of the genome. Moreover, noncoding regions, such as small RNAs, introns, enhancers, silencers, and insulators, influence gene expression through several mechanisms.

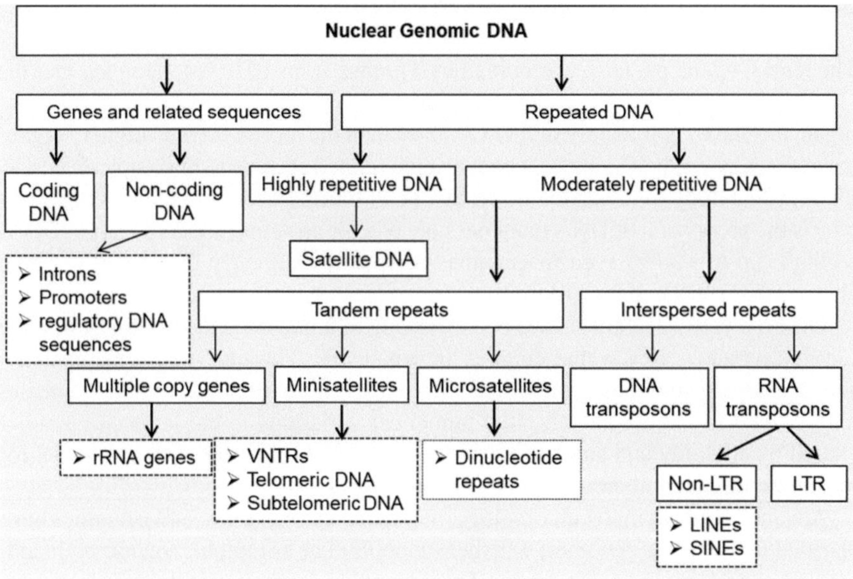

Fig. 8.2 Structural components of plant genome highlighting the higher amount of repetitive DNA sequence

Additionally, the variation in noncoding DNA may contribute to evolutionary plasticity, enabling organisms to adapt to changing environments. The impact of these variations can vary significantly among different species and their respective habitats. Changes in genome size can play a role in reproductive isolation and the process of speciation. Moreover, transposable elements add to genetic diversity by repositioning themselves within the genome, potentially creating new regulatory elements and influencing gene evolution.

Some experts have proposed that genic DNA holds precise information about an organism's development and functioning, while nucleotypic DNA sets thresholds within which genes can operate, influencing an organism's functional traits independently of the information encoded in the DNA. Analysis of single-celled amoebae genomes revealed that they are around 100 times larger than the human genome, suggesting that genomes can contain a significant amount of DNA other than genes and their regulatory sequences. Genome sequencing studies have also shown that a smaller number of regulatory genes contribute to developmental complexity, and organismal complexity appears to have an inverse relationship with gene number. This means that complex organisms may have fewer genes, while seemingly simple organisms may have more genes than expected. Genome size variations across different taxa are positively correlated with cell size and cell types in organisms. It has also been observed that the nuclei of large, slowly dividing cells inversely affect C-values in many cases, although the exact mechanisms for this are still uncertain. Finally, while some argue that the C-value paradox is a redundant issue, it is emphasized that future biological studies should focus on the evolutionary mechanisms and processes underlying DNA expansions or contractions and how the acquired DNA content has been repurposed in diverse groups (Lakhotia 2023).

8.4 Methods of Genome Size Estimation

Several techniques, such as flow cytometry, Feulgen densitometry, and sequencing-based methods, are available to measure the genome size of an organism. Each technique has its own principles, advantages, limitations, and applications (Table 8.3).

Flow cytometry is a relatively preferred choice for the determination of the C-value in plants due to its relatively fast and accurate results and requires relatively small amounts of tissue. However, the accuracy of flow cytometry relies on the quality of the sample preparation and the calibration of the instrument. The application of a well-characterized appropriate reference standard with a known genome size (C-value) is very crucial. The ploidy level of the sample must also be considered during the C-value measurement.

8.5 Key Steps to Determine the C-Value of Plant (or Any Organism) by Flow Cytometry

(i) *Collection of Sample:* Collect fresh, young, actively growing leaf tissue.
(ii) *Isolation of Nuclei:* The collected tissue is finely chopped in a buffer solution to release the nuclei. The resulting suspension is filtered to remove debris.
(iii) *DNA Staining:* Add a fluorescent dye like propidium iodide (PI) and 4′,6-diamidino-2-phenylindole (DAPI) to the nuclear suspension. These dyes bind specifically to DNA by intercalating between the DNA base pairs, making the nuclei fluorescent.
(iv) *RNAse Treatment:* Add RNAse to the solution to remove existing RNA, ensuring that only DNA is measured.

(v) *Flow Cytometry Analysis:* The stained nuclear suspension is passed through a flow cytometer, where a laser excites individual nuclei. The emitted fluorescence intensity is counted and measured by the instrument's detectors.

Table 8.3 Characteristic features of flow cytometry, Feulgen densitometry, and sequencing-based methods

Characters	Flow cytometry	Feulgen densitometry	Sequencing-based methods
Principle	It is a flow cytometry-based technique that measures the emitted fluorescence intensity from DNA-specific fluorescent-stained cells passing through a flow cytometer. The fluorescence intensity is proportional to the amount of DNA in the cell. The genome size can be estimated by comparing the fluorescence of the target sample to a reference sample with a known genome size	It is a microscopic technique that involves staining DNA with the DNA-specific Feulgen stain and measuring the intensity of Feulgen using a microscope with a photometer This provides a measure of the amount of DNA in the cell	Estimates genome size from next-generation sequencing data using k-mer (counting the frequency of short DNA sequences (k-mers) in sequencing reads) and read-depth analysis (by analyzing the depth of coverage of sequencing reads)
Advantages	Relatively fast and accurate and requires relatively small amounts of tissue	Allows measurement of DNA content in individual cells	Provides accurate estimates of genome size and additional information about the genome sequence
Limitations	Requires access to a flow cytometer and can be affected by factors such as the presence of secondary metabolites	More labor-intensive than flow cytometry and susceptible to staining variability	Requires access to sequencing facilities and bioinformatics expertise and can be affected by sequencing errors and repetitive sequences

(vi) *Data Interpretation:* The flow cytometry data is displayed as a histogram, with fluorescence intensity on the x-axis and the number of nuclei on the y-axis. The position of the peaks on the histogram indicates the amount of DNA in the nuclei.

A plant sample with a known genome size (C-value) is run along with the unknown sample as a *reference standard.* This allows for comparison and accurate calculations.

(vii) *Calculation:* The C-value of the unknown sample can be calculated by comparing the position of the peaks for the unknown sample to the peaks for a reference standard.

$$\text{C}-\text{value of unknown sample} = \begin{pmatrix} \text{Mean fluorescence of unknown /} \\ \text{Mean fluorescence of standard} \end{pmatrix} \times \text{C}-\text{value of standard}$$

8.6 Online Archives of Plant C-Value Estimates

Sparrow et al. (1972) highlighted the need for creating a comprehensive database to compile genome size variations for future reference. As more data accumulated, Bennett and Smith (1976) presented the first in a series of compilations of plant DNA C-values, which have been updated over the years (Bennett et al. 1982; Bennett and Smith 1991; Bennett and Leitch 1995, 1997, 2001; Murray 1998; Bennett et al. 2000; Voglmayr 2000). The first electronic version of the Angiosperm DNA C-values Database was made available during the meetings at Kew in 1997. The second version was released in 2003. The most recent update, version 7.1, was released in 2019 (Pellicer and Leitch 2020; https://cvalues.science.kew.org/), containing data for 12,273 species, including land plants and several algal lineages (10,770 species of angiosperms, 421 species of gymnosperms, 303 species of pteridophytes, 334 species of bryophytes, and 445 algae).

The Plant DNA C-values Database is crucial for biological research. It serves as a fundamental resource for researchers that provides essential data on plant genome complexities and their role in evolution, ecology, and conservation. Genome size is a key factor in plant biology, influencing evolutionary processes, ecological adaptations, and the feasibility of sequencing projects.

8.7 Significance of C-Value

The C-value is a crucial tool in understanding genome evolution, facilitating genomic research, and informing ecological and evolutionary studies. It helps researchers understand the relationship between genome size and biological traits, providing insights into how genome size has evolved across different plant lineages. Accurate C-value data facilitate planning sequencing efforts and ensure the correct assembly of genomes. It also contributes to understanding how genome size affects plant biodiversity and distribution. The database often includes information on ploidy levels and chromosome numbers, essential for cytogenetic studies. Understanding the genomic characteristics of endangered plant species is crucial for effective conservation efforts, as it provides valuable information for informed conservation planning.

References

Bennett MD, Leitch IJ (1995) Nuclear DNA amounts in angiosperms. Ann Bot 76:113–176

Bennett MD, Leitch IJ (1997) Nuclear DNA amounts in angiosperms—583 new estimates. Ann Bot 80:169–196

Bennett MD, Leitch IJ (2001) Nuclear DNA amounts in pteridophytes. Ann Bot 87:335–345

Bennett MD, Leitch IJ (2005) Nuclear DNA amounts in angiosperms: Progress, problems and prospects. Ann Bot 95(1):45–90

Bennett MD, Smith JB (1976) Nuclear DNA amounts in angiosperms. Philos Trans R Soc Lond, Series B 274:227–274

Bennett MD, Smith JB (1991) Nuclear DNA amounts in angiosperms. Philos Trans R Soc Lond, Series B 334:309–345

Bennett MD, Smith JB, Lewis Smith RI (1982) DNA amounts of angiosperms from the Antarctic and South Georgia. Environ Exp Bot 22:307–318

Bennett MD, Bhandol P, Leitch IJ (2000) Nuclear DNA amounts in angiosperms and their modern uses—807 new estimates. Ann Bot 86:859–909

Biscotti MA, Canapa A, Forconi M, Olmo E, Barucca M (2015) Transcription of tandemly repetitive DNA: functional roles. Chromosom Res 23(3):463–477

Deakin JE, Potter S, O'Neill R, Ruiz-Herrera A, Cioffi MB, Eldridge MDB, Fukui K, Marshall Graves JA, Griffin D, Grutzner F, Kratochvíl L, Miura I, Rovatsos M, Srikulnath K, Wapstra E, Ezaz T (2019) Chromosomics: bridging the gap between genomes and chromosomes. Genes (Basel) 10(8):627

Elliott TA, Gregory TR (2015) What's in a genome? The C-value enigma and the evolution of eukaryotic genome content. Philos Trans R Soc Lond Ser B Biol Sci 370(1678):20140331

Flavell RB (2021) Perspective: 50 years of plant chromosome biology. Plant Physiol 185(3):731–753

Gregory TR (2005) The C-value enigma in plants and animals: a review of parallels and an appeal for partnership. Ann Bot 95(1):133–146

Greilhuber J, Dolezel J, Lysák MA, Bennett MD (2005) The origin, evolution and proposed stabilization of the terms 'genome size' and 'C-value' to describe nuclear DNA contents. Ann Bot 95(1):255–260

Kuwada Y (1925) On the number of chromosomes in maize. Bot Mag (Tokyo) 39:227–234

Lakhotia SC (2023) C-value paradox: genesis in misconception that natural selection follows anthropocentric parameters of 'economy' and 'optimum'. BBA Adv 4:100107

Leitch IJ, Johnston E, Pellicer J, Hidalgo O, Bennett MD (2019) Plant DNA C-values database (release 7.1, Apr 2019). Royal Botanic Gardens, Kew. Available at https://cvalues.science.kew.org

Longley AE (1927) Supernumerary chromosomes in *Zea mays*. J Agric Res 35:769–784

Longley AE (1938) Chromosomes of maize from North American Indians. J Agric Res 56:177–195

Murray BG (1998) Nuclear DNA amounts in gymnosperms. Ann Bot 82(Suppl. A):3–15

Pellicer J, Leitch IJ (2020) The plant DNA C-values database (release 7.1): an updated online repository of plant genome size data for comprehensive studies. New Phytol 226:301–305

Pellicer J, Fay MF, Leitch IJ (2010) The largest eukaryotic genome of them all? Bot J Linn Soc 164(1):10–15

Šmarda P, Bureš P (2010) Understanding intraspecific variation in genome size in plants. Preslia 82(1):41–61

Sparrow AH, Price HJ, Underbink AG (1972) A survey of DNA content per cell and per chromosome of prokaryotic and eukaryotic organisms: some evolutionary considerations. In: Smith HH (ed) Evolution of genetic systems. Gordon and Breach, New York, pp 451–494

Swift H (1950) The constancy of desoxyribose nucleic acid in plant nuclei. Proc Natl Acad Sci USA 36(11):643–654

Thomas CA Jr (1971) The genetic organization of chromosomes. Annu Rev Genet 5:237–256

Vendrely R (1955) The deoxyribonucleic acid content of the nucleus. Nucleic Acids 2:155–180

Voglmayr H (2000) Nuclear DNA amounts in mosses (Musci). Ann Bot 85:531–546

Winkler H (1920) Verbreitung und Ursache der Parthenogenesis im Pflanzen- und Tierreiche. Jena: Gustav Fischer Verlag

Flow Cytometry: Foundations of Modern Cell Analysis

9

Abstract

Flow cytometry is an essential laboratory technique that revolutionizes cellular analysis by measuring various cellular properties as cells pass through a fluid stream and are excited by laser beams. This chapter explores the foundational principles, applications, and recent advancements of flow cytometry, highlighting its significance in biological research. The technique combines hydrodynamic focusing and light scattering to analyze individual cells or particles, making it invaluable across diverse fields such as immunology, cell biology, and cancer research. By detecting multiple physical, chemical, and fluorescent characteristics of thousands of cells per second, flow cytometry efficiently provides crucial data regarding cell size, granularity, and protein expression. The historical evolution of flow cytometry is traced from its inception in the 1930s, with key milestones including the invention of the Coulter Counter and innovations by Mack Fulwyler and Wolfgang Göhde. These developments laid the groundwork for fluorescence-based analysis and cell sorting methodologies. Modern flow cytometers are capable of analyzing up to 30 parameters simultaneously, revealing the complexity of cellular heterogeneity. Recent advancements, such as flow karyotyping and spectral flow cytometry, have broadened the applications of this technique, extending its utility to fields like plant breeding and ecological research. Understanding flow cytometry is essential for researchers investigating immune cell populations, cancer cell dynamics, and other critical biological questions, ultimately contributing to informed therapeutic strategies and clinical diagnostics.

9.1 Introduction

We examined classical and molecular methodologies for plant chromosomal analysis in the preceding chapters. This chapter focuses on the fundamental principles, applications, and recent advancements associated with a powerful, versatile, and reliable laboratory technique widely used in modern biological research called

T. B. Jha, M. Halder, *Classical and Molecular Plant Cytogenetics*,
https://doi.org/10.1007/978-981-95-8265-5_9

"flow cytometry." The term "flow cytometry" consists of three components: "flow," referring to fluid or motion; "cyto," meaning "cell" or "cells"; and "metry," indicating "measure" or "the process of measuring." Thus, flow cytometry is a laser-based technique that measures various cellular properties as cells are excited at a specific point by a laser beam while flowing through a fluid stream. This technique is also valuable for analyzing particles suspended in fluids (Goetz et al. 2018).

Flow cytometry utilizes the principles of hydrodynamic focusing and light scattering to analyze individual cells or particles within complex suspensions (samples) within a machine known as a flow cytometer (Fig. 9.1). This laser-based technique is powerful and reliable, capable of identifying, counting, and sorting cells or particles based on their physical and chemical characteristics. It has become an essential method for cellular analysis across various fields, including immunology, cell biology, and cancer research. A comprehensive understanding of this remarkable technology is crucial for those studying immune cell populations, analyzing cancer cells, developing new therapeutic strategies, or engaging in contemporary biological research and clinical diagnostics. The principles and applications covered in this chapter will provide researchers with a foundation to effectively leverage this powerful technology.

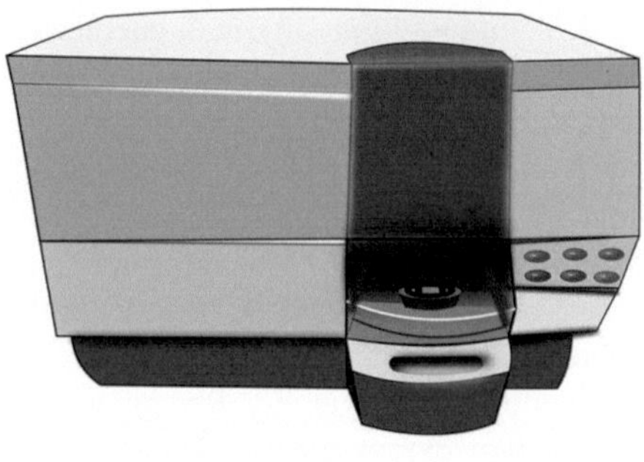

Fig. 9.1 Diagrammatic view of a flow cytometer

From a cytogenetic perspective, this process detects and analyzes multiple physical, chemical, and fluorescent properties of thousands of cells each second. This is achieved by passing individual cells through laser beams and collecting the light they emit or scatter. The technique rapidly provides cellular data through fluorescent labeling, including size, granularity, and the expression of specific proteins or other molecules of interest.

The journey begins with an overview of the technical foundations and core components of a flow cytometer. In-depth discussions on the critical aspects of sample preparation specific to plant tissues, an exploration of the principles of fluorescence and light scattering, and data analysis follow this. Additionally, we will examine the latest advancements in flow cytometry technology, including flow karyotyping and spectral flow cytometry. These innovations have expanded the applications of flow cytometry beyond traditional cellular analysis, paving the way for emerging uses in plant breeding, conservation biology, and ecological research.

9.2 History of Flow Cytometry

The fascinating journey of flow cytometry begins with the groundbreaking work of Moldavan (1934), who introduced an automated cell counting method using a photoelectric apparatus. This innovation laid the groundwork for the development of flow cytometry. In 1949, Wallace Coulter invented the "Coulter Counter," which enabled the counting of particles in suspension and provided a crucial principle for cell analysis in flow. Subsequently, Crosland-Taylor advanced the field in 1953 by applying the principles of laminar flow to design a chamber for optical cell counting, allowing for precise cell alignment and analysis (Coulter 1953; Crosland-Taylor 1953).

A pivotal moment in the evolution of this technology occurred in 1965 when Mack Fulwyler (1965) invented the first cell sorter, which integrated inkjet technology with the Coulter principle, marking a significant breakthrough in flow cytometry. Just a few years later, Dittrich and Göhde (1969) from Germany developed the first fluorescence-based flow cytometry device, which was commercialized by Partec in 1969. Initially known as pulse cytophotometry, the term "flow cytometry" was adopted in 1978.

The field continued to advance with the introduction of various instruments, including the Cytofluorograph in 1971 and later FACS (fluorescence-activated cell sorting). Today, modern flow cytometers can analyze up to 30 parameters simultaneously per cell, offering unprecedented insights into the complex world of cellular heterogeneity and function. This technological progression has empowered researchers to identify rare cell populations, investigate intricate cellular interactions, and monitor disease progression with remarkable precision.

Flow cytometry, also known as pulse cytophotometry, has been in use since the 1970s. Over the years, with various advancements, it has become a crucial tool for both animal and plant scientists in determining nuclear DNA content, estimating genome size, analyzing cell cycle dynamics, assessing ploidy levels, and conducting isolated chromosome sorting and analysis.

Spectral flow cytometry represents an advanced iteration of this technique, offering high-precision measurements of fluorescent spectra and light scattering properties across diverse cellular populations. At the dawn of the twenty-first century, innovations in detector technology began to enhance flow cytometry, providing the speed and resolution necessary for practical applications. Researchers have increasingly adopted sophisticated optics and detectors in spectral flow cytometers, which are characterized by their high gain and fast response times. This method employs a spectrograph and a multichannel detector (typically a CCD) to facilitate the multiparameter collection and analysis of data from individual cells or particles.

9.3 Challenges and Solutions in the Application of Flow Cytometry in Plant Sciences

Flow cytometry is mainly employed in biomedical applications, particularly for analyzing blood cells because they lack cell walls. While flow cytometry has demonstrated significant utility in plant science, its development in this area has encountered distinct challenges compared to animal systems. These obstacles stem from cell walls, the abundance of secondary metabolites, autofluorescence, and the complex structure of various plant tissues.

Over many decades, researchers have developed various strategies for analyzing plant tissues. These approaches include tissue-specific sample preparation protocols that incorporate modified chopping techniques using razor blades or specialized tissue homogenizers, adjusted incubation times, specific filtration steps, and revised buffer compositions. Modifications have been made to sampling gates and thresholds to improve plant-specific detection, alongside the creation of specialized algorithms for identifying nuclear aggregation and discriminating against debris. Furthermore, quality control measures tailored for plants involve selecting appropriate reference standards to estimate genome size, cell size, and nuclear measurements and modifications to existing protocols.

These strategies have broadened research possibilities and are currently applied to various plant species for tasks such as determining nuclear DNA content and genome size, analyzing cell cycle dynamics, and assessing ploidy levels. Flow cytometry provides researchers with quantitative data that would be impractical or impossible to obtain through traditional microscopy techniques. Pre-processing is crucial when working with plant materials, and researchers have long awaited standardized methods for preparing single-cell, protoplast, and nucleus suspensions within plant systems.

Additionally, the implementation of compensation protocols for managing autofluorescence—such as that caused by chlorophyll (with emission peaks at 670–680 nm), cell wall components like lignin and suberin, and various secondary metabolites—has further refined the methodology. Specialized filter combinations differentiate between natural fluorescence, fluorochrome signals, and cellular debris. These plant-specific modifications have significantly enhanced the applicability and reliability of flow cytometry in plant research, facilitating more accurate and comprehensive analyses of plant cellular characteristics and responses.

The application of flow cytometry in plants initially focused on isolated protoplasts (cells devoid of cell walls) from *Vicia faba*. However, the process of protoplast isolation proved to be time-consuming and complex due to challenges such as the low permeability of the plasma membrane, chlorophyll autofluorescence, and the off-center positioning of the nucleus. These factors hindered the generation of fluorescence histograms with adequate resolution (Loureiro et al. 2021).

This experience prompted researchers to explore the use of isolated nuclei for flow cytometry analysis in plants. Around this time, Puite and Ten Broeke (1983) successfully harvested intact nuclei after protoplast lysis in the presence of non-ionic detergents. Concurrently, Galbraith et al. (1983) developed a method for preparing a suspension of nuclei by chopping a small sample of fresh tissue in a hypotonic buffer supplemented with a non-ionic detergent. The protocol proposed by Galbraith et al. (1983) has been recognized for its simplicity, convenience, rapidity, and effectiveness in delivering high-quality histograms across various plant species. Since then, the use of isolated nuclei has gained traction not only in plant studies but also in a variety of eukaryotic organisms, including insects, mammals, fish, and fungi (Loureiro et al. 2021). The Plant DNA C-values Database has documented DNA estimations completed for approximately 12,500 plant species (https://cvalues.science.kew.org/).

Flow cytometry analysis in plants is primarily employed to estimate genome size and ploidy levels, with significant implications for future research in genomics and proteomics (Greilhuber et al. 2005, 2006; Galbraith 2010). It is essential to recognize that flow cytometry is an indirect method for ploidy estimation and should be corroborated with chromosome counting. If chromosome counts do not validate ploidy estimation, it is designated as "DNA-ploidy" (Sliwinska et al. 2022). The technique's non-destructive nature and high-throughput capabilities make it particularly valuable for studying rare or endangered plant species, where the conservation of samples is crucial.

Furthermore, advancements in flow cytometry have led to the development of fluorescence-activated cell sorting (FACS), which provides direct, precise information on chromosome sorting in higher plants. This method yields valuable insights into the organization of DNA sequences across various plant species. However, the data obtained through flow cytometry are relative and must be calibrated using chromosome counts derived from the same individual by traditional chromosome squashing or advanced EMA-based techniques (Suda et al. 2007). Consequently, flow cytometry cannot replace the indispensable information obtained from cost-effective plant chromosome analysis. Such analysis provides a comprehensive view of genomic structures by examining an individual's complete chromosome complement, offering a "bird's-eye view" of their genetic information. Additionally, the species-specific number of chromosomes serves as an essential indicator of ploidy.

The ability to swiftly analyze thousands of individual cells or nuclei within minutes has transformed our approach to fundamental questions in plant biology. This technique's versatility extends beyond basic research, playing a critical role in modern plant biology, plant breeding programs, biodiversity studies, and evolutionary research, thereby offering unprecedented insights into plant cell biology, genetics, and development. Recent technological advancements, particularly in multiparameter analysis and sorting capabilities, have broadened the scope of plant flow cytometry. Additionally, the integration of flow cytometry with other cutting-edge techniques, such as single-cell genomics and proteomics, has paved new pathways for exploring plant biology at an extraordinary level of resolution.

9.4 Components of a Flow Cytometer

The flow cytometer is comprised of three interconnected systems: (i) Fluidics, (ii) Optics, and (iii) Electronics (Fig. 9.2). These components work in unison to facilitate rapid, multi-parameter analysis of cells, making flow cytometers essential tools in various fields of research and medicine.

(i) *Fluidics System:* The fluidics system features a flow cell where a buffer solution containing samples (liquid suspensions of the desired cells or particles) is introduced. Upon injection, the cells or particles initially display a random distribution. However, a laminar flow of sheath fluid is created to achieve hydrodynamic focusing of the sample stream containing the suspended cells or nuclei. Hydrodynamic focusing is accomplished through the precise control of pressure differentials between the sample and the sheath fluid, which narrows and centers the sample stream within the sheath fluid. This process ensures optimal alignment of individual cells or particles within the single-cell sample stream during analysis. Such precise alignment is crucial for consistently delivering individual cells to the specific site where they encounter the optical laser beam, which intersects with the focused cells one at a time. This configuration allows for accurate measurements of each cell. Additionally, it enables multiple laser intercepts or interrogation points for comprehensive analysis. The sheath fluid is typically a balanced salt solution, though saline or other suitable fluids may also be utilized.

(ii) *Optics System:* A flow cytometer's optical system comprises excitation units (light sources), a lens unit, and light collection units (detectors). This system is essential for delivering excitation light and capturing the scattered and emitted light from the sample.

The excitation unit produces a focused beam of light using lasers such as argon and helium-neon. Modern flow cytometers frequently employ multiple lasers with varying wavelengths—typically 488 nm, 633 nm, and 405 nm—allowing for the simultaneous detection of multiple fluorescent markers. The lens unit directs this beam onto the cells as they flow through the system.

The light collection unit consists of an array of dichroic mirrors, optical filters, and photodetectors designed to capture the scattered light and fluorescence emissions from the sample. A series of dichroic filters directs fluorescent emissions to specific detectors by transmitting certain wavelengths while reflecting others at an angle. Bandpass filters are utilized to select specific light wavelengths for measurement. This method enables the detection and quantification of individual fluorochromes. Commonly used detectors in flow cytometry include photomultiplier tubes (PMTs), photodiodes (PDs), and avalanche photodiodes (APDs). While photodiodes, PMTs, and APDs operate on similar principles, it is worth noting that APDs tend to be more cost-effective than PMTs and provide enhanced sensitivity in the far-red regions of the spectrum.

As each cell traverses the flow cytometer, a laser beam interacts with individual cells, generating scattered light based on the structures both within and surrounding the cells. This scattered light is measured at two angles by two types of detectors known as a forward scatter detector (FSC) and a side scatter detector (SSC). The FSC detector captures light scattered in the forward direction at angles of 1 or 2 degrees along the laser path. In contrast, the SSC detector measures light scattered at a 90-degree angle perpendicular to the laser beam. Forward scatter provides information regarding the size of cells and particles, while side scatter reveals insights into the granularity or internal complexity of the cells. Additionally, multiple fluorescence detectors are employed to identify specific light emitted by fluorescently labeled cells or natural autofluorescence. These detectors are equipped with filters designed to isolate different wavelengths of fluorescence.

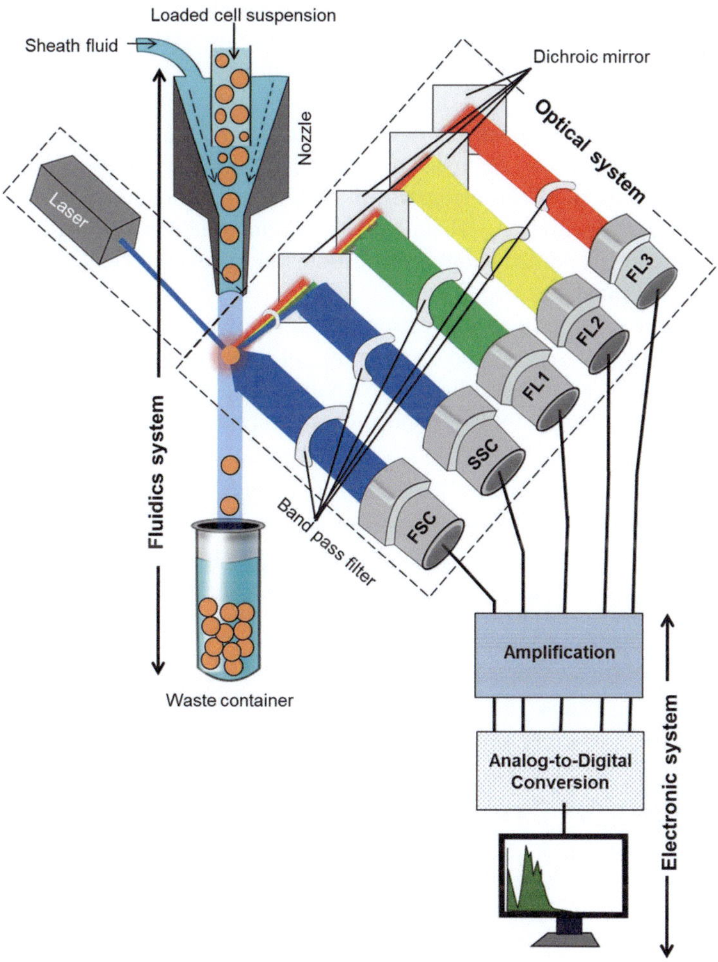

Fig. 9.2 Schematic diagram of a typical flow cytometer showing the three main components: fluidic systems, optical systems, and electronic systems of flow cytometry

(iii) *Electronics System:* The electronics system plays a crucial role in amplifying optical signals captured by the photodetectors, performing analog-to-digital conversion, acquiring data, and conducting data analysis. Typically, the optical signals generated from scattered light and fluorescence are comparatively weak. Electronic amplifiers enhance these weak optical signals received from the various detection channels (FSC, SSC, and fluorescence) of the flow cytometer, elevating them to measurable levels. This amplification allows for the identification and quantification of distinct cell populations within a sample, as well as enabling complex statistical analyses.

Once amplified, the analog signals (which represent light intensity) are converted into electronic signals and transformed into digital values using analog-to-digital converters (ADC). The signal processing units filter and analyze the shape, amplitude, height, area, and width of the signal pulses produced by each cell as it traverses the laser beam. Specialized flow cytometry software plays an essential role in managing the instrument and facilitating the collection, processing, and storage of digitized data within the connected computer system. This software generates reports in various formats (e.g., FCS files) and stores them for future reference or sharing. The analysis produces scatter density plots (dot plots) or histograms that reflect the characteristics of the cells. Scatter density plots are valuable for visualizing relationships between different parameters, while histograms illustrate the frequency distributions of various cell populations (Fig. 9.3).

Recent advancements in electronic technology have significantly improved the sensitivity, speed, and complexity of flow cytometry measurements. Additionally, electronic filtering techniques help to reduce noise and background interference, thereby enhancing measurement accuracy.

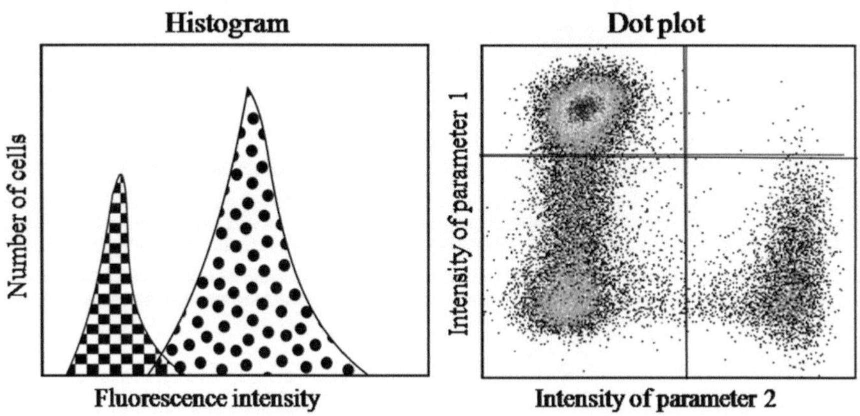

Fig. 9.3 Plotting patterns in flow cytometry

9.5 Isolation and Estimation of Plant Nuclear DNA Content

9.5.1 Requirements

(i) *Small apparatus or equipment*
- Micropipette and tips
- Ice pack
- Glass or Plastic Petri dishes
- Razor blade
- Nylon mesh of different sizes (20–50 μm, as required)

(ii) *Buffers for nuclei isolation*

9.5.2 Steps for Isolation of Plant Nuclei (Fig. 9.4)

(a) *Pre-isolation preparation:* The isolation of plant nuclei necessitates the use of an appropriate isolation buffer, as its composition is pivotal for obtaining intact nuclei and ensuring accurate, reliable, and high-resolution measurements of nuclear DNA content. Galbraith et al. (1983) introduced a method for isolating plant cell nuclei that involves finely chopping a small amount of fresh tissue in a hypotonic buffer solution supplemented with a non-ionic detergent. This approach effectively isolates the cell nuclei while simplifying the procedure, making it both convenient and reliable for a variety of plant materials. The isolation buffer serves to eliminate cytoplasmic debris, prevent the aggregation of nuclei, protect against DNA degradation, and maintain the stability of the nuclei.

Given the variability in tissue structure and chemical composition across different plant species, it is understandable that a universally optimal buffer does not exist. Nevertheless, certain isolation buffers have consistently yielded superior results, particularly when working with model species. Since the introduction of this technique in 1983, numerous scientists have proposed various isolation buffers, resulting in the description of over 30 distinct buffers for nuclear isolation or simultaneous isolation and staining. In the following section, we highlight some notable buffers, including their chemical compositions, commonly employed for flow cytometric analysis (Table 9.1). To achieve high-quality preparations of plant nuclei, it is crucial to carefully consider the composition of the buffer and optimize the isolation process. The selection of the appropriate buffer is essential, and adjustments to specific components and their concentrations may be required depending on the plant species, tissue type, and intended downstream application.

(b) *Plant Tissue Selection:* The young, healthy roots and leaves of freshly germinated seedlings or field-grown plants are considered the best plant materials for isolating plant nuclei. Although older roots, mature leaves, seeds, dry herbarium specimens, and silica-dried samples can be used, it is imperative to avoid senescent or damaged tissue.

(c) *Tissue Preparation:* Young roots or leaf tissues (at least 100 mg) are excised from the plant sample (Unknown 2C) and internal standard (Known 2C), cleaned thoroughly to remove any debris, and placed immediately on ice or in liquid nitrogen to minimize damage to cellular integrity.

(d) *Mechanical Disruption of Tissues by Chopping:* Excised tissues are placed in a small glass Petri dish (5 cm in diameter), and 1–2 mL of ice-cold chopping buffer solution is added. The tissues are then chopped gently into very fine pieces using a sharp single-edged razor blade until uniform maceration is achieved. The material is allowed to sit for two minutes. The composition of the isolation buffer varies depending on the plant species and intended downstream application.

(e) *Dealing with Secondary Metabolites:* Many plant species have secondary metabolites or mucilaginous compounds in their cytoplasm that may hinder the release of nuclei from the cells.

(f) *Filtration:* The buffer solution containing chopped tissue homogenate is filtered through a nylon mesh (20–100 μm of pore size) or gauze to eliminate large cell fragments and plant cell walls. The filtrate is collected in a clean tube and kept on ice to prevent warming.

(g) *Centrifugation:* The filtered homogenate is centrifuged at low speed to separate and pellet the nuclei. It is advisable to wash the nuclei pellet several times with a fresh isolation buffer to remove any cellular debris and contaminants.

Table 9.1 Composition of commonly used buffer solutions for plant nuclei isolation

Chemical composition	Galbraith's buffer (1983)	LB01 buffer (Doležel et al. 1989)	Tris-$MgCl_2$buffer (Pfosser et al. 1995)	GPB (Loureiro et al. 2007)	WPB (Loureiro et al. 2007)	Otto's buffer (1992)	Nuclear buffer (Doležel et al. 2007)
$MgCl_2.6H_2O$	45 mM	–	4 mM	–	4 mM	Otto I: 100 mM citric acid, 0.5% (v/v) Tween 20 Otto II: 400 mM $Na_2PO_4.12H_2O$	–
MgSO4	–	–	–	–	–		20 mM
Sodium citrate	30 mM	–	–	30 mM	–		–
Sodium metabisulfite	–	–	–	–	10 mM		–
MOPS	20 mM	–	–	20 mM	–		–
Tris-HCl	–	15 mM	200 mM	–	0.2 M		15 mM
$Na_2EDTA2H_2O$	–	2 mM	–	–	2 mM		2 mM
Spermine.4HCl	–	0.5 mM	–	0.5 mM	–		0.5 mM
KCl	–	80 mM	–	80 mM	–		80 mM
NaCl	–	20 mM	–	20 mM	86 mM		20 mM
PVP-10	–	–	–	–	1%		–
Triton X-100	0.1% (w/v)	0.1% (v/v)	0.5% (v/v)	0.5% (v/v)	1% (v/v)		
β-Mercaptoethanol	–	15 mM	–	–	–		–
pH	7.0	7.5–8.0	7.5	7.0	7.5	Otto I: (2–3); Otto II: (8–9)	7.5

Store the buffer at 4 °C for regular use or at −20 °C in 10 mL aliquots for future use

(h) *Purification (Optional):* To achieve greater purity, additional purification of the nuclei can be performed using density gradient centrifugation, such as the Percoll gradient, which separates the nuclei from other organelles based on their respective densities.

Function of Chemicals

- $MgCl_2$ or $MgSO_4$ helps in chromatin and membrane stabilization.
- KCl and NaCl provide ionic strength.
- Sodium citrate helps in nuclear stabilization.
- MOPS (4-morpholinepropane sulfonate) and Tris-HCl maintain the pH of the solution.
- Triton X-100 is a mild detergent that helps in membrane disruption and permeabilization.
- Tris-EDTA buffer maintains pH and chelates divalent cations.
- β-mercaptoethanol breaks disulfide bonds (S-S) in proteins,
- Spermine helps stabilize chromatin.
- Polyvinylpyrrolidone (PVP) is used to bind phenolic compounds.
- Ascorbic acid as an antioxidant.
- Enzymatic digestion protocols using (i) Pectolyase for middle lamella degradation, and (ii) Cellulase for cell wall breakdown.

(i) *RNase Treatment (Optional):* DNAase-free RNase (10–20 mg mL^{-1}) can be added at this stage.

(j) *Quality and Quantity Assessment of Isolated Nuclei:* High debris levels, clumping, low yield, and degradation of nuclei may interfere with the result obtained from the flow cytometry. Thus, prior to staining, the quality (integrity and purity) and quantity of isolated nuclei must be assessed by examining them under a microscope.

If debris levels are high, repeat the filtration steps. If clumping occurs, add more buffer. For low yield, it may be necessary to adjust the intensity of chopping. If degradation is observed, assess the freshness of the buffer. Additionally, verify the final volume and concentration.

(k) *Fluorescent Dyes and Nuclei Staining:* Staining is essential for obtaining accurate, high-resolution assessments of nuclear DNA content. Homogenized nuclei are stained using DNA-specific fluorochromes, such as 4′, 6-Diamidino-2-phenylindole (DAPI) and Propidium iodide (PI), before being injected into the flow cytometer. The basic characteristic features of DAPI and PI are discussed in Table 9.2. An appropriate volume of fluorochrome is added, gently mixed by inversion, and incubated in darkness for a specific duration ranging from 5 to 30 min at 4 °C. It is crucial to standardize both the concentration and duration of fluorochrome application for optimal staining results. Longer staining periods may diminish fluorescence and/or increase background noise. The recommended staining concentrations are 50 μg mL^{-1} for PI, while it is 4 μg mL^{-1} for

DAPI. Methods for preparing fluorescent dyes have been detailed in previous chapters. It is important to verify staining efficiency and protect samples from light exposure.

The entire procedure should be completed within 30 min, ensuring the maintenance of a cold chain throughout.

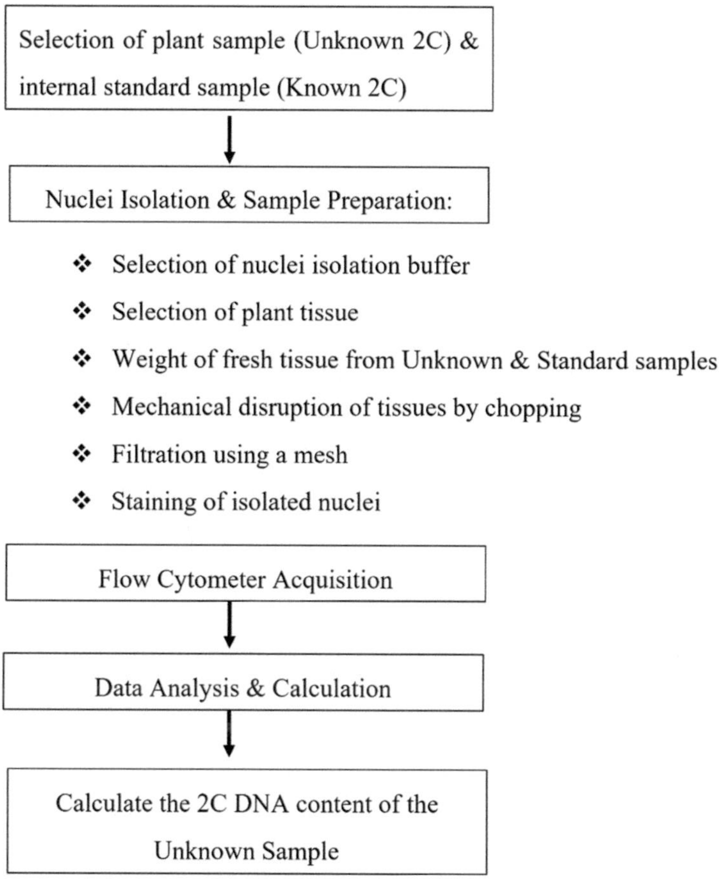

Fig. 9.4 Flowchart illustrating the step-by-step process for determining the 2C DNA content of a plant sample using a flow cytometer

Table 9.2 Difference between Propidium Iodide (PI) and 4′, 6-Diamidino-2-phenylindole (DAPI)

Feature	Propidium iodide (PI)	4′,6-Diamidino-2-phenylindole (DAPI)
Excitation maximum	535 nm	358 nm
Emission maximum	617 nm	461 nm
Binding affinity	Stoichiometric binding to DNA, no sequence specificity	Binds to AT-specific sequences
Applications	Genome size estimation, ploidy level determination, and cell cycle analysis	High-resolution DNA content analysis, chromosome studies, and nuclear architecture investigation
Advantages	High quantum yield, linear DNA binding, and stable fluorescence	High specificity for DNA, low background interference, and no RNase requirement
Limitations	(i) Requires RNase treatment, (ii) DNA intercalation may affect chromatin structure, (iii) potential interference with some plant metabolites	(i) AT preference affects stoichiometry, (ii) UV excitation requirement, and (iii) limited use in some comparative studies

Fluorochromes Used in Plant Flow Cytometry:

- *DNA-Specific Fluorochromes:* Propidium Iodide (PI) and 4′, 6-Diamidino-2-phenylindole (DAPI).
- *Fluorochromes used for viability and membrane integrity studies:* Fluorescein Diacetate (FDA) and Calcein-AM.
- *Organelle-Specific Fluorochromes:* MitoTracker Dyes, ER-Tracker Dyes.
- *Compatible Fluorochrome Combinations for Multiparameter Analysis:*
  - (i) DNA/Viability Analysis: PI + FDA, DAPI + Calcein-AM, Hoechst + PI.
  - (ii) Organelle/Membrane Studies: MitoTracker + ER-Tracker, PI + MitoTracker, SYTO dyes + organelle markers.

Key Points to Remember:

- The quality of water is essential. Always utilize double-distilled water when mixing high-quality chemicals.
- Verify the pH level; if necessary, adjust the pH of the buffers with either 1 M NaOH or 1 N HCl. Store the solution after steam or cold sterilization using a 0.22 μm filter.
- To prepare a 1 M solution of any chemical, refer to the molecular weight indicated on the container. Dissolve the required amount (gm) of the chemical in 1000 mL of water.

9.6 Fluorescence-Activated Cell Sorting (FACS)

Fluorescence-activated cell sorting (FACS) is a sophisticated and specialized technique commonly used in biological research. It is an extension of flow cytometry principles. This analytical process assesses the fluorescent intensity of individual cells and physically sorts or isolates specific cells into multiple homogeneous subpopulations from a heterogeneous mixed sample with remarkable accuracy (99.99%). Cells can be categorized based on their fluorescent characteristics and scattering intensity, which are linked to their physical and chemical properties, such as size, surface markers, density, volume, electrical impedance, and functional and immune-phenotypic traits.

FACS functions similarly to a flow cytometer, integrating fluidic, optical, and electrostatic systems, as well as a cell sorter unit that separates individual cells based on their characteristics for further analysis. The fluidic system employs a vibrating nozzle to break the flowing stream into uniform, small droplets containing individual cells. These droplets then detach from the main liquid stream at a precise moment. An electric charge is subsequently applied to deflect the droplets containing the cells of interest. The sorted cells can then be used for more detailed examinations of cell populations.

9.6.1 Steps of Fluorescence-Activated Cell Sorting

(a) *Fluorescent Labelling of Cells:* Cells are labelled with fluorescent dyes targeting certain surface marker proteins (CD45, CD14, and HLA-DR2) or DNA sequences. These dyes are usually attached to antibodies that bind selectively to target markers, making each cell uniquely identifiable.

(b) *Flow Cytometry:* Once labeled, these cells flow into a flow cytometer, where lasers excite the fluorescent dyes. FSC and SSC channels measure the intensity of emitted fluorescence from each cell as they reach the sensing zone (Fig. 9.5).

(c) *Cell Sorting:* As the cells flow downstream, they reach a break-off point where a voltage pulse is applied to the liquid jet. This causes the formation of small droplets, each containing a single cell from the intact liquid stream. The droplets that contain the cells of interest are then sorted and directed towards a specific location using an electric field generated between two deflection plates (Fig. 9.5).

(d) *Collection of Sorted Cells:* The sorted droplets are collected in tubes or vessels placed beneath the deflection plates (Fig. 9.5). The cells are categorized into different groups based on their fluorescence intensity and size for further analysis.

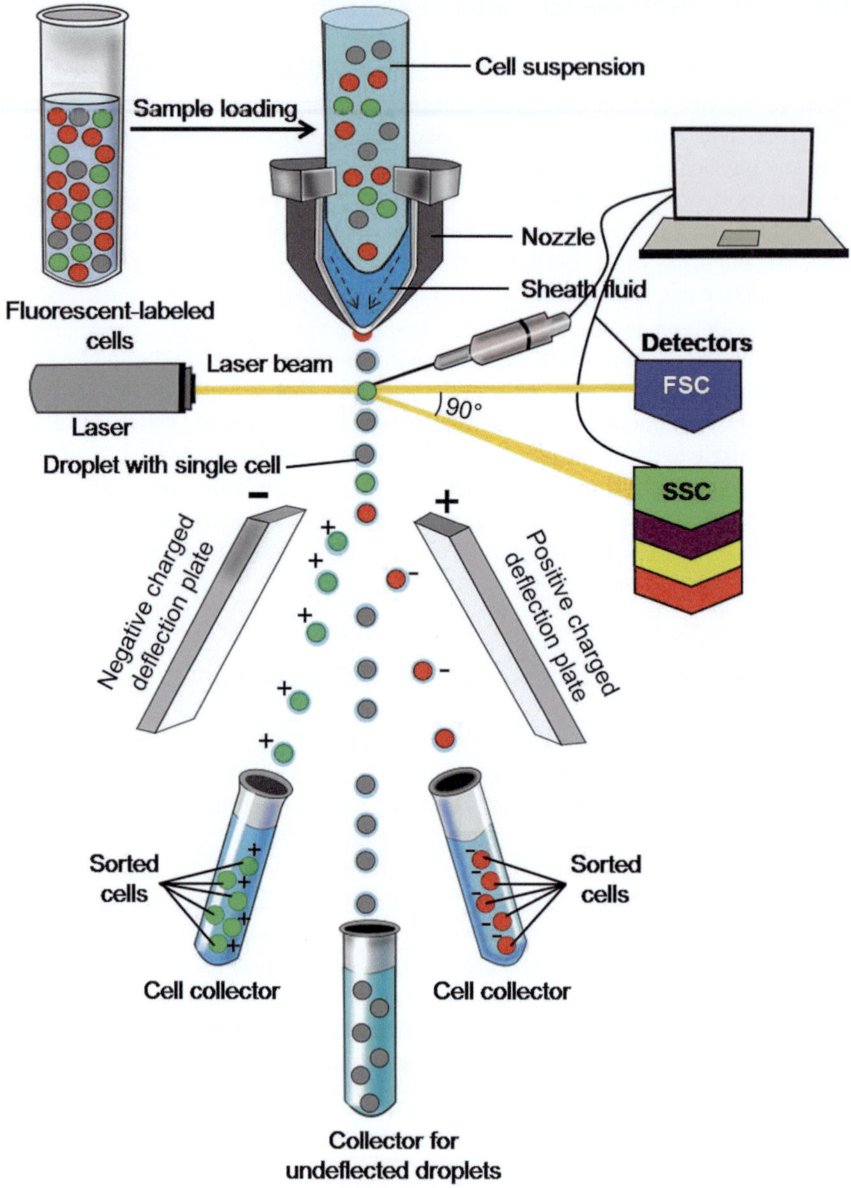

Fig. 9.5 Schematic representation of fluorescence-activated cell sorting process

9.7 Flow Karyotyping

In plant sciences, the isolation and analysis of individual fluorescent-stained chromosomes through FACS have ushered in a new era for modern cytogenetic techniques. This FACS-based approach to chromosome isolation and analysis is known as flow karyotyping. Developed over several decades, this technique highlights the crucial role of chromosomes in genomics, enabling the quantitative characterization of chromosome complements.

The isolated and stained chromosomes from Chinese hamster cells were individually classified by flow cytometry. Since then, flow cytometric sorting of chromosomes, or flow karyotyping, has emerged as an alternative quantitative tool for assessing numerical and structural chromosomal aberrations, gene mapping, and preparing chromosome-specific DNA libraries. Notably, Van Dilla and Deaven (1990) documented the flow sorting of human chromosomes and the subsequent preparation of DNA libraries.

The advancement of flow karyotyping for plant chromosomes was delayed due to necessary modifications in methodologies. The first successful preparation of a chromosome-specific DNA library in plants was reported in wheat chromosomes by Wang et al. (1992). Doležel et al. (1994) emphasized that the introduction of flow cytometry created opportunities to complement or even replace traditional visual chromosome analysis with quantitative chromosome analysis. While it cannot entirely supplant the advantages of classical and molecular karyotyping in plant chromosome studies, flow karyotyping simplifies the genetics and genomics of plant species with large genomes. This technique has emerged as an effective method for dissecting genomes into individual chromosomes by isolating and purifying them. Additionally, it is valuable for analyzing chromosome structure and evolution, as well as isolating chromosomes with specific genes of interest. Flow karyotyping enables the rapid classification of isolated chromosomes based on their relative DNA content and optical properties.

Plant chromosomes sorted by flow cytometry have been utilized for further analysis in various applications, including molecular cytogenetics, molecular biology, genome sequencing, gene cloning, genomics, and proteomics (Vrána et al. 2016; Doležel et al. 2021). However, fluorescence-activated cell sorting (FACS) remains a more expensive technique than traditional and molecular cytogenetics, primarily due to its reliance on advanced instrumentation.

The process begins with the induction of cell cycle synchrony to accumulate cells at the metaphase stage, followed by the extraction of intact chromosomes from mitotic cells in liquid suspensions. These chromosomes are stained with fluorescent dyes, and their analysis and sorting are performed using flow cytometry. During this procedure, chromosomes are classified based on their fluorescence and light scatter properties as they move rapidly through a liquid stream.

In flow karyotyping, the fluorescence intensity of an individual chromosome is directly related to its size. Each chromosome contains a specific number of base pairs that linearly correspond to its size. Mitotic chromosomes can be purified through flow cytometry, utilizing light scatter and fluorescence parameters due to their distinct optical properties when isolated in an aqueous suspension. This method allows for rapid analysis of large populations, resulting in representative distributions of chromosomal DNA content, known as flow karyotypes.

9.8 Protocol for Isolation of Plant Chromosomes by FACS

(a) *Material:* The root apical meristems of seedlings are considered the most suitable tissue for chromosome preparation and their analysis. However, isolation of plant chromosomes for flow cytometric analysis began with cell suspension cultures (Laat and Blaas 1984; Wang et al. 1992) due to their ease of handling and their ability to achieve a high degree of metaphase synchrony and maintain karyological stability. On the other hand, Doležel et al. (1992) initiated the use of root tip meristems of young seedlings for the isolation of plant chromosomes by mechanical homogenization. Nowadays, chromosome isolation for flow karyotyping has been well standardized from (i) suspension-cultured cells, (ii) protoplasts isolated from leaf mesophyll, (iii) root tips of young seedlings, and (iv) root tips of hairy root cultures (Gualberti et al. 1996; Zwyrtková et al. 2020; Doležel et al. 2023).

(b) *Germination of Seeds:* Water-soaked seeds are placed on moist filter paper inside a container or Petri dish for germination (Fig. 9.6a).

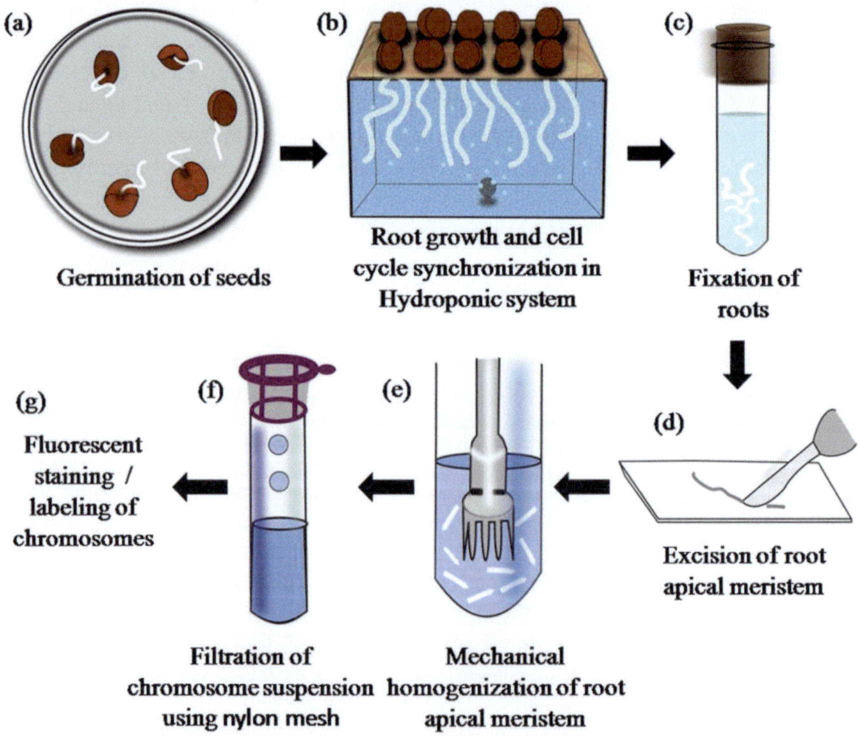

Fig. 9.6 Steps for chromosome preparation and fluorescence-staining or labeling for flow karyotyping

(c) *Cell Cycle Synchronizing and Root Harvesting:* Germinated seedlings with a maximum root length of 2 cm are transferred to a hydroponic system filled with a well-aerated nutrient solution. Hydroxyurea is added to this solution and incubated at 25 °C to synchronize the cell cycle (Fig. 9.6b).
(d) *Fixation:* The root tips are fixed in formaldehyde (Fig. 9.6c).
(e) *Preparation of Chromosomal Suspension:* The root apical meristems are excised (Fig. 9.5d) and mechanically homogenized to release the chromosomes into the selected buffer solution (Fig. 9.5e). The mixture is then filtered through a nylon mesh (Fig. 9.6f).
(f) *Fluorescent Staining or Labeling of Chromosomes:* The filtered chromosomal suspension is then stained with an appropriate fluorochrome (Fig. 9.6g).
(g) *Flow of Chromosomes:* The suspension containing isolated fluorescent-labelled chromosomes is loaded into the flow cytometer. As the chromosomes flow in a fluid stream through a flow cell, they pass through a laser beam that excites fluorescence and scatters light from the cells (Fig. 9.7).
(h) *Detection of Emitted Signals:* Detectors capture FSC, SSC, and fluorescent signals emitted by each cell.
    (i) *Chromosome sorting:* As the sample stream reaches a designated break-off point, a selective electric charge or voltage pulse (positive or negative) is applied to the liquid jet to produce droplets containing chromosomes encapsulated within small liquid droplets. As droplets with chromosomes of interest exit the stream, they are deflected towards high-voltage deflection plates by the applied external electric field. The direction of the deflection of droplets depends on their charge. Uncharged droplets are directed to waste, while charged droplets are collected in designated collection tubes positioned below the deflection plates. This process allows sorting of multiple cell populations from a single sample (Fig. 9.7).

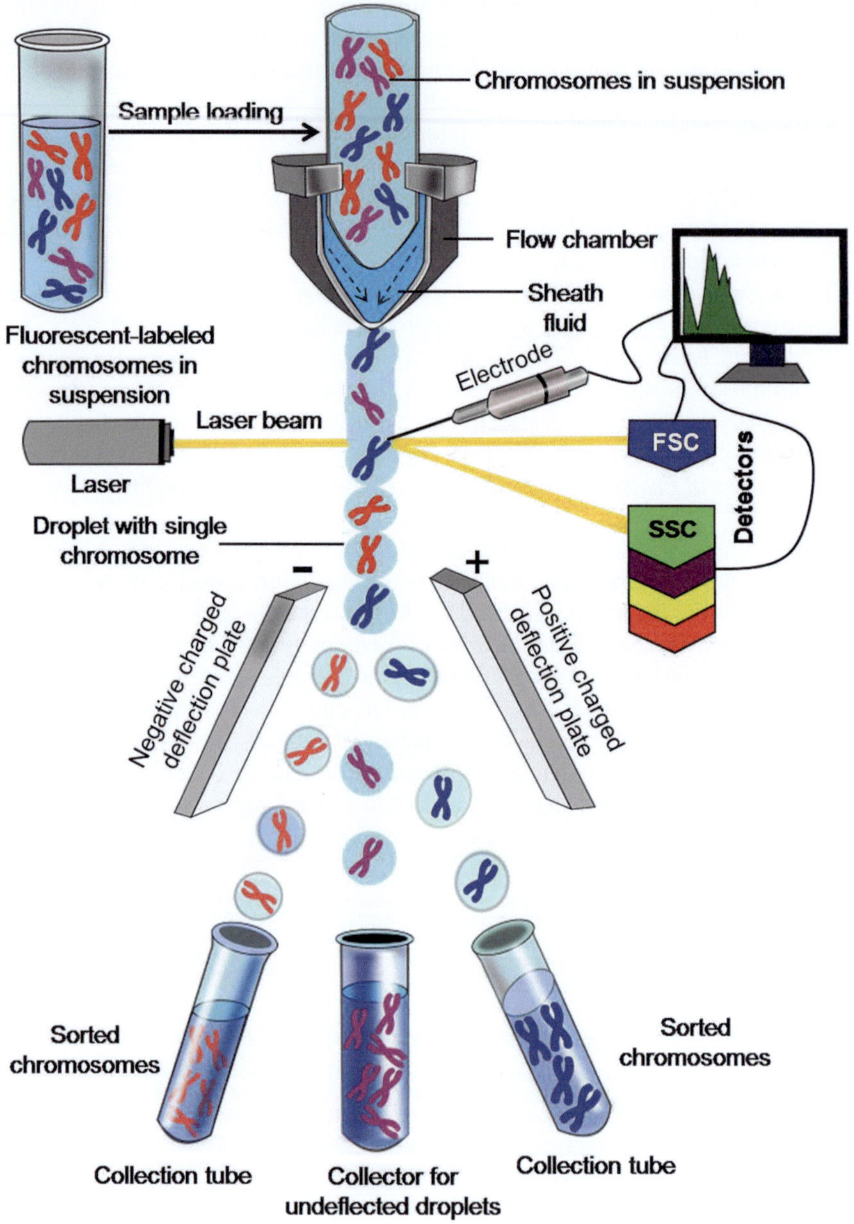

Fig. 9.7 Schematic representation of the flow karyotyping process

- *Bivariate flow karyotyping* enhances chromosome resolution, enabling more precise differentiation of individual chromosomes compared to univariate methods. It typically employs two fluorescent dyes, such as Chromomycin A3 and Hoechst 33258, to selectively stain A: T and G: C regions. Furthermore, this method may incorporate fluorescently labeled microsatellite probes to visualize GAA clusters on DAPI-stained chromosomes.

9.9 Spectral Flow Cytometry (SFC)

Spectral flow cytometry (SFC), an innovative advancement of classical flow cytometry, was introduced by Robinson et al. (2004). This technique employs a grating to disperse light across a 32-channel multi-anode PMT array, enabling high-speed single-cell spectral analysis (Nolan 2022). A defining feature of this groundbreaking technology is its capacity to capture the complete emission spectrum of fluorophores, rather than being restricted to discrete wavelengths. This is achieved through the use of an array of multichannel detectors (typically a CCD) along with optical components, such as gratings or prisms.

SFC allows for the simultaneous detection and measurement of multiple parameters related to the spectral signatures of various fluorochromes, even when their emission profiles overlap significantly, all in unmixed conditions at the level of individual cells or particles. In contrast to classical flow cytometers that can analyze up to 30 parameters per cell simultaneously, spectral flow cytometers can analyze more than 40 fluorescent dyes concurrently, even with considerable overlap in their emission spectra. Additionally, SFC effectively distinguishes autofluorescent signals in stained specimens, ranging from mammalian cells to chlorophyll-containing organisms like cyanobacteria.

This advanced technique significantly enhances the sensitivity and resolution of conventional flow cytometry, facilitating deeper single-cell analysis, improved multiplexing capabilities, and greater flexibility in experimental design. While there are several distinctions between conventional and spectral flow cytometry (Table 9.3), both techniques operate on the same foundational principles, analyzing approximately 10,000 events per second and focusing on individual cells.

9.10 Applications of Classical Flow Cytometry (FCM), Spectral Flow Cytometry (SFC), and Fluorescence-Activated Sorters (FACS)

Classical flow cytometry (FCM), spectral flow cytometry (SFC), and fluorescence-activated cell sorting (FACS) are powerful technologies with diverse applications across various fields, including immunology, molecular biology, virology, molecular cytogenetics, genome sequencing, gene cloning, proteomics, pathology, and cancer biology. Flow cytometry has two main advantages. First, it allows the evaluation of a large number of particles quickly and facilitates rapid, multiparametric analysis per particle at rates of up to 100,000 per second, which ensures statistically robust results. Second, it can physically sort single cells from mixed populations at rates of up to 70,000 cells per second, enabling the isolation of rare events (e.g., 1:100,000) while preserving cell viability (Robinson and Grégori 2007). Figure 9.8 shows a comparison between spectral and conventional cytometry.

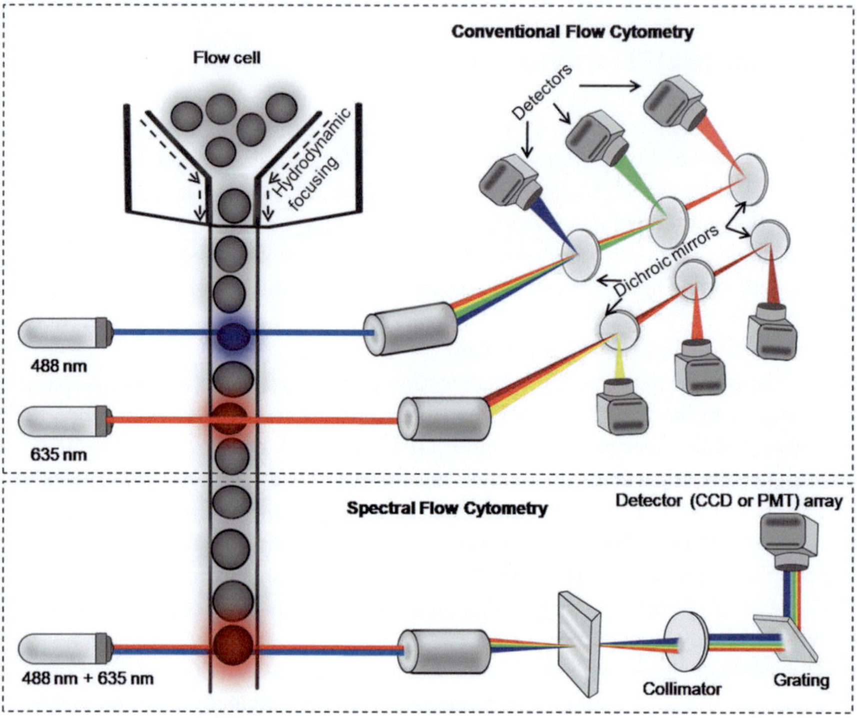

Fig. 9.8 Comparison between spectral and conventional cytometry

Key applications of classical FCM, SFC, and FACS encompass immunophenotyping, cell sorting, cell proliferation assays, cell cycle analysis through DNA content measurement, apoptosis detection, transfection efficiency evaluation, exosome detection, CAR T-cell therapy, stem cell analysis, gene expression analysis, disease diagnosis, and monitoring (Vorobjev et al. 2023). Researchers can efficiently analyze thousands of cells per second within a continuously flowing single-cell suspension (Galbraith 2010).

These technologies have been effectively integrated into plant biology, enabling rapid and multiparametric analysis of single cells or nuclei. They support fundamental research in areas such as biosystematics, ecology, taxonomy, population biology, genomics, and conservation, among others (Sliwinska et al. 2022). FCM is particularly valued for measuring plant genome sizes and has significant applications in agriculture, including ploidy detection (haploid /polyploid), breeding, and seed quality testing. Applications of FCM and FACS are summarized below:

(i) *Genome Size Estimation:* The most valuable application of FCM in plant systems is measuring nuclear DNA content (genome size) by staining nuclei with DNA-specific fluorochromes and comparing the amount of fluorescence emitted to a reference standard (Bennett and Leitch 2011; Doležel and Bartoš 2005).
(ii) *Ploidy Level Assessment:* FCM is used to determine the ploidy level of plant cells by estimating the nuclear C-value (Choi et al. 2020). Thus, it is used to determine ploidy status in commercial seed lots.
(iii) *Estimation of DNA Base Composition:* In all living organisms, DNA is made up of adenine (A), guanine (G), cytosine (C), and thymine (T) bases. DNA base composition, i.e., percentage content of GC and AT bases, is a basic species characteristic. FCM helps to estimate the base composition of a genome (Sliwinska et al. 2022).

Table 9.3 Comparison of conventional and spectral flow cytometry

Feature	Conventional flow cytometry	Spectral flow cytometry
Detector	It uses a single detector, either a photodiode (PD) or a photomultiplier tube (PMT)	It uses a linear array of detectors of charge-coupled devices (CCDs) or multi-anode PMTs
Detection method	It utilizes a combination of mirrors and optical filters to detect fluorophore-emitted signals	It utilizes dispersive optics, such as spectrographs or prisms, to detect fluorophore-emitted signals
Detection spectra	It utilizes and detects only a selected fraction of the fluorophore emission spectra (maximum peak emission) within defined wavelength ranges that pass through bandpass, short pass, and long pass filters that selectively reflect, block, or transmit a photon based on its wavelength	It detects complete spectral of the fluorophore emission spectra
Multi-color panel capability	It can analyze up to 30 parameters simultaneously for each cell	It can simultaneously analyze over 40 fluorescent dyes
Resolution of highly similar fluorochromes	Dichroic mirrors and selective bandpass filters cannot distinguish fluorophores with similar emission spectra	It can distinguish fluorophores with similar emission spectra using a set of prisms that separate them before sending them to 32-channel PMT arrays
Autofluorescence extraction	Autofluorescence can hamper data analysis, particularly in larger and more granular cells	Autofluorescence is treated as another color
Fluorochrome choice	Dependent on the laser and filter configuration	Dependent on laser configuration
Individual fluorochrome contribution determination	Compensation	Unmixing

(iv) *Flow Karyotyping* represents a significant advancement in the application of flow cytometry (FCM). Using fluorescent-activated cell sorting (FACS), this technique allows for the isolation and classification of chromosomes extracted from mitotic cells. The chromosomes are stained with fluorescent dyes and sorted individually based on their fluorescent intensity, enabling rapid identification of individual chromosomes. This method is effective in detecting both numerical and structural chromosome aberrations. Flow-sorted chromosomes have been utilized for PCR with specific primers to integrate physical and genetic maps, physical mapping of DNA sequences via FISH and PRINS, chromosomal protein analysis, and the creation of subgenomic DNA libraries (Dolezel et al. 2007).

The first reports of isolating individual chromosomes for flow karyotyping were made by Laat and Blaas (1984) and Wang et al. (1992), who worked with *Haplopappus gracilis* ($2n = 4$) and a cell suspension culture of bread wheat, respectively. Meanwhile, the pioneering work on isolating plant chromosomes from root apical meristems of young seedlings through mechanical homogenization was conducted by Doležel et al. in 1992. More recently, Zwyrtková et al. (2020) noted that chromosome isolation for flow karyotyping is now feasible from several sources, including (i) suspension-cultured cells, (ii) protoplasts obtained from leaf mesophyll, (iii) root tips of young seedlings, and (iv) root tips from hairy root cultures. Additionally, Doležel et al. (2023) provided a review of the advancements in flow karyotyping across various plant species.

(v) *Detection of Specific DNA Sequence:* Flow-FISH is a cytogenetic technique that combines flow cytometry with FISH to map particular DNA sequences within cell populations' genomic DNA.

(vi) *Cell Cycle Analysis:* FCM estimates the distribution of cell populations in different phases of the cell cycle (in percentage), as cells in each phase have a distinct nuclear DNA content and DNA-specific dye stoichiometrically binds with the cell's DNA. FCM has the ability to measure the intensity of fluorescence and produce cell count versus linear fluorescence, creating a histogram of the DNA content distribution across the cell cycle steps to determine their stage in the cell cycle.

(vii) *Detection of Endoreduplication:* Endoreplication is a process in which somatic cells undergo successive rounds of nuclear genome replication without cytokinesis. FCM can effectively detect endoreduplication in plant cells by analyzing the DNA content of nuclei. The frequency of cells within each peak can then be used to calculate accurate endoreduplication levels for different genotypes or tissues.

Endoreduplication Index: The level of endoreduplication can be quantified using an "Endocycle index" (E), using the formula $E = [(1 \times f8\text{C}) + (2 \times f16\text{C}) + \ldots$ where f is nuclei frequency (%)

(viii) *Gene Mapping and Preparation of Chromosome-Specific DNA Libraries:* FCS helps in gene mapping and preparation of chromosome-specific DNA libraries (Wang et al. 1992).

FCM and FCS have benefited molecular cytogenetics, molecular biology, genome sequencing, gene cloning, and proteomics (Doležel et al. 2021).

(ix) *Applications in Plant Breeding and Agriculture:* FCM can be applied to the identification and production of polyploid lines possessing desired agricultural traits, quality assessment of commercial cultivars, quality control of euploidy status for commercial seed lots, identification of individuals having novel ploidy levels, characterization of interspecific hybridization, classification of ploidy distributions within germplasm collections, resolving many phylogenetic ambiguities of taxonomically disputed groups, sex determination in dioecious plants, and identification of hybrids formed between wild species.

In plant research, particularly in areas such as genome size and ploidy estimation, epigenetic analysis, cell cycle assessment, and metabolic activity evaluation, SFC offers enhanced capabilities and in-depth analytical power. This technology is essential for advancing plant breeding, monitoring plant health, and optimizing plant cell cultures, all of which are vital for ensuring food security and supporting biotechnological advancements. Additionally, SFC is instrumental in the screening and selection of elite cell lines and the detection of biomarkers.

References

Bennett MD, Leitch IJ (2011) Nuclear DNA amounts in angiosperms: targets, trends and tomorrow. Ann Bot 107(3):467–590

Choi IY, Kwon EC, Kim NS (2020) The C-and G-value paradox with polyploidy, repeatomes, introns, phenomes and cell economy. Genes Genom 42:699–714

Coulter WH (1953) Means for counting particles suspended in a fluid, US Patent no 2,656,508

Crosland-Taylor PJ (1953) A device for counting small particles suspended in a fluid through a tube. Nature 171:37–38

Dittrich W, Göhde W (1969) Impulsfluorimetrie bei Einzelzellen in Suspensionen. Z Naturf 24b:360–361

Doležel J, Bartoš JAN (2005) Plant DNA flow cytometry and estimation of nuclear genome size. Ann Bot 95(1):99–110

Doležel J, Binarova P, Lucretti S (1989) Analysis of nuclear DNA content in plant cell by flow cytometry. Biol Plant 31(2):113–120

Doležel J, Číhalíková J, Lucretti S (1992) A high-yield procedure for isolation of metaphase chromosomes from root tips of *Vicia faba* L. Planta 188(1):93–98

Doležel J, Lucretti S, Schubert I (1994) Plant chromosome analysis and sorting by flow cytometry. Crit Rev Plant Sci 13(3):275–309

Doležel J, Greilhuber J, Suda J (2007) Flow cytometry with plant cells: analysis of genes, chromosomes and genomes. John Wiley & Sons

Doležel J, Lucretti S, Molnár I, Cápal P, Giorgi D (2021) Chromosome analysis and sorting. Cytometry A 99(4):328–342

Doležel J, Urbiš P, Said M, Lucretti S, Molnár I (2023) Flow cytometric analysis and sorting of plant chromosomes. Nucleus 66(3):355–369

Fulwyler MJ (1965) Electronic separation of biological cells by volume. Science 150:910–911

Galbraith WD (2010) Flow cytometry and fluorescence-activated cell sorting in plants: the past, present, and future. Bioméd 30:65–70

Galbraith DW, Harkins KR, Maddox JM, Ayres NM, Sharma DP, Firoozabady E (1983) Rapid flow cytometric analysis of the cell cycle in intact plant tissues. Science 220(4601):1049–1051

Goetz C, Hammerbeck C, Bonnevier J (2018) Flow cytometry basics for the non-expert (Vol. 1). Springer

Greilhuber J, Doležel J, Lysák MA, Bennett MD (2005) The origin, evolution and proposed stabilization of the terms 'genome size' and 'C-value' to describe nuclear DNA contents. Ann Bot 95(1):255–260

Greilhuber J, Borsch T, Müller K, Worberg A, Porembski S, Barthlott W (2006) Smallest angiosperm genomes found in Lentibulariaceae, with chromosomes of bacterial size. Plant Biol 8(06):770–777

Gualberti G, Doležel J, Macas J, Lucretti S (1996) Preparation of pea (*Pisum sativum* L.) chromosome and nucleus suspensions from single root tips. Theor Appl Genet 92(6):744–751

Laat AM, Blaas J (1984) Flow-cytometric characterization and sorting of plant chromosomes. Theor Appl Genet 67(5):463–467

Loureiro J, Rodriguez E, Doležel J, Santos C (2007) Two new nuclear isolation buffers for plant DNA flow cytometry: a test with 37 species. Ann Bot 100(4):875–888

Loureiro J, Kron P, Temsch EM, Koutecký P, Lopes S, Castro M, Sílvia Castro S (2021) Isolation of plant nuclei for estimation of nuclear DNA content: overview and best practices. Cytometry 99:318–327

Moldavan A (1934) Photo-electric technique for the counting of microscopical cells. Science 80(2069):188–189

Nolan JP (2022) The evolution of spectral flow cytometry. Cytometry A 101(10):812–817

Otto F (1992) Preparation and staining of cells for high-resolution DNA analysis. In: Radbruch A (ed) Flow cytometry and cell sorting. Springer-Verlag, Berlin, pp 101–104

Pfosser A, Amon A, Lelley T, Heberle-bors E (1995) Evaluation ofsensitivity of flow cytometry in detecting aneuploidy in wheat using disomic and ditelosomic wheat-rye addition lines. Cytometry 21:387–393

Puite KJ, Ten Broeke WR (1983) DNA staining of fixed and non-fixed plant protoplasts for flow cytometry with Hoechst 33342. Plant Sci Lett 32(1–2):79–88

Robinson JP, Grégori G (2007) Principles of flow cytometry. In: Dolezˇel J, Greilhuber J, Suda J (eds) Flow cytometry with plant cells analysis of genes, chromosomes and genomes. John Wiley & Sons, pp 19–40

Robinson JP, Rajwa B, Grégori G, Jones J, Patsekin V (2004) Collection hardware for high speed multispectral single particle analysis. Cytometry 59:12

Sliwinska E, Loureiro J, Leitch IJ, Šmarda P, Bainard J, Bureš P, Chumova Z, Horova L, Koutecký P, Lučanová M, Trávníček P, Galbraith WD (2022) Application-based guidelines for best practices in plant flow cytometry. Cytometry A 101(9):749–781

Suda J, Kron P, Husband BC, Trávníček P (2007) Flow cytometry and ploidy: applications in plant systematics, ecology and evolutionary biology. In: Dolezˇel J, Greilhuber J, Suda J (eds) Flow cytometry with plant cells. Wiley-VCH, pp 103–130

Van Dilla MA, Deaven LL (1990) Construction of gene libraries for each human chromosome. Cytometry: J Int Soc Anal Cytol 11(1):208–218

Vorobjev IA, Kussanova A, Barteneva NS (2023) Development of spectral imaging cytometry. Methods Mol Biol 2635:3–22

Vrána J, Cápal P, Šimková H, Karafiátová M, Čížková J, Doležel J (2016) Flow analysis and sorting of plant chromosomes. Curr Protoc Cytom 78(1):5–3

Wang ML, Leitch AR, Schwarzacher T, Heslop-Harrison JS, Moore G (1992) Construction of a chromosome-enriched Hpall library from flow-sorted wheat chromosomes. Nucleic Acids Res 20(8):1897–1901

Zwyrtková J, Šimková H, Doležel J (2020) Chromosome genomics uncovers plant genome organization and function. Biotechnol Adv 46:107659

Concept and Applications of Fluorescent In Situ Hybridization (FISH)

10

Abstract

Fluorescence In Situ Hybridization (FISH) is a powerful molecular cytogenetic technique utilized for the visualization and localization of specific DNA sequences within chromosomes and interphase chromatin. This method has evolved significantly since its inception in the 1960s, with the introduction of non-radioactive probes that have enhanced safety and efficacy. The technique employs fluorescently labeled nucleic acid probes that, after denaturation, anneal to complementary target DNA at lower temperatures. Subsequent washing steps ensure the removal of non-specifically bound probes, allowing precise visualization of hybridization signals via fluorescence microscopy. FISH offers a versatile approach for detecting various genomic features, including centromeres, telomeres, gene loci, and chromosomal rearrangements. The success of FISH largely depends on the choice of probes, which can be categorized into traditional and advanced types. Traditional probes, such as Bacterial Artificial Chromosomes (BAC) and ribosomal DNA, are larger and cloned using diverse vectors, while advanced probes consist of short synthetic oligonucleotides that provide enhanced sensitivity and specificity. This technique has broadened the field of cytogenetics, facilitating the detailed study of chromosomal structure and function in both plant and animal sciences. Its applications span genome mapping, evolutionary studies, clinical diagnostics, and the investigation of chromosomal abnormalities, making FISH a cornerstone of modern molecular cytogenetics. This chapter provides a comprehensive overview of FISH, detailing its principles, probe types, and significant applications, reflecting its central role in advancing our understanding of genomic organization and dynamics.

10.1 Introduction

The skill-based cytogenetic technique that employs microscopy is instrumental in revealing the fundamental chromosomal numbers and morphology of individual species. This method also facilitates the examination of basic structural variations in

T. B. Jha, M. Halder, *Classical and Molecular Plant Cytogenetics*,
https://doi.org/10.1007/978-981-95-8265-5_10

genome organization at the chromosomal level. Since the 1970s, advancements in molecular biology, plant chromosome preparation, and microscopy have led to the emergence of molecular cytogenetics. This evolution has expanded the scope of plant cytogenetics to encompass the study of heterochromatin as well as specific DNA and RNA sequences.

In 1969, Gall and Pardue (1969), along with John et al. (1969), developed a molecular cytogenetic technique known as in situ hybridization (ISH), which allows for the detection of specific nucleic acid sequences directly on chromosomes. The original method employed radioactive RNA or DNA probes in conjunction with autoradiographic techniques to locate hybridized nucleic acids within the cells or chromosomes of both plants and animals. However, the dangers posed by radioactive materials to human health led scientists to seek alternative approaches. In 1981, researchers began using a naturally occurring, non-radioactive molecule, biotin (vitamin H or B_8), to label probes, along with avidin, for detecting in situ hybridization sites (Schwarzacher et al. 2023). Biotin, a water-soluble vitamin produced by plants, bacteria, and some fungi, is essential for all forms of life, while avidin is a protein that binds biotin and is found in egg whites.

Langer-Safer et al. (1982) were pioneers in utilizing this non-radioactive probe in combination with fluorophores to visualize specific DNA sequences on mitotic or meiotic chromosomes, interphase nuclei, and extended chromatin fibers under a microscope. The incorporation of biotin, avidin, and fluorophores led to the development of a groundbreaking molecular cytogenetics technique known as fluorescence in situ hybridization (FISH). This technique has transformed the field of molecular cytogenetics, evolving significantly over the decades and playing a crucial role in advancing modern molecular cytogenetics in both animal and plant sciences (Pinkel et al. 1986).

10.2 Basic Principles of Fluorescence In Situ Hybridization (FISH)

Fluorescence In Situ Hybridization (FISH) is an advanced molecular cytogenetic technique that visually identifies and locates specific DNA sequences within target samples, including metaphase chromosomes or interphase chromatin in cells fixed on a microscope slide. This technique utilizes fluorescently labelled nucleic acid probes. Both the labeled probes and target DNA are denatured by heat to form single strands, which then anneal during hybridization at lower temperatures. To remove any non-specific hybridization and unbound probes, a washing step is performed. A fluorescence microscope is employed to visualize FISH signals generated by the specifically hybridized fluorescent probes that bind to complementary nucleic acid sequences in situ. The linear arrangement of DNA along the longitudinal axis of the chromosomes enhances the direct hybridization of probe DNA for identifying specific genes. Various probes, each labeled with different fluorophores, enable the detection of multiple signals in a single experiment.

10.3 Probes of Fluorescence In Situ Hybridization

Fluorescently labeled single-stranded or double-stranded DNA, cDNA, or RNA serves as probes in FISH. The lengths of these probes can vary widely, typically ranging from 20 base pairs (bp) to 1500 bp. They are employed to directly detect complementary DNA sequences on chromosomes or interphase chromatin. The success of a FISH experiment largely depends on the types of probes used. Probes may target centromeres, telomeres, chromosome bands, genes, translocation breakpoints, and interspersed repeated sequences, as well as whole chromosome-specific and whole genome-specific DNA sequences (Luke et al. 1997).

Many researchers have categorized FISH probes into traditional cloned DNA and advanced probes. Traditional probes include Bacterial Artificial Chromosomes (BAC), ribosomal DNA (rDNA; 5S and 35S), tandem repeats, and distributed repetitive sequences (Harun et al. 2023). Depending on their size, traditional DNA probes are cloned using various vectors, such as plasmids (5–6 kb), phages (3–20 kb), cosmids (40 kb), P1 artificial chromosomes (150 kb), yeast artificial chromosomes (YAC) (100–200 kb), and bacterial artificial chromosomes (BAC). In contrast, synthetic single-stranded oligonucleotides are classified as advanced probes. These probes are short, synthetic DNA or RNA sequences, generally ranging from 20 to 50 nucleotides in length. Oligo probes provide significantly improved sensitivity and specificity compared to traditional probes. Based on their target sites, probes are sometimes categorized as chromosomal probes, "arm-specific" probes, satellite probes, telomeric probes, subtelomeric probes, genome probes, and more (Lakatošová and Holečková 2007).

10.4 Labeling of Probes

Before hybridization, it is essential to label probes. Several techniques can be employed for this labeling, including nick translation, random primed labeling, and PCR. The two main labelling strategies used in FISH are direct and indirect labeling. Indirect labeling involves attaching a non-fluorescent hapten to modified nucleotides, which can subsequently be recognized through an enzymatic or immunological detection system. In contrast, direct labeling attaches fluorophores directly to the nucleotides. Although the indirect labeling process is more time-consuming, it offers the benefit of signal amplification by employing multiple layers of antibodies, potentially resulting in a brighter signal relative to background levels.

The synthesis of oligonucleotide probes requires careful computational design to facilitate their creation. These oligo sequences are labelled with a fluorochrome (such as Cy3, Cy5, or 6-Carboxyfluorescein), biotin-dUTP, or digoxigenin-dUTP through PCR amplification. High-quality oligo probes with various labels are now readily available from numerous multinational biotech companies.

Probes are crucial for constructing karyotypes, identifying chromosomes, and determining interspecific homoeologous relationships among different plant species (Harun et al. 2023).

10.5 Requisites

10.5.1 Equipment

(i) Fluorescence microscope equipped with filters, CCD camera, computer, and software
(ii) Incubator
(iii) −20 °C and 4 °C refrigerator
(iv) Heat block or modified thermo cycler
(v) Micropipettes
(vi) Glass slides and cover slips
(vii) Coplin jars

10.5.2 Solutions and Chemicals

(i) *20× Saline-Sodium Citrate (SSC) Buffer Used as Hybridization Buffer*
- Sodium chloride (MW: 58.44 g mol^{-1})
- Sodium citrate (MW: 294.10 g mol^{-1})

SSC buffer is often used as a hybridization buffer in FISH experiments. To prepare 1 L of 20× SSC buffer (3 M NaCl and 0.3 M sodium citrate, pH 7), take 800 mL of double-distilled water in a flask or beaker and add 175.3 g of sodium chloride (3 M) and then 88.3 g of sodium citrate (0.3 M). Add double-distilled water to reach 1 L. Check the desired pH of 7.0 before use

(ii) *Ribonuclease or RNase* is a type of enzyme that is used to remove RNA from DNA solutions. Working solution: 100 μg mL^{-1} in 2× SSC.
(iii) *Pepsin*: It is an enzyme involved in protein digestion—working solution: 40 units mL^{-1} in 10 mM HCl.
(iv) 4% Paraformaldehyde electron microscopy-grade (EM grade) in PBS is a ready-to-use methanol-free fixative for cells or tissues.
(v) *Hybridization Mixture*:
- 50% Formamide (v/v): 50 mL of formamide is measured, and then the total volume is increased to100 mL by adding deionized water
- 10% Dextran sulfate: Dextran sulfate is a white to off-white powder easily soluble in water. 10 g of dextran sulfate powder is dissolved in 90 mL of dd water, and then the volume is 100 mL by adding the required amount of deionized water
- 0.1% SDS (Sodium dodecyl sulphate): It is also known as lauryl sulfate. SDS is an ionic detergent and a powerful inhibitor of both RNase and DNase activity. 0.1 g of SDS is dissolved in 100 mL of dd water to make 0.1% SDS
- 0.5–1.5 ng μL^{-1} labeled probe
- 300 ng mL^{-1} salmon sperm DNA in 2× SSC

(vi) *Washing Buffer*: 20% formamide in 0.1× SSC is prepared.

(vii) *Detection Buffer*: 0.2% Tween 20 (Tween 20 is a non-ionic surfactant used as a detergent and emulsifier) is mixed in 4× SSC buffer.

(viii) *Blocking Buffer*: 5% bovine serum albumin (BSA) is mixed in 4× SSC buffer to prepare the blocking buffer. It may contain Streptavidin-Cy3 (antibody) as a detection compound. It plays an essential role in reducing nonspecific binding, a common source of background noise.

(ix) *Fluorescent Dye*: A 2 μg mL^{-1} solution of DAPI is added to the antifade mounting medium.

10.6 Steps of Fluorescence In Situ Hybridization

Since the introduction of the non-radioactive FISH technique in the late 1980s, this molecular-cytogenetic method has undergone continuous evolution, resulting in various adaptations of classical FISH. Typically, a FISH experiment consists of the following steps: (i) Slide preparation, (ii) pre-hybridization, (iii) hybridization, (iv) post-hybridization, and (v) detection or visualization (Fig. 10.1). This section will specifically emphasize protocols tailored for plants rather than animals.

(i) *Slide Preparation*

In plants, the EMA method of chromosome preparation is commonly applied to obtain cytoplasm-free metaphases, interphase nuclei, or extended chromatin fibers fixed on the slide surface. Chapter 4 describes the details of plant chromosome preparation using the EMA method.

(ii) *Pre-hybridization Steps*

In the next step, prepared slides are passed through the following steps:

- Add 200–300 μL of RNase to each slide, cover it with a cover glass, and incubate at 37 °C for 1 h.
- Carefully remove the cover glass and wash the slides twice in 2× SSC for 5 min each.
- Dehydrate the slides in an ethanol series (70%, 80%, 95%, and 100%) for 5 min each and then air-dry at room temperature for a minimum of 30 min.

(iii) *Hybridization*

During hybridization, single-stranded DNA or RNA probes with known sequences bind to their complementary sequences located in a denatured target sequence and form a stable double-stranded hybrid molecule. Here, we illustrated the steps with metaphase chromosomal DNA as the target material; however, this may vary.

- *Composition of 100 μL of hybridization mixture*: 50 μL of 50% deionized formamide, 20 μL of 50% dextran sulfate, 10 μL of 20× SSC, 1.5–3 μL sonicated salmon sperm DNA (10 mg mL^{-1}) and 1–4 μL of DNA probe (40 ng μL^{-1}), and add sterile distilled water to make the volume 100 μL.
- DNA probes containing hybridization solution are kept in a water bath or heat block at 70–72 °C for 5–10 min to denature the DNA probe and then immediately transferred to ice for at least 5 min.

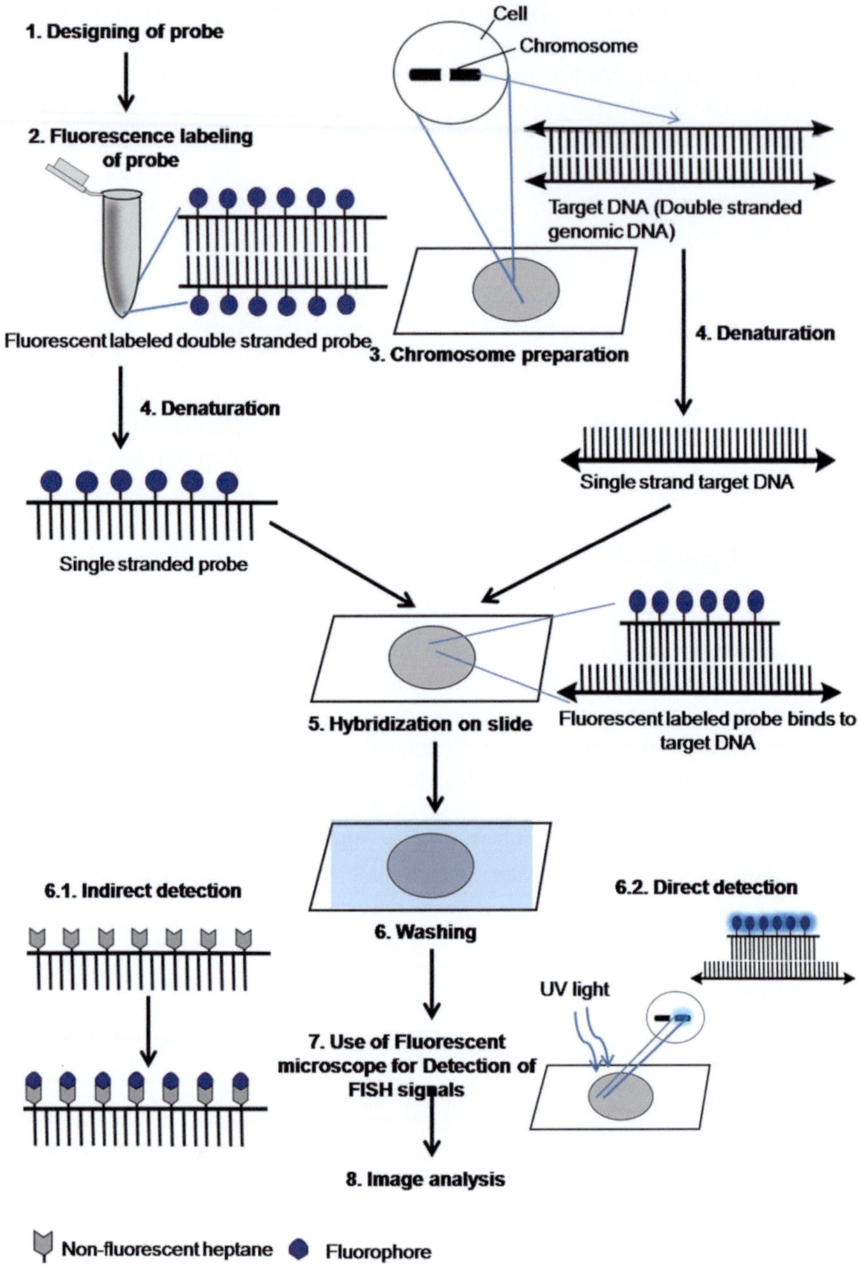

Fig. 10.1 Steps of fluorescence in situ hybridization (FISH)

- The slides containing chromosomes are placed in a 70% formamide bath containing 2× SSC at 70 °C for 2 min for denaturation. After this, the slide is transferred immediately to ice-cold ethanol for 5 min and then air-dried with a blower.
- 40–50 μL of hybridization solution is added to the slides and covered with a cover glass.
- Then, slides are placed directly on a thermal cycler with a flat aluminum bed and incubated at 72 °C for 10 min, followed by 5 min at 55 °C. Finally, the slides are kept in a humidity chamber at 37 °C for 6–48 h.

(iv) *Post-hybridization*

- In the post-hybridization steps, the buffer solutions (0.1× SSC, 2× SSC, 4× SSC) are preheated in a water bath to 42 °C.
- The cover glass on the slide is carefully removed, and the preheated buffer solutions of 2× SSC for 5–10 min (2–4 times), 20–50% formamide in 2× SSC for 5–10 min (2 times), 0.1× SSC for 5–10 min, 2× SSC for 5–10 min, and 4× SSC for 5–10 min, are subsequently applied to the slides for washing.
- 100 μL of blocking buffer is added to the slides, covered with cover glass, and incubated for 5 min at room temperature in the dark.
- Remove the cover glass, add 50 μL antibody or detection compound, and incubate at 37 °C for 30–60 min.
- Remove the cover glass and wash the slides in 4× SSC 3 times for 5 min each.
- Then counterstaining is done using DAPI solution for 5–10 min.
- The slides are washed in 2× SSC solution to remove the cover glasses, mounted in the antifade mounting medium, and kept in the dark at 40 °C.

(v) *Detection or Visualization*

Following the operation steps of the fluorescence microscope, set the desired visualization settings, including selecting the appropriate filter and fluorescent wavelengths to detect specific fluorescent signals from the target DNA sequences. The documentation and analysis of fluorescence signals can be performed using a cooled CCD camera and software designed for the fluorescent microscope.

The application of multiprobe FISH for identifying the ten pairs of somatic chromosomes in maize ($2n = 20$) using nine fluorochrome-labeled DNA probes was detailed by Kato et al. in 2005 (Fig. 10.2). An additional FISH image (Fig. 10.3) illustrates the distribution of centromeric repeated sequences in green, utilizing the Cent repeat probe, while red coloration indicates the presence of knobs or centromeric retrotransposons of maize (CRM).

10.7 Applications

Fluorescence in situ hybridization (FISH) is a labor-intensive and expensive technique utilized in cytology; however, it has proven to be an invaluable cytogenetic tool. Detection of ribosomal RNA gene (rDNA) sites using FISH offers molecular cytogenetic information and is essential for chromosome identification and the determination of phylogenetic relationships among species (Fukui et al. 1994; Murata et al. 1997; Yamamoto et al. 2016). FISH facilitates the visualization and localization of the 18s, 26s, and 5s rRNA genes located in the nucleolar organizer region (NOR), as well as the identification of various structural changes directly on metaphase chromosomes. This technique has notably advanced modern plant cytogenetics. Its initial application in plants was reported in wheat and has since been extended to numerous other species, including *Triticum, Aegilops, Secale, Oryza, cucumber,* and *melon* (Mukai 1996; Jiang 2019; Smajkan 2021). Moreover, RNA fluorescence in situ hybridization (FISH) enables the visualization of specific messenger RNA transcripts within cultured cells, tissue sections, or whole mounts in different organisms (Young et al. 2020). Some of the significant applications of FISH in plants are discussed below.

- In plants, chromosome numbers and sizes exhibit significant variability. One important and reliable method in molecular cytogenetics for identifying various types of chromosomes within a karyotype is FISH.

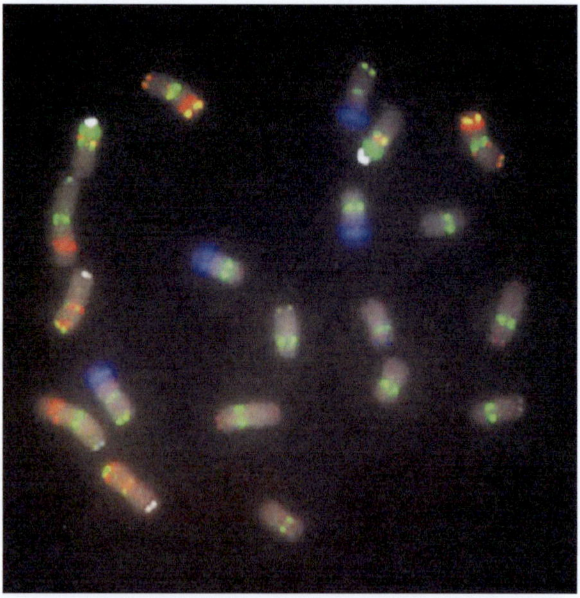

Fig. 10.2 Identification of the ten pairs of somatic chromosomes of maize inbred line B37 using nine fluorochrome-labelled DNA probes. Each homologous chromosome pair in the karyotype is numbered. The nine probes were knob 180-bp repeat (revealed as blue), 5S rDNA (yellow); NOR on chromosome 6 short arm, CentC, subtelomeric 4-12-1 (revealed as green), Cent4, microsatellite TAG repeat and pMTY9ER telomere-associated sequence (revealed as red), and TR 1 knob repeat (white). (Photograph taken from the publication of Kato et al. (2005) with permission from ELSEVIER)

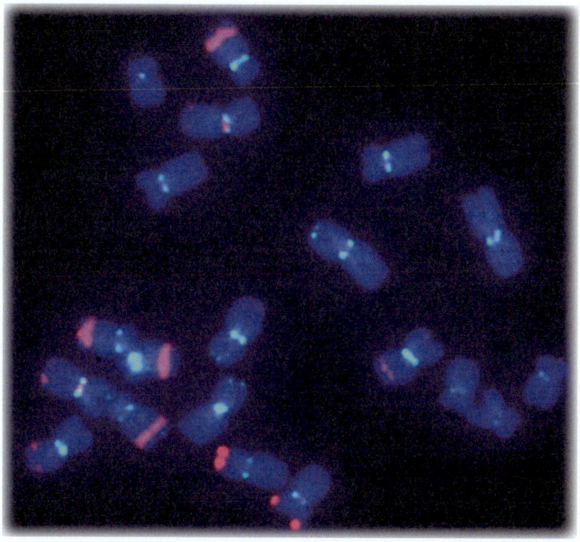

Fig. 10.3 FISH of maize chromosomes ($2n = 20$). Centromeres (Cent) are green, while knobs (CRM) are in red. Permission granted by Dr. Kelly, Dept. of Genetics, University of Georgia)

- FISH-based karyotyping provides an evolutionary and phylogenetic perspective on related plant species (Lim et al. 2000; Hizume et al. 2002). Repetitive DNA sequences likely evolved under distinct evolutionary pressures compared to functional genes. As a result, phylogenetic frameworks derived from repeat-based comparative FISH karyotyping serve as an independent test of molecular-based phylogenies (Lim et al. 2000).
- FISH is instrumental in identifying and validating satellite repeats in plant genomes.
- The technique is also proficient in detecting chromosomal translocations with precision. For example, a chromosomal translocation between chromosome 9 and chromosome 11 in *Oryza sativa* has been identified through oligo-FISH mapping (Harun et al. 2023).
- Chromosome painting is an advancement of FISH that utilizes oligo probes, which have been successfully applied in various crops, including rice, wheat, maize, potato, rye, barley, and *Brassica* (Harun et al. 2023).
- FISH can facilitate detailed chromosomal mapping of plant species, providing insights into genome structural organization and evolutionary processes.
- Furthermore, FISH can elucidate gene arrangements within related species.
- The sRNA-FISH protocol is now available for plants (Huang et al. 2019), allowing for the detection of mRNAs and small RNAs (sRNAs) localization in plant tissues.

References

Fukui K, Ohmido N, Khush GS (1994) Variability in rDNA loci in the genus *Oryza* detected through fluorescence *in situ* hybridization. Theor Appl Genet 87(8):893–899

Gall JG, Pardue ML (1969) Formation and detection of RNA DNA hybrid molecules in cytological preparations. Proc Natl Acad Sci USA 63(2):378–383

Harun A, Liu H, Song S, Asghar S, Wen X, Fang Z, Chen C (2023) Oligonucleotide fluorescence in situ hybridization: an efficient chromosome painting method in plants. Plants 12(15):2816

Hizume M, Shibata F, Matsusaki Y, Garajova Z (2002) Chromosome identification and comparative karyotypic analyses of four *Pinus* species. Theor Appl Genet 105(4):491–497

Huang K, Patricia Baldrich P, Meyers BC, Jeffrey L, Caplan JL (2019) sRNA-FISH: versatile fluorescent in situ detection of small RNAs in plants. Plant J 98(2):359–369

Jiang J (2019) Fluorescence in situ hybridization in plants: recent developments and future applications. Chromosom Res 27(3):153–165

John H, Birnstiel ML, Jones KW (1969) RNA-DNA hybrids at the cytological level. Nature (London) 223:582–587

Kato A, Vega JM, Han F, Lamb JC, Birchler JA (2005) Advances in plant chromosome identification and cytogenetic techniques. Curr Opin Plant Biol 8(2):148–154

Lakatošová M, Holečková B (2007) Fluorescence in situ hybridisation. Biol 62(3):243–250

Langer-Safer PR, Levine M, Ward DC (1982) Immunological method for mapping genes on *Drosophila* polytene chromosomes. Proc Natl Acad Sci USA 79(14):4381–4385

Lim KY, Matyasek R, Lichtenstein CP, Leitch AR (2000) Molecular cytogenetic analyses and phylogenetic studies in the *Nicotiana* section Tomentosae. Chromosoma 109(4):245–258

Luke S, Belogolovkin V, Varkey JA, Ladoulis CT (1997) Fluorescence in situ hybridization. In: Jiang G (ed) Analytical morphology: theory, applications and protocols. Birkhäuser, Boston, pp 139–173

Mukai Y (1996) In situ hybridization. In: Fukui K, Nakayama S (eds) Plant chromosomes: laboratory methods, 1st edn. CRC Press

Murata M, Heslop-Harrison JS, Motoyoshi F (1997) Physical mapping of the 5S ribosomal RNA genes in *Arabidopsis thaliana* by mulricolor fluorescence *in situ* hybridization with cosmid clones. Plant J 12(1):31–37

Pinkel D, Straume T, Gray JW (1986) Cytogenetic analysis using quantitative, high-sensitivity, fluorescence hybridization. Proc Natl Acad Sci USA 83:2934–2938

Schwarzacher T, Liu Q, Heslop-Harrison JS (2023) Plant cytogenetics: from chromosomes to cytogenomics. In: Heitkam T, Garcia S (eds) Plant cytogenetics and cytogenomics: methods and protocols. Springer, New York, pp 3–21

Smajkan A (2021) Applications of fluorescent in situ hybridization (FISH). Bioeng Stud 2(2):12–20

Yamamoto M, Moriya S, Yamamoto T (2016) Detection of ribosomal RNA genes in apple (*Malus* × *domestica*) using fluorescence *in situ* hybridization. Chromosome Sci 19(1–4):33–36

Young AP, Jackson DJ, Wyeth RC (2020) A technical review and guide to RNA fluorescence in situ hybridization. PeerJ 8:e8806

Biostatistics

11

Abstract

This chapter on biostatistics introduces foundational concepts crucial for interpreting data in the biological sciences. Originating from the Latin word "statista," statistics encompasses not only the collection but also the analysis and interpretation of data, serving as a scientific toolbox for discerning patterns within diverse fields. Biostatistics, a specialized branch, focuses on analyzing data related to various organisms, providing quantitative insights that support decision-making in research. The chapter emphasizes the importance of understanding populations and samples. A population refers to the entirety of individuals or data points relevant to a study, while a sample is a representative subset of that population, chosen for practical reasons, such as time and cost constraints. Thus, sampling is a critical process that ensures the selected subset accurately reflects the characteristics of the larger group. Through descriptive and inferential biostatistics, researchers can estimate central tendencies and variability, and make predictions based on sample data. The chapter outlines essential steps in statistical analysis, including study design, hypothesis generation, data collection, and the application of techniques. It also highlights various sampling methods, with random sampling often regarded as the ideal choice for its fairness in representation. Overall, this chapter aims to equip students with the analytical tools necessary for conducting chromosome analysis and encourages further exploration of biostatistical methodologies applicable in broader biological research settings.

11.1 Introduction

The word "statistics" originated from the Latin word "statista," which refers to a person, who deals with affairs such as censuses, taxation, and war planning. Statistics is not just about gathering information but also about interpreting it to better understand the event from which the data has been collected. It is like a scientific toolbox filled with mathematical methods. It offers a critical lens through which data from various disciplines can be analyzed.

T. B. Jha, M. Halder, *Classical and Molecular Plant Cytogenetics*,
https://doi.org/10.1007/978-981-95-8265-5_11

It involves the collection, organization, presentation, analysis, and interpretation of data (Fig. 11.1). Any existing information or knowledge collected by an investigator in any form (numbers, letters, and figures) for a definite purpose is termed as data. The data is called primary data when an investigator collects it from the source for the first time using surveys, experiments, or clinical trials, through observation, measurement, or counting. While secondary data are not initially collected by an investigator from the source, they are collected from published or unpublished data of other investigators. Raw data are unprocessed or unorganized data collected by researchers. According to the need, the collected raw data may be organized for better interpretation of data or processed to remove outliers for accurate analysis (Pagano et al. 2022). This data is called processed or organized data.

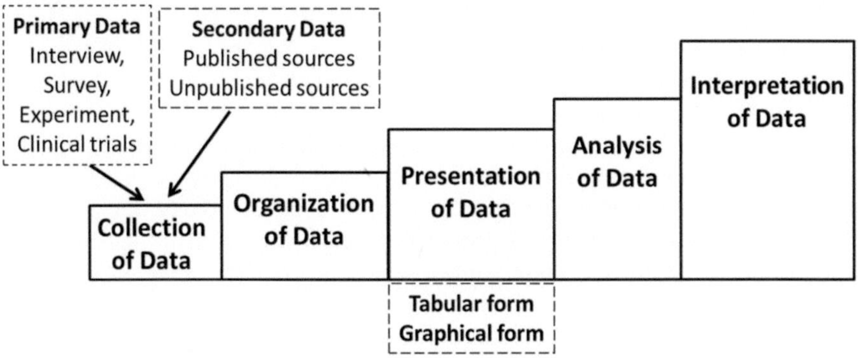

Fig. 11.1 Steps of statistical analysis

Statistics do not prove or disprove anything. They organize data efficiently to find patterns and trends, analyze data to get a quick snapshot of the key characteristics of the data, and draw conclusions. Understanding statistics is crucial for making informed decisions, as it offers insights into data that drive strategic actions, whether for students learning about statistics or professionals analyzing business data.

Biostatistics is a branch of applied statistics that focuses on analyzing and interpreting data from the biological sciences to support decision-making. It provides a quantitative lens to understand variations in data collected from individual samples or populations of any species, including plants, animals, human beings, and other organisms. Biostatistics generally involves several key steps, such as study design (formulating research questions and selecting appropriate methods), hypothesis generation, relevant data collection, and applying statistical techniques to draw meaningful conclusions (Mackay et al. 2019).

Descriptive biostatistics organizes, summarizes, and interprets data to understand patterns and explain the distribution of population measurements by estimating central tendency (mean, mode, and median) and variability (standard deviation and correlation coefficient). Inferential biostatistics helps make predictions and generalizations about a population based on sample data. It includes hypothesis testing and assessing the significance of relationships or effects in a study.

This chapter's primary objective is to teach students about biostatistical analyses commonly associated with chromosome analysis. However, several other biostatistical methods exist that are very useful for other experiments. Students are recommended to consult a comprehensive biostatistics book as needed.

11.2 Population, Sample, and Sampling

Understanding concepts such as population, sample, and sampling in biostatistics is fundamental to conducting effective research and making inferences about larger groups from smaller subsets (Elston and Johnson 2008). A population in biostatistics represents the totality of all individuals, items, data points, or observations that are the focus of a study. It represents the "whole" from which data is collected, and it can be finite or infinite. However, when the population is large, it may not be feasible to study/collect data from all individuals due to cost, time, or logistical constraints.

To address this challenge, researchers usually select a smaller, non-biased subset of the whole population to estimate and draw conclusions about the population. The subset of the population chosen for study is termed a sample (Fig. 11.2). The process of selecting a representative subset (sample) from the entire population is called sampling (Fig. 11.2). Remember, the sample should represent the population's actual characteristics. The sample size may be small, medium, or large, but it is always less than the total size of the population (Fig. 11.2).

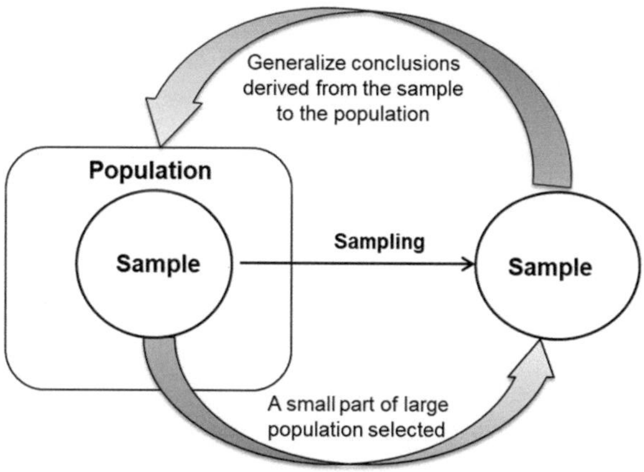

Fig. 11.2 Concept of population, sample, and sampling

The sampling process involves several steps: (i) Clearly define the population, (ii) choose an appropriate sampling method, (iii) decide the sample size, (iv) collect data from the selected sample, and (v) analyze and interpret the results obtained to conclude the population based on sample data.

Various sampling methods are available, including random sampling, stratified sampling, cluster sampling, systematic sampling, convenience sampling, judgmental sampling, etc. (Fig. 11.3). In biological studies, random sampling is often preferred as it gives every member an equal chance of being selected. A standard sampling error occurs when the chosen sample does not represent the whole population. A small sample size has a greater risk of sampling error, whereas using a sufficiently large sample and random selection minimizes sampling error. Researchers also calculate a margin of error during their analysis, which indicates the permissible amount of error for a miscalculation between the sample and the actual population. These strategies enhance the accuracy and reliability of statistical interpretations regarding the population by minimizing sampling errors.

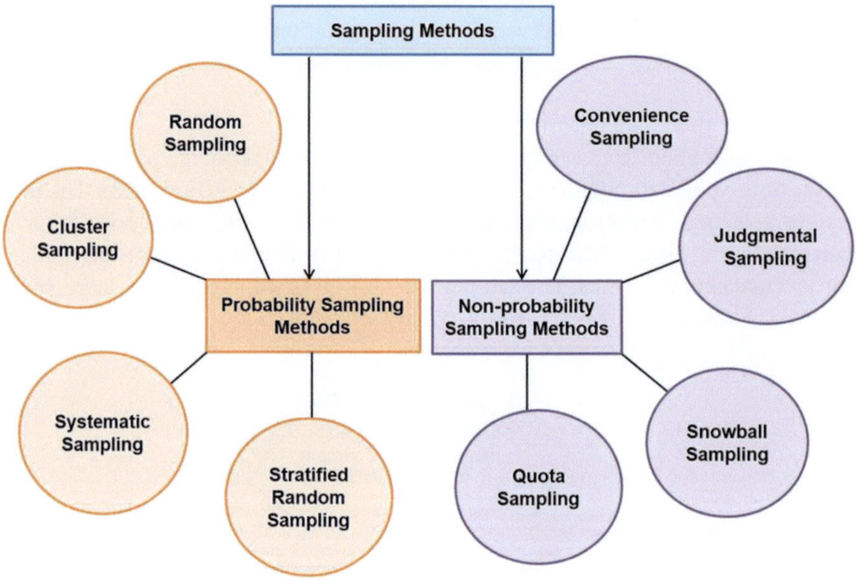

Fig. 11.3 Types of sampling methods

From a cytogenetic perspective, healthy root tips of germinated seedlings or potted plants are randomly selected for chromosome analysis. Once chromosome preparations are made and stained, thousands of cells are found on the slide. Since studying every individual cell is impossible, data for mitotic index studies are collected from randomly selected fields. Similarly, randomly selected, well-scattered metaphase plates are examined for karyotyping to determine the chromosome number. Cells that are non-dividing, non-metaphasic, or have uncountable chromosomes in metaphase are excluded, considering them as outliers. They may introduce errors in the analysis and affect interpretation. It is essential to distinguish this process from the purposeful selection because, during the selection of metaphase plates, the investigator does not prefer metaphase plates that have a specific chromosome number. Instead, only countable metaphase plates are included in the sample and eliminated. Well-scattered, countable metaphase plates form the foundational visual data for karyotype analysis. Integrating biostatistics into chromosome research is vital, as it transforms results into more meaningful, accurate, and reliable information. This approach mitigates the risks of misinterpreting artifacts from the staining process and minimizes the potential impact of random variations on the final interpretation.

For karyotype analysis using biostatistical methods, it is essential to consider at least 5 well-scattered metaphase plates showing distinct chromosome morphology. A larger sample size is also preferred, as it enhances the reliability and confidence in the results and their interpretation. Random sample selection from the population is most commonly practiced in biostatistics. However, in cytogenetic studies, specific preferences may influence the random selection process; for instance, metaphase plates with scattered chromosomes are typically chosen for analysis.

11.3 Central Tendency

A central tendency in statistics is the central or representative value of a statistical series. Among the different statistical averages in statistics, the mean, median, and mode are the three popular central tendency calculation methods used for analyzing data effectively and drawing inferences. The mean, or arithmetic mean represents the mathematical average, whereas the median and mode represent the positional average.

11.3.1 Mean or Arithmetic Mean

The sample mean or arithmetic mean ($\bar{x}$) is the most commonly used central tendency measure. It represents the average value in a dataset. It is calculated by the summation of all the variants or observations ($\sum x$ or "sigma of x") in a dataset or sample and is divided by the number of observations or sample size (N).

$$\text{Sample mean}\,(\bar{x}) = \Sigma x / N$$

$$\text{Population mean}\left(\bar{\mu}\right) = \Sigma \mu / N$$

where, $\sum$ = summation, "x" = individual observation, "μ" = individual observation, and "N" = total sample size (number)

11.3.2 Median

The median (denoted by the symbol "M") is another important measure of central tendency. It represents a positional average and is the series's middle or centrally located value when the data are organized in ascending or descending order. In other words, the median is the value that divides the arranged series into two equal parts, comprising all values greater than the median and all values less than the median.

$$\textbf{Median value } (M) = \text{value of } (N+1)/2^{\text{th}} \text{ observation, when } N \text{ is an odd number}$$

Note: If N of a series or dataset is an even number, $(N + 1)/2$ will be a fraction. In such a case, the median value will be the arithmetic mean of the two middle values of the series.

Steps to Calculate the Median Value of a Given Dataset

(i) Rearrange the given data in an ascending or descending order of magnitude.
(ii) Calculate the median using the formula, median (M) = the value of $(N + 1)/2^{\text{th}}$ observation.

- *Calculate the median from the following data: 25, 16, 19, 13, 15, 21, 17, 20, and 28.*
  Rearrangement of the given data in ascending order generates the data series as follows:
  13, 15, 16, 17, 19, 20, 21, 25, 28

  $$\text{Here, the } N = 9, \textit{so} \text{ the median}(M) = \text{value of } (N+1)/2^{th} \text{ observation}$$
  $$= \text{value of } (9+1)/2^{th} \text{ observation}$$
  $$= \text{value of } 5^{th} \text{ observation}$$
  $$= 19$$

- *Calculate the median from the following data: 190, 210, 318, 213, 285, 320, 290, and 350.*

  Rearrangement of the given data in ascending order generates the data series as follows: 190, 213, 210, 286, 290, 318, 320, and 350

  $$\begin{aligned}\text{The Median} &= \text{value of } (N+1)/2^{\text{th}} \text{ observation}\\ &= \text{value of } (8+1)/2^{\text{th}} \text{ observation}\\ &= \text{value of } 9/2^{\text{th}} \text{ observation}\\ &= \text{value of } 4.5^{\text{th}} \text{ observation}\end{aligned}$$

  Here, $N = 8$, which is an even number. According to the formula, the median value is the value of the 4.5^{th} observation, which does not exist. The median value will be calculated in such a situation by averaging the middle two scores.

So, the Median (M) = the average of value of 4th and 5th observations

$$=(\text{value of } 4^{\text{th}} \text{ observation} + \text{value of } 5^{\text{th}} \text{ observation})/2 = (286+290)/2 = 288$$

11.3.3 Mode (*Z*)

Mode (denoted by the symbol "*Z*") is another important measure of central tendency. Mode is the value that occurs most frequently in a series or dataset—that is, it has the greatest frequency in a distribution. Mode is used when the values are widely varying.

For example, in many vegetatively propagated plants, we obtained different chromosome numbers in the same and different root tips. Then, the chromosome number of a plant is described by its most frequently occurring chromosome number, i.e., its mode.

11.4 Dispersion

In statistical analysis, the mean, median, and mode serve as measures of central tendency, indicating the central value of a given variable. However, they do not provide insight into the distribution of observations around the mean. For instance, three varieties may have the same average chromosome length, but this does not imply that their karyomorphological features are identical; variability may exist among them, as illustrated in Table 11.1. Consequently, after determining the central tendency, it is crucial to examine additional statistical parameters, such as dispersion.

Table 11.1 Calculation of average chromosome length of three hypothetical plant varieties having $2n = 6$ chromosomes

	Variety A		Variety B		Variety C	
Sl. No.	Individual chromosome length in μm (x)	Average of chromosome length in μm $(\bar{x} = \Sigma x / N)$	Individual chromosome length in μm (x)	Average of chromosome length in μm $(\bar{x} = \Sigma x / N)$	Individual chromosome length in μm (x)	Average of chromosome length in μm $(\bar{x} = \Sigma x / N)$
1	7.5	45/6 = 7.5	9.5	45/6 = 7.5	1.5	45/6 = 7.5
2	7.5		9.5		1.5	
3	7.5		7.5		8	
4	7.5		7.5		8	
5	7.5		5.5		13	
6	7.5		5.5		13	

The measurement of data dispersion is essential in data analysis in biostatistics, and cytogenetics. Dispersion, often referred to as scatter, spread, or variation, indicates the degree to which individual observations in a dataset deviate from a central value (such as the mean, mode, or median) or from one another. It quantifies how much the observation varied. Depending on the specific requirements, dispersion can be calculated using various methods (Fig. 11.4). A dispersion value of zero indicates that there is no variation among the observations, meaning all values are identical. In contrast, high dispersion signifies substantial differences between the observed values and the average, while low dispersion indicates that these differences are minimal.

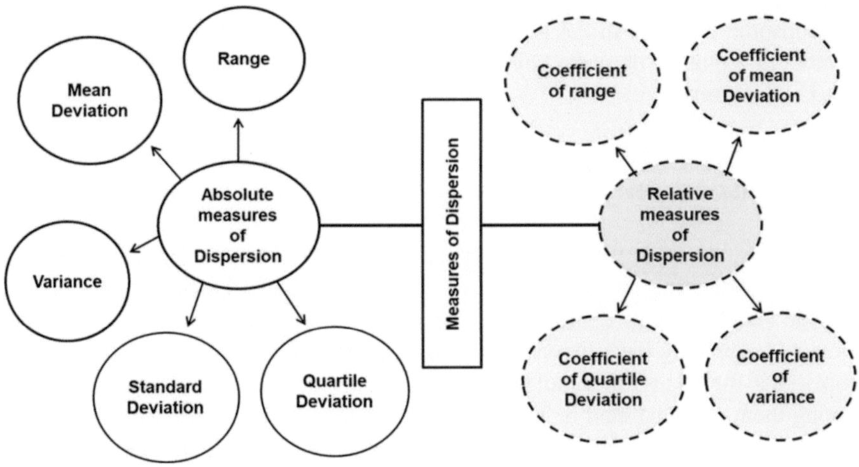

Fig. 11.4 Different methods of dispersion calculation

11.4.1 Range (*R*)

Among the various methods for studying dispersion, the range is the simplest to understand and easiest to calculate. The range (R) is defined as the difference between the maximum observed value (x_{max}) and the minimum observed value (x_{min}) within a given series or dataset. Calculating the range takes only a few minutes. However, it is the least reliable measure of dispersion, as it does not consider every observation in the dataset and is highly sensitive to sampling fluctuations. In cytogenetics, the chromosome range is an important parameter, calculated alongside the average chromosome length (ACL).

$$\text{Range}(R) = x_{\text{max}} - x_{\text{min}} = \text{H} - \text{L}$$

The coefficient of range is calculated using the formula-

$$\text{Coefficient of range} = \frac{\text{H} - \text{L}}{\text{H} + \text{L}} \times 100$$

11.4.2 Mean Deviation (MD)

Mean deviation is defined as the average or mean of the deviations of the values from the mean.

Steps for Calculating Mean Deviation

(i) Calculate the arithmetic mean $(\bar{x} = \sum x / N)$ of the given dataset
(ii) Find the deviation of each observation $(d_x = x - \bar{x})$ from the mean
(iii) Calculate the total deviation $(|\sum d_x| = d_{x1} + d_{x2} + d_{x3} + d_{x4} + \ldots.. + d_{xn})$ by ignoring the negative sign of deviation
(iv) Apply the formula:

$$\text{Mean Deviation from mean (MD)} = |\Sigma d_x| / N = \Sigma |x - \bar{x}| / N$$

$$\text{Mean Deviation from median} = MD_{Me} = |\Sigma d_x| / N = \Sigma\ |x - M| / N$$

- Suppose you have counted $2n = 12$ chromosomes in a metaphase plate with size variations. After measuring the size of the individual chromosomes as given below, you median, mode, and range of chromosome length.

Chromosome number ($2n = 12$)	Individual chromosome length (x)	Total chromosome length in µm (Σx)	Average chromosome length in µm ($\bar{x} = \Sigma x/N$)	Chromosome range in µm ($R = H–L$)	Median (value of $\frac{(N+1)}{2}$th observation)	Mode
1	6.5	64	64/12 = 5.33	3.7–7.5	Value of $\frac{(12+1)}{2}$th observation= value of 6.5th observation = $\frac{(6.2+4)}{2}$ = 5.1 [decending order of chromosome length is 7.5, 7.5, 6.5, 6.5, 6.2, 6.2, 4, 4, 4, 4, 3.8, 3.8]	4
2	3.7					
3	6.5					
4	4.0					
5	7.5					
6	7.5					
7	6.2					
8	6.2					
9	4.0					
10	4.0					
11	4.0					
12	3.7					

11.4.3 Variance

Variance is a statistical measure that quantifies the spread or dispersion of a set of data points in relation to their mean (average). It evaluates the distance of each data point from the mean by calculating the average squared deviation of each point. Variance is typically denoted by σ^2 for population variance and s^2 for sample variance. A higher variance indicates a greater dispersion among the data points, while a lower variance suggests that the points are more closely clustered around the mean. Population variance measures dispersion across all members of a population, whereas sample variance estimates the population variance based on a subset (sample) of the data. This concept is closely linked to the idea of variability within a data set, reflecting the extent of dispersion or spread. Greater variability results in a larger variance around the mean.

The formula for the calculation of the two types of variances is given below

$$\text{Population Variance } \left(\sigma^2\right) = \frac{\Sigma(x-\mu)^2}{N}$$

$$\text{Sample Variance } \left(s^2\right) = \frac{\Sigma(x-\overline{x})^2}{n-1}$$

$$\text{Coefficient of variance} = \frac{\text{SD}}{\overline{x}} \times 100$$

Where x = individual value in the data set, $\overline{x}$ = sample mean value, μ = population mean value, N = the population size, n = total number of observations, s^2 = sample variance, and σ^2 = population variance

11.4.4 Standard Deviation (SD)

Standard deviation, commonly abbreviated as SD, is a fundamental statistical analysis tool that provides a more precise and in-depth understanding of the variability within a data set relative to its mean value. It measures the extent to which data points are spread out or deviate from the mean. The symbols "σ" and "s" are used to represent the standard deviation of a population and a sample, respectively. Standard deviation plays a vital role in comparing the dispersion of two distinct datasets that have approximately the same mean value, as well as in identifying outliers. Data points that lie significantly outside the standard deviation may indicate errors or highlight unusual cases.

Steps to Calculate Population Standard Deviation

(i) *Calculate the population means* ($\mu = \sum x/N$): Summation of all the observed values in the dataset and divide by the population size.
(ii) *Calculate deviations from the population mean* ($d_\mu = x - \mu$): Subtract the population mean (μ) from each individual data point (x).
(iii) *Square each deviation* (d_μ^2): Square each deviation you calculated in step (ii). This ensures positive and negative deviations contribute equally since squaring makes them both positive.
(iv) *Find the Variance* ($\sigma^2 = \sum d_\mu^2/N$): Add all the squared deviations you obtained in step (iii). Divide this sum by the total number of data points in your population (N) for population SD.
(v) *Calculate the population standard deviation* $\left(\sigma = \sqrt{\sigma^2}\right)$: Take the square root of the calculated population variance (σ^2) in step (iv).

Steps to Calculate Sample Standard Deviation (S)

(i) Calculate the sample means $\left(\overline{x} = \sum x / N\right)$
(ii) Calculate deviations from the sample mean for each data $\left(d_{x1} = x - \overline{x}\right)$
(iii) Square each deviation from the mean (d_{xi}^2)
(iv) Summation of the squared deviations ($\sum d_{xi}^2$)
(v) Find the sample variance (s^2) by dividing the sum of squared deviations by the number of data points of sample (N) minus 1
(vi) Calculate the sample standard deviation $\left(s = \sqrt{s^2}\right)$ by taking the square root of the calculated population variance (s^2).

The formula for the calculation of the two types of Standard Deviation is given below:

$$\text{Sample Standard Deviation } (s) = \sqrt{s^2} = \sqrt{\frac{\Sigma\left(x-\overline{x}\right)^2}{N-1}}$$

$$\text{Population Standard Deviation } (\sigma) = \sqrt{\sigma^2} = \sqrt{\frac{\Sigma\left(x-\overline{\mu}\right)^2}{N}}$$

Where x = individual value in the data set, $\overline{x}$ = sample mean value, $\overline{\mu}$ = population mean value, N = the population size, n = total number of observations, s^2 = sample variance, and σ^2= population variance

While the calculation of standard deviation may be done manually, the process can be time-consuming. On the other hand, software applications, such as MS Excel, SPSS, and PAST streamline calculations, enhancing data analysis efficiency and accuracy.

A data set characterized by a smaller standard deviation exhibits a narrower spread, or compact clustering of measurements around the mean, and generally contains comparatively fewer high or low values. An item randomly selected from a data set with a low standard deviation has a better chance of being close to the mean than one drawn from a data set with a higher standard deviation. It is important to remember that the standard deviation is more sensitive to extreme values. A single extreme value can significantly influence the calculation of the standard deviation.

In cytogenetics, a low standard deviation of chromosome length means the data are clustered closely around the mean value, indicating that individual chromosome lengths are more or less the same. Conversely, a high standard deviation of chromosome length suggests the data are scattered far from the mean value, with some chromosome lengths significantly greater or smaller than the mean value.

11.4.5 Standard Error (SE)

The standard error (SE) is a fundamental concept in statistics that is based on the standard deviation of a sample population. It serves as a statistical measure that emphasizes the variability of sample means derived from a population, rather than the individual data points. The SE estimates the extent to which the sample mean is expected to fluctuate among different samples drawn from the same population, offering valuable insights into the accuracy of the sample mean as a representation of the true population mean. This assessment relies on standard deviation, with the deviation being interpreted as the error. The concept of standard error is crucial in inferential statistics, as it aids in estimating the range within which the true population mean is likely to reside.

Steps to Calculate Standard Error (SE)

(i) Calculate the population standard deviation ($\sigma = \sqrt{\sigma^2}$). following the steps mentioned previously.

(ii) Calculate standard error $\left(\mathrm{SE} = \frac{\sigma}{\sqrt{n}}\right)$ by dividing the population standard deviation (σ) by the number of observations (n)

$$\text{Standard error}(\mathrm{SE}) = SE = \frac{\sigma}{\sqrt{n.}}$$

Where, SE = standard error, σ = population standard deviation, and n = number of observations

Interpretation of Standard Error

A larger standard error suggests that the sample means are more spread out, indicating that the sample mean may not accurately represent the true population mean. On the other hand, a low standard error indicates that the sample mean is a more accurate estimate of the population mean. The data points within the sample are likely close together, providing a more reliable picture of the whole population.

Standard error is crucial in calculating confidence intervals, which express the range of values within which the true population mean is likely to fall with a specified level of confidence. Understanding standard error helps interpret the results derived from studies and surveys. It allows researchers to assess the degree of confidence in the conclusions drawn from a particular sample.

It has been noticed that a larger sample size generally leads to a lower standard error, thereby enhancing the precision of the sample mean as it pertains to the population. The formula associated with standard error quantifies the extent to which sample statistics may deviate from the actual population parameters. This allows for the prediction of uncertainty regarding the reliability of the estimates obtained.

Imagine that you have randomly selected multiple metaphase plates from a large population to estimate the chromosome length of a particular plant species. After calculating the average chromosome length (ACL) for this species, it is important to note that another random sampling of several metaphase plates from the same population may yield slightly different ACL values. This variation reflects the standard error.

11.4.6 Computer-Aided Statistical Analysis

In the ever-evolving field of biostatistics, professionals frequently turn to statistical software to analyze complex biological data. Among the wide array of software options available, Microsoft Excel is particularly well-suited for introductory analyses and handling smaller datasets. For students in the biological sciences, cultivating proficiency in Excel is immensely valuable due to its accessibility, flexibility, versatility, and robust computational capabilities. Furthermore, a strong command of Excel serves as an essential asset, laying a solid foundation for understanding biostatistical principles and conducting fundamental analyses. Acknowledging its broader applicability—especially in chromosomal analysis—we have made a concerted effort to illustrate straightforward use of this tool in the following sections.

Before entering any data into Excel, it is crucial to gather the karyomorphological data. This data may include the total number of chromosomes on selected metaphase plates and the lengths of individual chromosomes. The measurements to be recorded for each chromosome include the short arms (p), long arms (q), and any secondary constrictions, if present. Next, launch MS Excel by clicking the Excel icon on your computer, then create a new worksheet. Organize your data in the spreadsheet by placing headers in the first row and entering the corresponding measurements in the subsequent rows, ensuring that you complete the data entry accurately and without errors.

Step 1: Open MS Excel by clicking on the MS Excel icon on the computer (Figs. 11.5a, b).

Step 2: Use the first row for headers and the subsequent rows for the entry of the measured data (the chromosome number, and the lengths of long, short, and secondary constrictions of individual chromosomes are entered in A2-A7, B2-B7, C2-C7, and D2-D7) into the spreadsheet (Fig. 11.5c).

Step 3: Calculation of individual chromosome lengths:

- In a new column adjacent to your existing data, label it as "Total Length." *in the cell (E1)* (Fig. 11.5d)
- Click on the empty cell (E2 Fig. 11.5e) corresponding to the first chromosome's long, short, and secondary constriction length and either enter the formula "=B2 + C2 + D2" into the formula bar (Fig. 11.5f), then press enter to get the result (Fig. 11.5g) or utilize the "Sum" function via the formula bar, which includes the following steps:
  - Click the fx in the formula bar (Fig. 11.5h); a new window for the Insert Function will pop up (Fig. 11.5i).
  - Select the "SUM" function and click on Ok (Fig. 11.5j).
  - A new window of function arguments will be opened (Fig. 11.5k); click on the specific symbol as shown in fig. 11.5k, and a new window will open (Fig. 11.5l).
  - Select three cells that represent the lengths of the long, short, and secondary arms by dragging (Fig. 11.5m).
  - Press Enter, then click OK (Fig. 11.5n) to display your result in the selected cell (Fig. 11.5g).
- Drag this formula down to apply it to the remaining rows, ensuring each chromosome's total length is calculated.

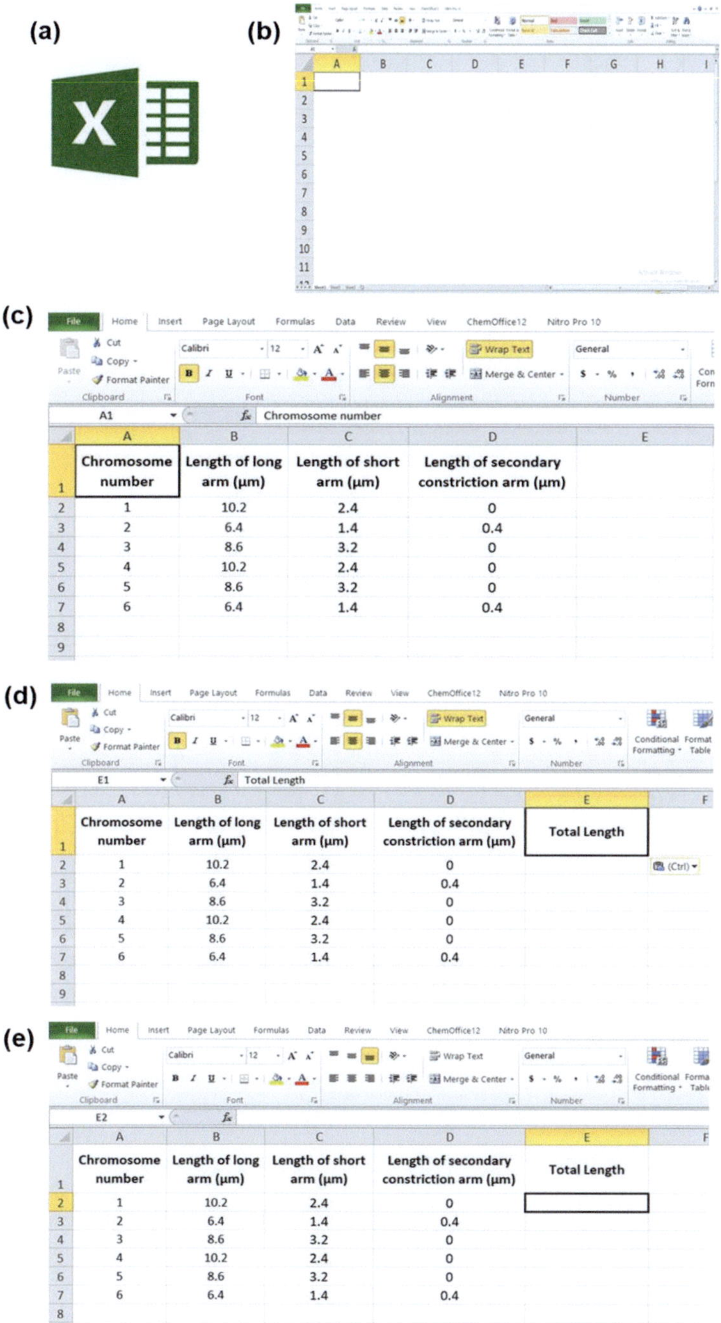
(a)
(b)
(c)
A1
Chromosome number
Chromosome number
Length of long arm (µm)
Length of short arm (µm)
Length of secondary constriction arm (µm)
1 10.2 2.4 0
2 6.4 1.4 0.4
3 8.6 3.2 0
4 10.2 2.4 0
5 8.6 3.2 0
6 6.4 1.4 0.4
(d)
E1
Total Length
Chromosome number
Length of long arm (µm)
Length of short arm (µm)
Length of secondary constriction arm (µm)
Total Length
1 10.2 2.4 0
2 6.4 1.4 0.4
3 8.6 3.2 0
4 10.2 2.4 0
5 8.6 3.2 0
6 6.4 1.4 0.4
(e)
E2
Chromosome number
Length of long arm (µm)
Length of short arm (µm)
Length of secondary constriction arm (µm)
Total Length
1 10.2 2.4 0
2 6.4 1.4 0.4
3 8.6 3.2 0
4 10.2 2.4 0
5 8.6 3.2 0
6 6.4 1.4 0.4

(f)

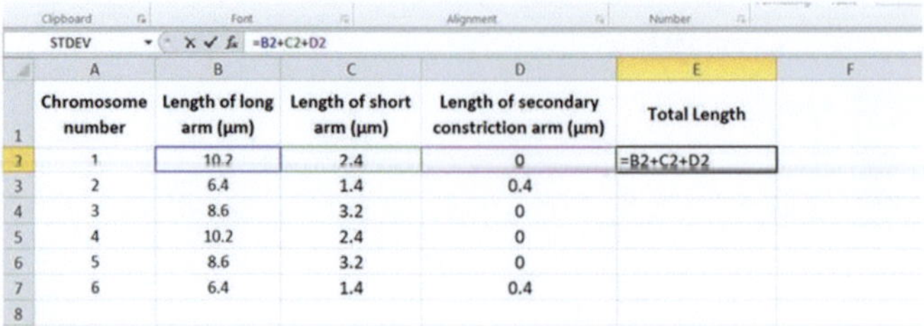
STDEV
=B2+C2+D2
Chromosome number
Length of long arm (µm)
Length of short arm (µm)
Length of secondary constriction arm (µm)
Total Length
1 10.2 2.4 0 =B2+C2+D2
2 6.4 1.4 0.4
3 8.6 3.2 0
4 10.2 2.4 0
5 8.6 3.2 0
6 6.4 1.4 0.4

(g)

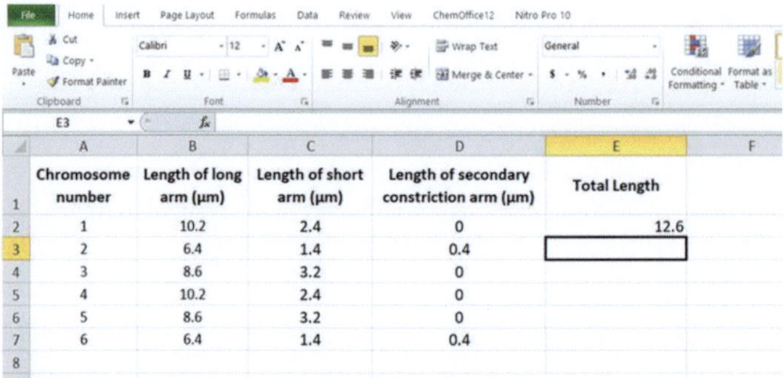
E3
Chromosome number
Length of long arm (µm)
Length of short arm (µm)
Length of secondary constriction arm (µm)
Total Length
1 10.2 2.4 0 12.6
2 6.4 1.4 0.4
3 8.6 3.2 0
4 10.2 2.4 0
5 8.6 3.2 0
6 6.4 1.4 0.4

(h)

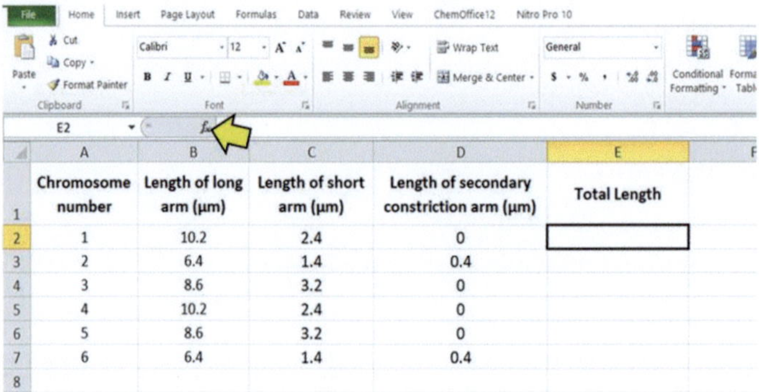
E2
Chromosome number
Length of long arm (µm)
Length of short arm (µm)
Length of secondary constriction arm (µm)
Total Length
1 10.2 2.4 0
2 6.4 1.4 0.4
3 8.6 3.2 0
4 10.2 2.4 0
5 8.6 3.2 0
6 6.4 1.4 0.4

(i)

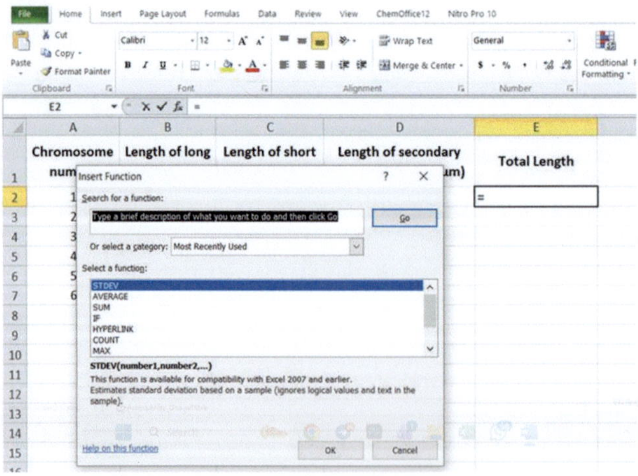
E2
=
Insert Function
Search for a function:
Type a brief description of what you want to do and then click Go
Go
Or select a category: Most Recently Used
Select a function:
STDEV
AVERAGE
SUM
IF
HYPERLINK
COUNT
MAX
STDEV(number1,number2,...)
This function is available for compatibility with Excel 2007 and earlier.
Estimates standard deviation based on a sample (ignores logical values and text in the sample).
Help on this function
OK
Cancel

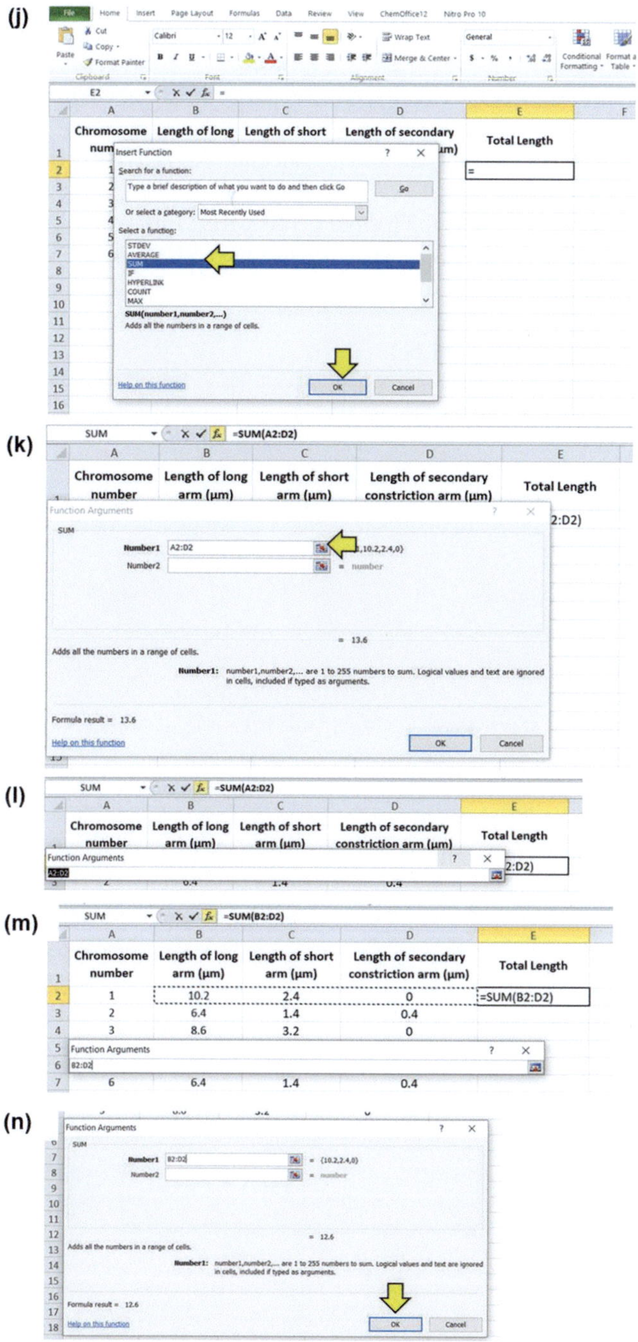

Fig. 11.5 (**a–n**) Steps of calculation of individual chromosome length analysis using MS-Excel

Step 4: Calculation of Total Chromosome Length (TCL) of Individual Metaphase Plate:

- Label the next column adjacent to existing data (e.g., F1) as "Total Chromatin Length" (Fig. 11.6a).
- Click on the empty cell (F2) where you want to calculate the *TCL value* under column F.
- Click on the formula bar, and type the formula "=SUM(E2:En)" into the formula bar and then press enter (Fig. 11.6b), or use the "SUM" function as described in step 3 to calculate the total. (Replace "n" with the last row number of your data).
- Press the Enter to get result (Fig. 11.6c).

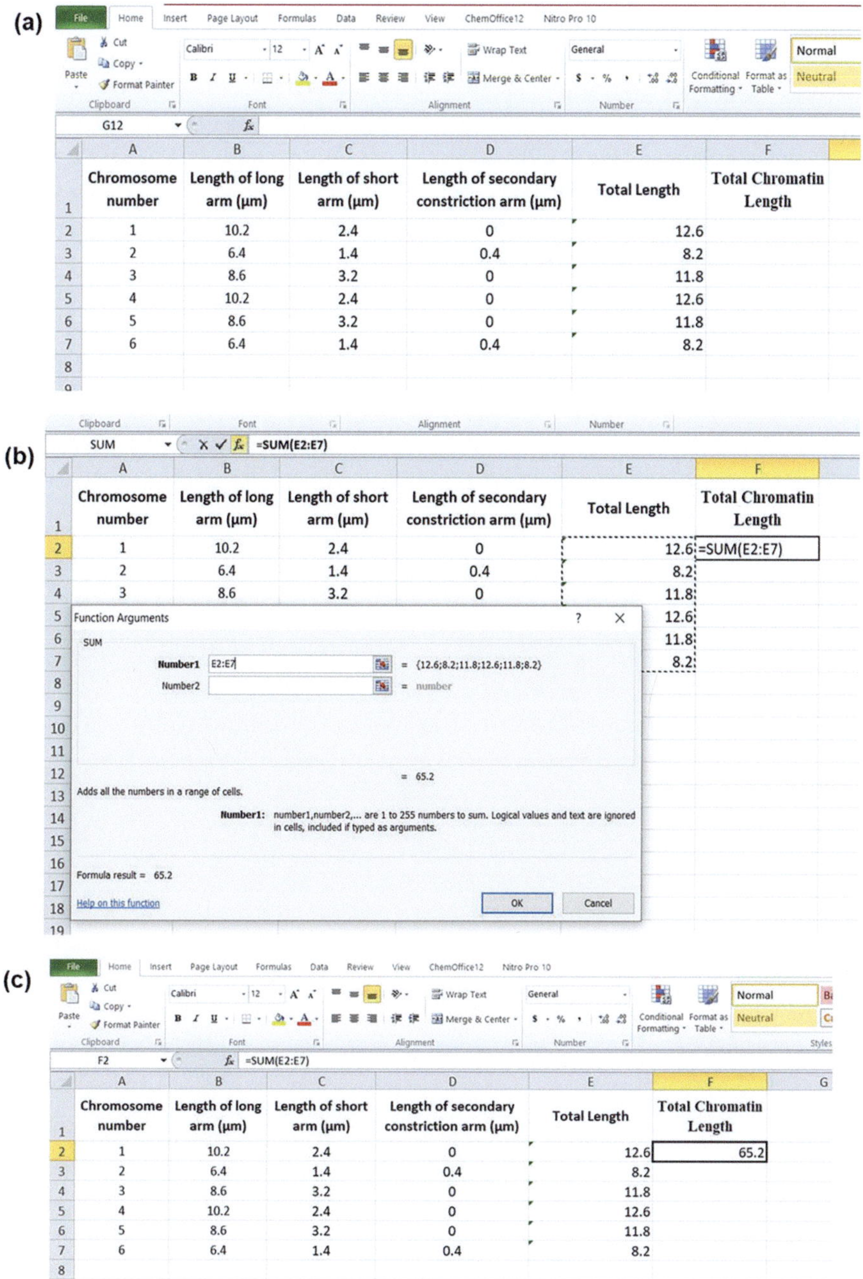

Fig. 11.6 (a–c) Steps of calculation of total chromatin length (TCL) of individual metaphase plate using MS-Excel

Step 5: Calculation of ACL of Individual Metaphase Plate:

- Move to a new column adjacent to existing data and label the cell (e.g., G1) as "Average Chromatin Length" (Fig. 11.7a).
- Click on the empty cell (G2) where you want to calculate the ACL value under column F (Fig. 11.7b).
- Click on the formula bar, and type "=" and click on the cell of TCL value, then insert divide and enter the number of chromosomes (Fig. 11.7c).
- Press the Enter to get result (Fig. 11.7d).

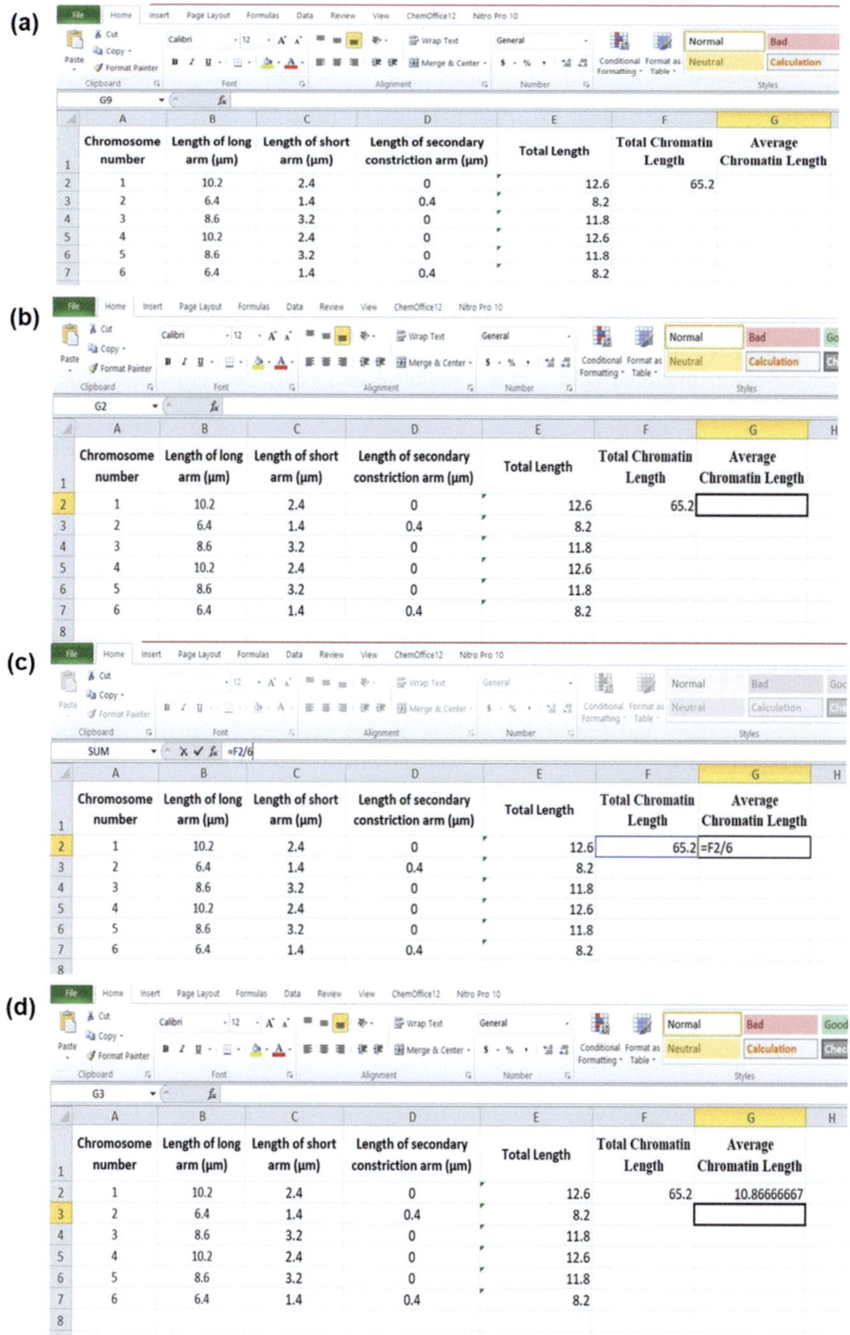

Fig. 11.7 (**a**–**d**) Steps of calculation of average chromatin length (ACL) of individual metaphase plate using MS-Excel

Calculation of the Mean of ACL and SD

Step 1: Open the Excel by clicking on the MS-Excel icon on your computer (Fig. 11.8a, b)

Step 2: Use the first row for headers and the subsequent rows for the entry of the measured data (the metaphase plate numbers are entered in A2-A6 cells, and their corresponding calculated ACL values are entered in B2-B6) into the spreadsheet (Fig. 11.8c)

Step 3: Calculation of the Mean of ACL:

- Move to a new column adjacent to existing data and label the cell (e.g., C1) as "mean of ACL" (Fig. 11.8d).
- Click on an empty cell (C2) where you want to calculate the mean ACL value under column C (Fig. 11.8e).
- Click on the *fx* in the formula bar, and a new window of Insert Function will pop up (Fig. 11.8f).
- Select the *AVR* function and click on Ok (Fig. 11.8g).
- A new window of function arguments (Fig. 11.7h) will open; click on the symbol, and a new window will open (Fig. 11.8i).
- Insert the numbers by selecting all the cells that represent individual ACL values by dragging (if the data is located in a single column or row without interruption) or by pressing Ctrl and selecting the cells using mouse clicking (Fig. 11.8j).
- Press Enter, and then click on Ok to get your result in the selected cell (Fig. 11.8k).

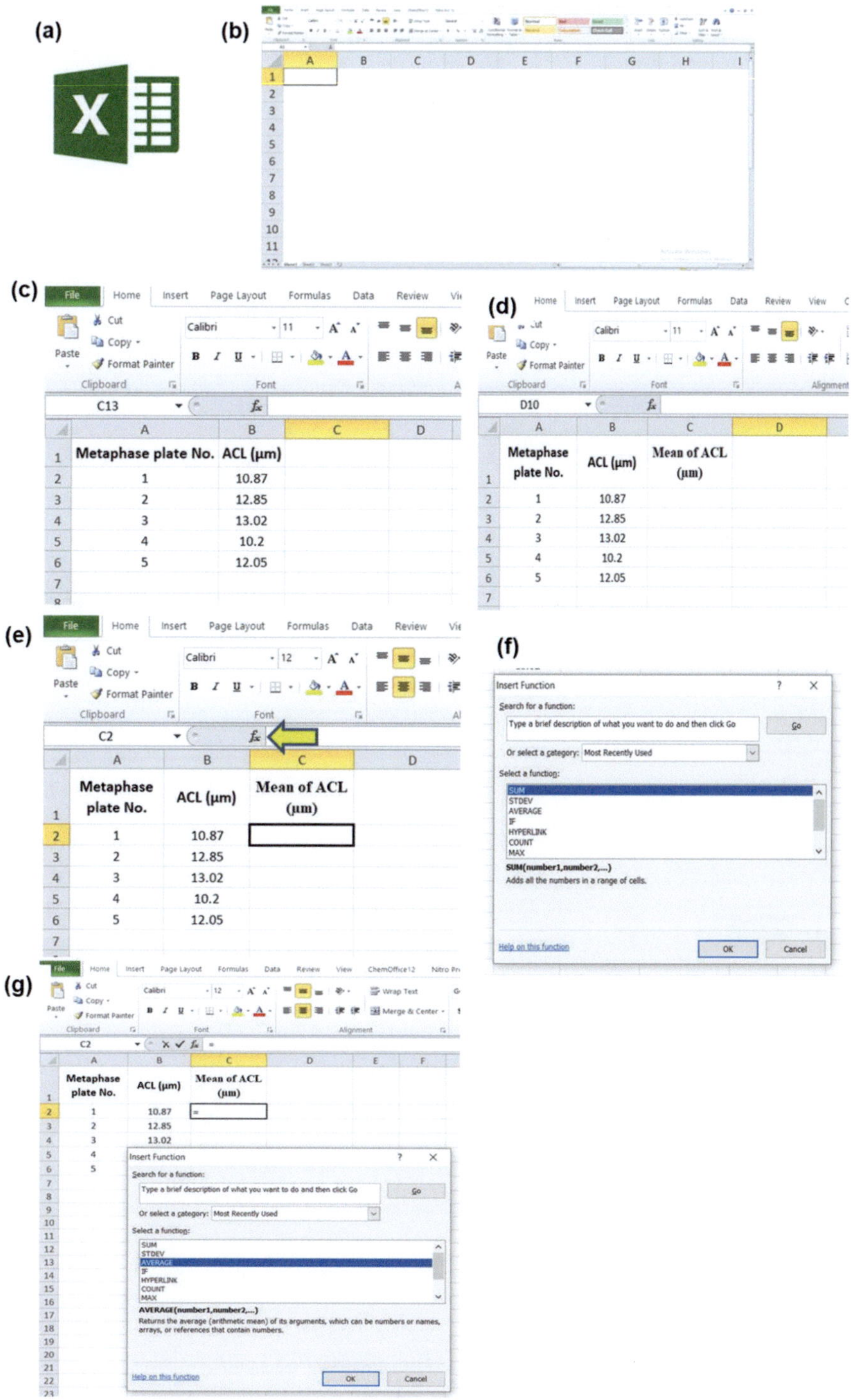

(a)
(b)
(c)
File
Home
Insert
Page Layout
Formulas
Data
Review
Cut
Copy
Format Painter
Paste
Clipboard
Calibri
11
Font
C13
Metaphase plate No.
ACL (μm)
1 10.87
2 12.85
3 13.02
4 10.2
5 12.05
(d)
D10
Metaphase plate No.
ACL (μm)
Mean of ACL (μm)
Alignment
(e)
C2
(f)
Insert Function
Search for a function:
Type a brief description of what you want to do and then click Go
Go
Or select a category: Most Recently Used
Select a function:
SUM
STDEV
AVERAGE
IF
HYPERLINK
COUNT
MAX
SUM(number1,number2,...)
Adds all the numbers in a range of cells.
Help on this function
OK
Cancel
(g)
Wrap Text
Merge & Center
AVERAGE(number1,number2,...)
Returns the average (arithmetic mean) of its arguments, which can be numbers or names, arrays, or references that contain numbers.

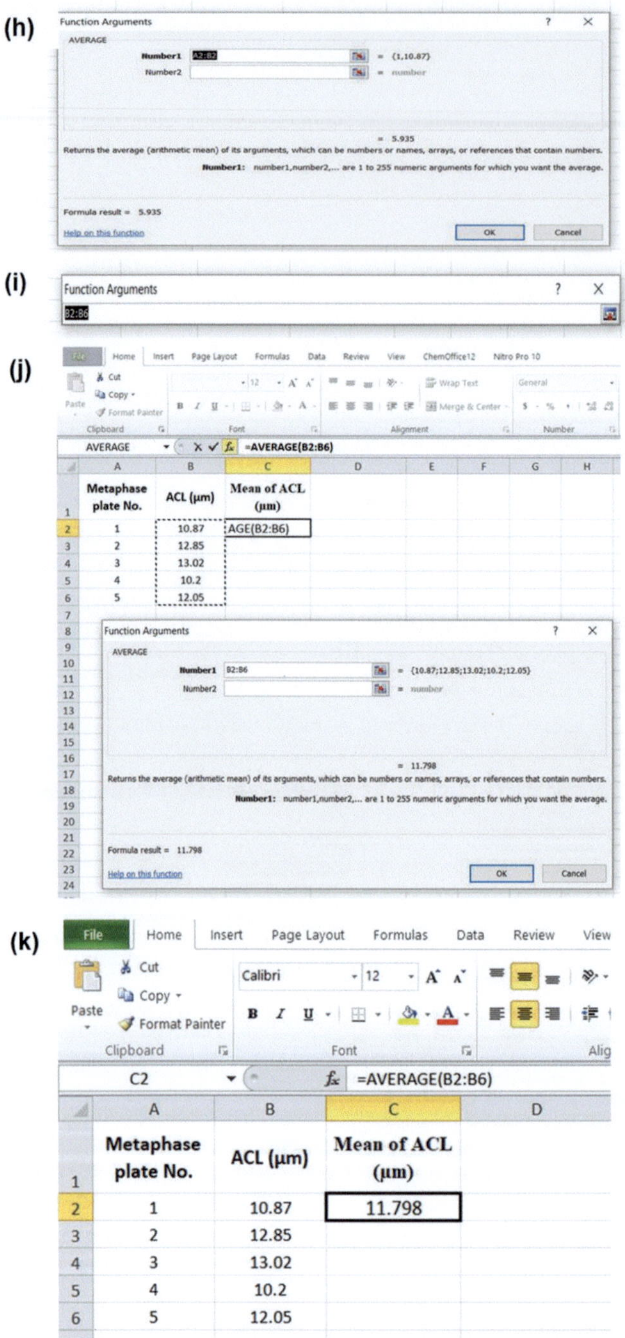

Fig. 11.8 (**a–k**) "Steps of calculation of mean ACL using MS-Excel"

Step 4: Calculation of Standard Deviation

- Move to a new column adjacent to existing data and label the cell (e.g., D1) as "Standard Deviation" in the cell.
- Click on an empty cell where you want to calculate the standard deviation under column D.
- Click on the fx in the formula bar, and a new window of Insert Function will pop up.
- Select the STDEV function and click on Ok.
- A new window of function arguments will be opened; click on the symbol, and a new window will open.
- Insert the numbers by selecting all the cells that represent individual ACL values by dragging (if the data is located in a single column or row without interruption) or by pressing Ctrl and selecting the cells using mouse clicking.
- Press Enter, and then click on Ok to get your result in the selected cell.

11.4.7 Correlation Coefficient

The correlation coefficient (r) is a statistical measure that quantifies the strength and direction of the relationship between two variables. It is expressed as a numerical value that always lies between −1 and +1. A larger correlation coefficient signifies a stronger association, which can be either positive or negative in nature. The maximum value of the correlation coefficient is 1, which indicates a perfect positive correlation, where an increase in one variable results in a proportional increase in the other variable. In a scatter plot, all data points lie on a straight line with a positive slope, rising from left to right, indicating that the values of the x-axis and y-axis are high or low together. Conversely, a correlation coefficient of −1 indicates a perfect negative correlation, where an increase in one variable leads to a proportional decrease in the other. In a scatter plot, all data points align on a straight line with a negative slope extending from left to right, indicating high values on the y-axis correspond to low values on the x-axis. A correlation coefficient of 0 indicates the absence of a linear relationship between the two variables, suggesting that changes in one variable do not predict changes in the other. The corresponding scatter plot will display a random distribution of points without any discernible pattern. A value closer to +1 indicates a strong positive correlation, meaning that as one variable increases, the other tends to increase as well. A value near −1 denotes a strong negative correlation, indicating that as one variable increases, the other tends to decrease. A correlation coefficient closer to 0 suggests a weak or no linear correlation, which signifies that there is no clear pattern in which the variables change together.

It is important to recognize that a correlation between two variables does not necessarily imply a cause-and-effect relationship. The observation that two variables just change at the same time does not mean one causes the other. Additionally, a weak correlation may still be statistically significant if the sample size is sufficiently large. By understanding correlation coefficients, one can gain valuable insights into the interrelations among various factors, which can be crucial in various fields like research, business, and social science.

Interpreting the Correlation from the Scatter Plot
The distribution of the dots in a scatter plot helps to interpret the nature and strength of the correlation between the two variables.

(a) *Positive Correlation*: When the dots show a general upward trend from left to right, it represents a positive correlation, as the X variable increases; the Y variable tends to increase (Fig. 11.9a).
(b) *Negative Correlation*: When the dots show a general downward trend from left to right, it represents a negative correlation as the X variable increases; the Y variable tends to decrease (Fig. 11.9b).
(c) *No Correlation*: When dots appear randomly scattered with no clear trend, this indicates a lack of significant linear correlation between the variables (Fig. 11.9c).

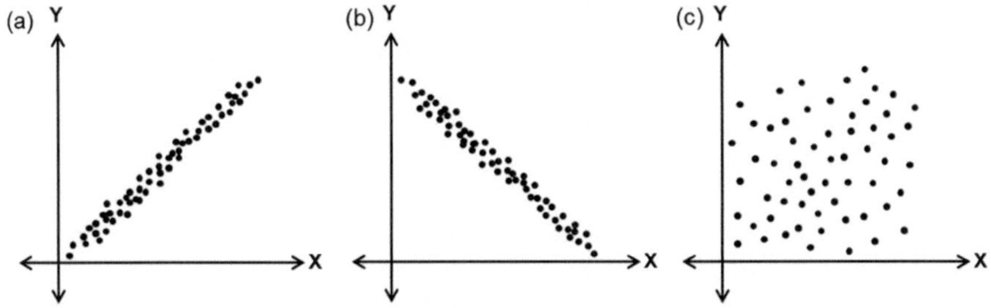

Fig. 11.9 Scatter plots showing (**a**) positive correlation, (**b**) negative correlation, and (**c**) no correlation

11.5 Analysis of Variance (ANOVA)

The analysis of variance (ANOVA) is a powerful statistical method used to validate the null hypothesis by comparing the means of more than two independent groups. It is one of the most common statistical tests used to check whether the variability within and among the groups is statistically significant (Turner and Thayer 2001). It helps determine whether observed differences between these groups result from chance or reflect genuine distinctions. In addition, it aids in designing experiments with appropriate sample sizes and controlling factors and supports stronger conclusions based on sound statistical evidence.

11.5.1 Concept of ANOVA

ANOVA does not directly compare group means; instead, it compares the variance between them. Variance indicates the spread of data points within each group. If the means are indeed the same across all groups, the data within each group are expected to be similarly spread out. However, if there are real differences between the means, the data points in each group will likely be spread out differently. The ANOVA procedure divides the variance in the data into two parts: variance between groups and variance within groups. ANOVA calculates an F-statistic, which compares the variance between groups (indicating potential differences) to the variance within groups (indicating randomness). If the F-statistic is large, it suggests that there is a significant difference between the group means. The p-value associated with the F-statistic is the probability of observing a test statistic as extreme or more extreme than the calculated one, assuming the null hypothesis is true. A small p-value (usually less than 0.05) indicates that the null hypothesis should be rejected, signifying a statistically significant difference between the means of the groups. A hypothesis test is performed to see if the F-statistic is large enough to reject the null hypothesis, which states that there are no significant differences between the group means.

11.5.2 Types of ANOVA

Depending upon the number of independent variables, ANOVA can be classified into different types:

One-way ANOVA The one-way ANOVA test involves the comparison of the means of *three or more groups* with one independent variable. Some of the assumptions of one-way ANOVA are independent and identically distributed data, homoscedasticity (the variances of the dependent variable should be equal across all groups)

Two-way ANOVA Two-way ANOVA compares the means of multiple groups with *two independent variables* and their interaction effect. A two-way ANOVA test is used to check the main effect of each independent variable and to determine whether any interaction effect between them exists. Each factor is considered separately, as done in a one-way ANOVA, to examine the main effect. Furthermore, to check the interaction effect, all factors are considered at the same time.

Certain assumptions need to be met for a two-way ANOVA test to be valid, including independent sampling from the population, an approximately normal distribution of the population, equal sample sizes for all groups, and equal population variances.

Nowadays, ANOVA analysis is performed through different software like SPSS, instead of the classical method, due to more accurate results in less time. SPSS offers a user-friendly interface for various types of ANOVA analysis, such as one-way ANOVA, two-way ANOVA, factorial ANOVA, and repeated measures ANOVA.

11.5.3 Steps to Perform One-Way ANOVA in SPSS

(i) *Data Entry*: To perform One-Way ANOVA in SPSS, the user should enter the data of the dependent variable and the independent variable (with group labels) in two different columns in SPSS software after opening the SPSS software.
(ii) *Access to one-way ANOVA*: Go to the Analyze tab, then click on the "Compare Means" option, and finally select the "One-Way ANOVA" option.
(iii) *Define Variables*: To define the variables, click Add to transfer the dependent variable to the "Dependent List" box and the independent variable to the "Factor" box.
(iv) *Options*: Click on the "Options" tab and select one or more desired statistics options, such as Descriptive statistics, Homogeneity of variance test, or Means plot, by right-ticking the corresponding box.
(v) *Post Hoc Tests*: To perform pairwise comparisons (e.g., Tukey, Bonferroni) between groups, click on the "Post Hoc" tab and select the specific analysis method as required.
(vi) *Continue and OK*: Click "Continue" to return to the main ANOVA dialogue box, then click "OK" to run the analysis.
(vii) *Display of Result*: The result will be displayed according to your pre-selection options. It may include descriptive statistics (mean, standard deviation, and sample size for each group), a homogeneity of variance test, an ANOVA table with the F-statistic, degrees of freedom, p-value, and post hoc tests.

In cytogenetics, ANOVA can be used to determine whether differences in TCL, ACL, etc. are significant across different germplasms, varieties, or landraces.

11.6 Scatter Plot (Scatter Diagram)

A scatter plot, also known as a scatter diagram, graphically illustrates the correlation or relationship between two variables. It consists of three main components: the X-axis, the Y-axis, and the data points. The X-axis typically represents one variable—often the independent variable—while the Y-axis depicts the other variable, usually the dependent variable. Each data point is represented by a dot on the graph, positioned according to the values of both variables (Fig. 11.9a).

The visual representation of dot patterns in scatter plots is essential for quickly assessing (i) the relationship between the two variables (Fig. 11.9a–c), (ii) the strength of that relationship (Fig. 11.10b, c), and (iii) the identification of outliers (Fig. 11.10a). The degree to which the dots cluster closely around a trend line indicates the strength of the correlation (Fig. 11.9a–c). Points that deviate significantly from the central cluster are considered outliers and may warrant further investigation (Fig. 11a).

It is crucial to recognize that scatter plots primarily capture linear correlations between two variables; hence, this method cannot detect non-linear relationships (such as U-shaped or curved patterns). Moreover, scatter plots are not designed to simultaneously represent interactions involving multiple variables.

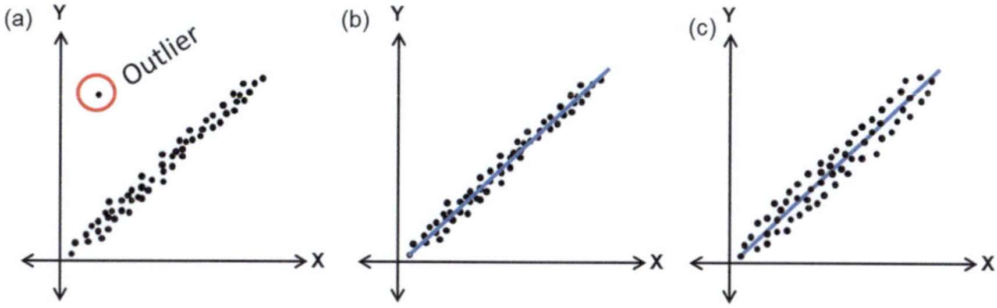

Fig. 11.10 (**a**) Detection of an outlier from scatter plot, (**b**) scatter plot showing strong correlation, and (**c**) scatter plot showing weak correlation

11.7 Cluster Analysis

Cluster analysis is an advanced multivariate statistical technique used to uncover hidden patterns or structures within exploratory data based on measured variables. It is particularly valuable for data visualization, prototyping, sampling, and segmentation, especially when the relationships between variables are not well understood. The primary aim of cluster analysis is to group similar data points, individuals, or observations into clusters based on shared characteristics. This method can result in either a single cluster or multiple clusters, which are distinct groups of similar data produced through the segmentation process.

Cluster analysis identifies underlying similarity patterns among individuals or data without prior knowledge or assumptions about relationships. It demonstrates high internal homogeneity within clusters and significant external heterogeneity between different clusters within the studied subjects or objects. However, it does not differentiate between relevant and irrelevant variables; thus, the selection of variables for inclusion in a cluster analysis should be guided by sound conceptual considerations.

11.7.1 Methods of Cluster Analysis

In today's landscape, various types of cluster analysis exist, each employing specific algorithms to organize datasets or taxa into distinct clusters. Once data has been collected, an appropriate distance or similarity metric is selected to quantify the resemblance between data points. It is crucial to choose a suitable type of cluster analysis based on the characteristics of the data, the number of clusters desired, the intended output (such as cluster shape), and the analytical objectives. Some of the most prevalent methods of cluster analysis include centroid-based (partition) clustering, model-based (distribution-based) clustering, density-based clustering, and hierarchical clustering, among others. A range of metrics is utilized to evaluate the quality of clustering results. It is essential to recognize that the choice of distance metric, the existence of outliers, and any missing values can significantly impact the outcomes of the clustering process.

Distance-based clustering remains a fundamental approach in which the similarity or dissimilarity between data points is measured using a distance metric. A pairwise distance matrix serves as input for analysis by a particular clustering algorithm (Johnson and Wichern 1992). Different clustering algorithms leverage this distance information to form clusters. Hierarchical clustering is the most common method for constructing a tree-like diagram known as a dendrogram, which visually represents the arrangement of clusters at varying levels of similarity or distance as data points are progressively grouped.

The biological sciences, particularly plant sciences, have greatly benefited from the application of phylogenetics. As a result, aspiring biologists must cultivate the skills and knowledge necessary to comprehend phylogenetic trees and their significance in modern evolutionary theory. Students will gain valuable insights by learning about the various cluster analysis methods and their applications within the biological sciences.

11.7.2 Software and Programming Languages for Cluster Analysis

- Displayer
- R (a popular open-source programming language and environment for statistical computing and data analysis)
- Python (a versatile programming language with a rich ecosystem of libraries and tools for data analysis and machine learning)
- MATLAB
- SPSS (a software package for statistical analysis)

11.7.3 Visualization of Cluster Analysis

Effective visualization of clustering is crucial for improving comprehension of the structure, separation, and characteristics of clusters derived from analyzed datasets (Sokal and Rohlf 1987; Scitovski et al. 2021). The choice of visualization techniques is influenced by the data volume and the clustering method employed. Common visualization methods include scatter plots, dendrograms, and heatmaps.

(i) *Scatter Plots*: Scatter plots are among the simplest methods for visualizing data points, particularly effective for 2D or 3D datasets. Each data point is represented as a dot, with different clusters distinguished by various colors or shapes based on cluster assignments. This presentation aids in interpreting group patterns and the separations between clusters.
(ii) *Heatmaps*: For high-dimensional data, heatmaps offer a color-coded matrix that illustrates the similarity or distance between data points and clusters. This

type of visualization is instrumental in identifying which data points are strongly associated with a particular cluster.

(iii) *Dendrograms*: Dendrograms are tree-like diagrams utilized for hierarchical clustering, illustrating the hierarchical relationships and clustering patterns among datasets (both non-biological data and various objects or taxa) based on their similarity (Fig. 11.11a). The term "dendrogram" derives from the Greek words "dendron," meaning tree, and "grámma," meaning drawing. Each branch of a dendrogram represents a category or class, with objects grouped together according to their similarities. Dendrograms can be rooted or unrooted, consisting of several branches and nodes. Each horizontal branch indicates a lineage, while the tips of the branches represent individual organisms, species, cultivars, or specific genes (Fig. 11.11a). Branches are interconnected by nodes (Fig. 11.11a), with each node symbolizing the last common ancestor of the two lineages descending from it. Internal branches or internodes connect pairs of nodes, while external branches link a tip to a node.

The vertical axis of the dendrogram represents the degree of similarity or distance between clusters. Longer vertical lines indicate significant separations between clusters, while shorter lines suggest less distinct separations, implying that the clusters are more similar. Dendrograms can be generated using various hierarchical clustering algorithms, such as agglomerative clustering (which begins with individual data points and merges them into clusters) and divisive clustering (which starts with a single cluster encompassing all data points and progressively divides it into smaller sub-clusters until each sub-cluster includes only one data point).

11.7.4 Types of Dendrograms

Based on the data types, dendrograms can be classified into three types: phenograms, cladograms, and phylograms.

- *Phenograms*: Phenograms are tree-like dendrograms created based on the overall degree of phenetic (phenotypic characters) similarity or dissimilarity to depict taxonomic relationships among taxa. During phenogram development, evolutionary history or the evolutionary significance of specific characters is not considered.
- *Cladograms*: Cladograms are tree-like specialized diagrams in evolutionary biology that highlight hypothetical relationships among different groups of taxa called "clades," based on their shared derived phenotypic or molecular characteristics. Cladograms can reconstruct the evolutionary history (phylogeny) of taxa. They represent only the branching pattern of evolutionary relationships among taxa. Cladograms are always unscaled, meaning the branch lengths do not reflect the evolutionary distance or evolutionary time between taxa or operational taxonomic units (Fig. 11.11b).

- *Phylograms*: Phylograms or phylogenetic trees are dendrograms that diagrammatically represent the evolutionary relationships among different species, organisms, or genes derived from a common ancestor (Baum and Offner 2008). They illustrate the branching pattern and the amount of evolutionary divergence (evolutionary time or distance). It should be remembered that phylogenetic trees provide hypothetical evolutionary relationships based on the available data and our present understanding of the data. The actual evolutionary history may not be identical to or different from these representations. In contrast to cladograms, phylograms are scaled (Fig. 11.11c), which means that branch lengths are proportional to the amount of evolutionary divergence.

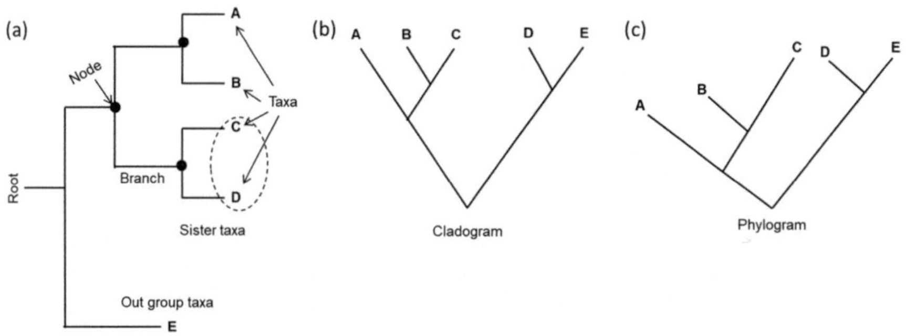

Fig. 11.11 (**a**) Different components of dendrogram, (**b**) cladogram, and (**c**) phylogram

11.7.5 Preparations of Dendrogram Based on Karyomorphological Features

A dendrogram can be constructed using various features, such as morphological, anatomical, karyomorphological, and molecular sequence data in plants. From a cytogenetic perspective, dendrograms of various plant taxa may be constructed based on karyomorphological features or specific coding and non-coding DNA sequences. Researchers have effectively constructed UPGMA dendrograms based on features like chromosome number and asymmetry indices, which have helped identify distinct clusters of plant species.

The following outlines the steps necessary to prepare a dendrogram based on karyomorphological features/DNA sequence:

(i) *Selection of plant sample*: Select the plant species, variety, or cultivar to be analyzed.
(ii) *Chromosome preparation*: Use classical or molecular chromosome techniques followed by appropriate staining methods to obtain metaphase plates having scattered, countable chromosome numbers with distinct morphology.
(iii) *Collection of data*: Different karyomorphological data, such as chromosome number, ACL, TCL, chromosome types, karyotype asymmetry indices, etc., were calculated from the documented metaphase plates for each plant sample.

(iv) *Data organization*: Organize the collected data into a matrix format, a table, or a spreadsheet. Each row represents a different plant species/cultivar, and each column represents a karyomorphological feature. If needed, preprocess the data to eliminate missing values, normalize scales, or standardize features to ensure comparability. To correct for multiple substitutions, it is common to use distances corrected by a model of molecular evolution, such as the Jukes-Cantor model (Jukes and Cantor 1969).

(v) *Calculate distance matrix*: Calculate the distance or similarity matrix among the species/cultivars based on their karyomorphological features or molecular data. Standard distance measures methods include Euclidean distance (suitable for numeric data), Jaccard index (for binary presence/absence data), Manhattan distance, Nei's genetic distance (for molecular markers), Pearson correlation (if applicable to the data), and other suitable metrics appropriate for the data type.

(vi) *Apply hierarchical clustering*: Select suitable hierarchical clustering algorithms such as the Unweighted Pair Group Method with Arithmetic Mean (UPGMA), neighbor-joining, Single-linkage clustering, and Complete-linkage clustering to group plant samples based on their calculated similarity distances. UPGMA and the neighbor-joining algorithms are among the most commonly utilized distance-based methods. Additional algorithms, such as single-linkage or complete-linkage clustering, may be employed depending on the specific characteristics of the data. Methods for constructing molecular phylogenies can be categorized as either distance-based or character-based. Distance-based methods condense all sequence information into a single metric, whereas character-based methods aim to infer phylogeny by analyzing individual characters (nucleotides or amino acids). Character-based techniques include parsimony, maximum likelihood, and Bayesian methods, while distance-based methods comprise UPGMA and neighbor-joining.

UPGMA (Unweighted Pair Group Method with Arithmetic Mean) is a foundational and widely utilized agglomerative hierarchical clustering algorithm that generates rooted dendrograms by clustering individual elements based on the average distances between clusters (Sneath and Sokal 1973). This method produces an ultrametric tree in which the distances from the root to all branch tips remain equal, operating under the molecular clock hypothesis, which assumes a constant rate of evolution. Consequently, the dendrogram reflects that divergence occurs at the same constant rate throughout the tree.

The algorithm commences with each element in its separate cluster and progressively merges them into larger clusters based on average distances. UPGMA is particularly prevalent in bioinformatics and phylogenetic studies. While it is characterized by its simplicity, speed, and relative insensitivity to outliers when compared to other clustering methods, a significant limitation arises from its assumption of uniform evolutionary rates across all lineages, which can potentially result in inaccurate tree topologies. Additionally, it generally shows greater resilience to outliers compared to single-linkage or complete-linkage methods, as it employs average distances to form clusters.

Neighbor-joining algorithm: The neighbor-joining algorithm, developed by Saitou and Nei in 1987, constructs a tree that accommodates varying evolutionary rates across different lineages without imposing a fixed root. Unlike the UPGMA method, which relies on pairwise distances, the neighbor-joining approach adjusts these distances by subtracting the distance to all other nodes from the pairwise distance. This adjustment effectively addresses situations in which the two closest nodes may not be immediate neighbors in the "true" tree. The neighbor-joining algorithm is widely recognized for its effectiveness and is commonly employed in phylogenetics, although modifications to its cubic time complexity can significantly enhance performance for larger datasets.

Single linkage: Clusters are formed based on the shortest distance between points in different clusters.

Complete linkage: Clusters are established by considering the longest distance between points in different clusters.

(vii) *Constructing the dendrogram*: Today, there are numerous software tools—both online and offline—available for constructing dendrograms. Prominent among these are commonly used programs such as R (e.g., the hclust function) and Python (via the SciPy library), as well as specialized software for phylogenetic studies such as MEGA (Molecular Evolutionary Genetics Analysis) or PAST. When creating a dendrogram, it is crucial to select the appropriate software, enter the distance matrix, and specify the clustering method. The resulting dendrogram will visually illustrate how cultivars cluster based on their karyomorphological or molecular similarities.

When choosing software for constructing dendrograms derived from karyomorphological data, it is important to consider several factors: the capability to input karyomorphological features (such as chromosome number, asymmetry indices, and chromosome lengths), support for distance calculation methods suited to karyomorphological data, flexibility in available clustering algorithms (such as UPGMA, single linkage, and complete linkage), a user-friendly interface with versatile data input options, visualization capabilities, and options for exporting the final dendrogram.

(viii) *Visualization*: Rectangular dendrograms are the most preferred form of visualizing the dendrogram to interpret relationships among plant samples. In this type of visualization, branches are drawn at right angles, either vertically or horizontally. Leaves at the bottom represent individual samples, while branches depict clusters at various levels of similarity. Other forms include triangular dendrograms and circular or radial dendrograms.

(ix) *Interpretation*: Analyze the dendrogram to identify clusters of similar cultivars based on their karyomorphological traits. The height at which branches merge indicates the level of similarity; shorter branches signify closer relationships. Determine the number of clusters by deciding where to "cut" the dendrogram horizontally. Determine the number of clusters by cutting the dendrogram at an appropriate height. Determine evolutionary relationships or diversity patterns by examining branch lengths and cluster arrangements.

References

Baum DA, Offner S (2008) Phylogenics and tree-thinking. Am Biol Teach 70(4):222–229

Elston RC, Johnson W (2008) Basic biostatistics for geneticists and epidemiologists: a practical approach. John

Gurumani N (2015) An introduction to biostatistics. MJP publisher

Johnson RA, Wichern DW (1992) Applied multivariate statistical analysis, 3rd edn. Prentice-Hall, Englewood Cliffs

Jukes TH, Cantor CR (1969) Evolution of protein molecules. In: Munro HN (ed) Mammalian protein metabolism. Academic, New York

Mackay I, Piepho HP, Garcia AA (2019) Statistical methods for plant breeding. In: Handbook of statistical genomics: two volume set, vol 29. Wiley, pp 501–520

Pagano M, Gauvreau K, Mattie H (2022) Principles of biostatistics. Chapman and Hall/CRC

Saitou N, Nei M (1987) The neighbor-joining method: a new method for reconstructing phylogenetic trees. Mol Biol Evol 4:406–425

Scitovski R, Sabo K, Martínez-Álvarez F, Ungar Š (2021) Cluster analysis and applications. Springer, Cham

Sneath PH, Sokal RR (1973) Unweighted pair group method with arithmetic mean. Numer Taxon 1973:230–234

Sokal RR, Rohlf FJ (1987) Biostatistics, vol 10. Francise & Co, New York, pp 2088–2092

Turner JR, Thayer JF (2001) Introduction to analysis of variance: design, analyis & interpretation. SAGE Publications Inc

12 Instruments

Abstract

This chapter provides a comprehensive overview of the essential instruments and concepts involved in plant chromosome analysis through microscopy. Microscopy serves as a critical scientific discipline, enabling the observation of microscopic objects that are beyond the resolution of the human eye. Due to its limitations in resolving fine structures, microscopy facilitates exploration of the micro-scale universe, significantly enhancing our understanding in various fields such as biology, medicine, and material science. The evolution of microscopy, beginning with simple magnifying glasses and progressing to complex optical devices, underscores humanity's quest for knowledge. Early pioneers like Hans and Zacharias Janssen, who invented the compound microscope, and Antonie van Leeuwenhoek, who first observed microorganisms, laid the groundwork for modern cytogenetics. Robert Hooke's discovery of cells using an improved compound microscope marked a significant milestone, further enhancing the functionality of these instruments through innovations in lens design and illumination techniques. The contributions of Carl Zeiss and Ernst Abbe in the nineteenth century led to advancements in optical quality, allowing for higher magnifications and improved resolution. As microscopy continues to evolve, it integrates state-of-the-art technologies, including digital imaging and computational tools, making it an indispensable tool for scientific research and discovery. This chapter not only outlines the technical aspects of these instruments but also emphasizes their broader significance in scientific inquiry and exploration.

12.1 Introduction

The foundational concepts and proper handling of essential instruments are integral components of a hands-on training program in plant chromosome analysis. Any cytogenetics laboratory should be equipped with a sufficient number of student microscopes. Ideally, the laboratory should include stereo and advanced bright-field microscopes with CCD cameras linked to dedicated computers. In addition, major and minor instruments and apparatus, such as refrigerators, incubators, precision

T. B. Jha, M. Halder, *Classical and Molecular Plant Cytogenetics*,
https://doi.org/10.1007/978-981-95-8265-5_12

balances, pH meters, and various smaller tools, are essential. To align with students' interests, this chapter incorporates conceptual insights into the operation of various key instruments.

12.2 Microscope and Microscopy

12.2.1 Introduction

The human eye is a remarkable optical device that enables us to see a vast spectrum of the world. However, it has inherent limitations for observing microscopic objects, such as bacteria, viruses, and nanoparticles, due to their small size. Even for visible objects and biological samples, the human eye cannot resolve structures smaller than about 0.1 mm (100 microns) due to its resolution limit (Fig. 12.1). Magnification is necessary to overcome these limitations and examine objects smaller than this threshold with precision. This is where microscopy is an essential tool that extends our vision and allows us to explore the fascinating microscopic world more comprehensively.

Microscopy is the scientific discipline investigating objects too small to be observed directly with the naked human eye, primarily using a microscope, with or without being equipped with various other instruments. Microscopes are optical devices used to examine and/or capture highly magnified, high-resolution images or videos of objects under inspection that the naked human eye cannot observe (Nechyporuk-Zloy 2022). The magnification and resolution power of the microscope have fundamentally reshaped our understanding of the world around us by revealing a previously invisible universe at the micro-scale. These microscopy capabilities have been critical to scientific progress in biology, medicine, and material science. Its applications extend beyond these core disciplines to areas such as forensics, environmental science, and archaeology. Nowadays, microscopy has become an indispensable tool for scientific exploration, discovery, and innovation.

12.2.2 Evolution of Microscope

The history of microscopy is a story of human curiosity and technological innovation. It starts with the desire to see beyond the limitations of our natural vision. The earliest whispers about the microscope date back to about 1000 AD, when it was used with polished lenses for enhanced magnification. The evolution of microscopes began with simple magnifying glasses and progressed to complex and powerful instruments that revolutionized the field of science. Like a magnifying glass, a simple microscope uses only one lens to refract light and produce fundamental magnification. In contrast to the single lens of a simple microscope, an optical compound microscope utilizes a system of two (objective lens and eyepiece lens) or more lenses to refract light and thus result in a magnified image. Credit was given to Dutch spectacle-makers Hans and Zacharias Janssen in 1590 for the invention of the first compound microscope, which marked a significant milestone in scientific discovery. However, it had a low resolution with a maximum magnification of 9×.

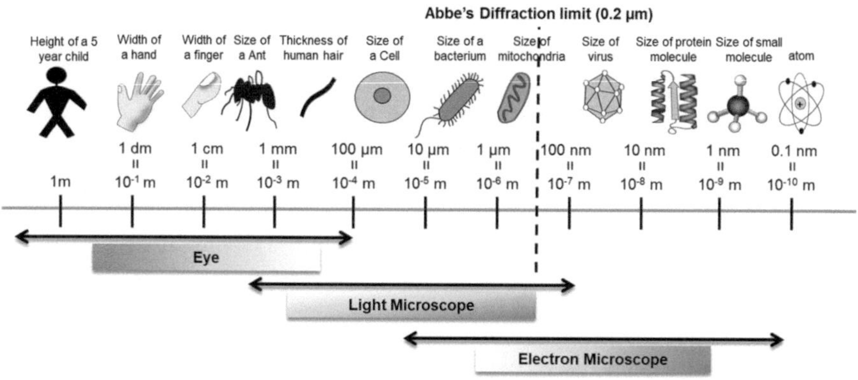

Fig. 12.1 Comparison of the resolving power of the human eye, light microscope, and electron microscope

Antonie van Leeuwenhoek (1670), a Dutch painter with a scientific mind, is considered a true pioneer in this arena. In the early seventeenth century, he was the first to observe bacteria, yeast, protozoa, muscle cells, and even sperm cells using his single-lens microscope, which reached a maximum magnification of 270×. His detailed observations challenged prevailing ideas about the origin of life and laid the foundations of modern microbiology.

Robert Hooke (1665), an English polymath of the seventeenth century, observed small, compartment-like structures in cork using a compound microscope and named them "cells". Hooke significantly improved the compound microscope by adding a stage to hold the specimen, an illuminator, and coarse- and fine-focus controls. His book, "Micrographia," published in 1665, presented his pioneering observations and illustrations made possible by the compound microscope. This discovery has opened a new scientific frontier for studying and understanding the universe at the microscopic scale.

Carl Zeiss (1816–1888) and Ernst Abbe (1840–1905) developed better lenses and introduced the substage condenser, which resulted in improved resolution and higher magnifications. Over time, different microscopes have been designed to meer various requirements and have evolved into more complex optical devices with multiple lenses, filters, polarizers, beam splitters, sensors, illumination sources, etc. These advanced microscopes have higher magnification and resolution that allow scientists to study the cellular structure of living organisms, investigate the composition of materials, and delve deeper into the mysteries of nature. Ultimately, it led to a transformative era of scientific discovery that forever changed our understanding of the microscopic world.

Max Knoll and Ernst Ruska (1931) invented the first electron microscope, a transmission electron microscope (TEM). They developed a new system that uses electron beams instead of light and electromagnets as lenses, replacing glass lenses. This technique is used to study topographical, morphological, compositional, and crystalline details of cells or other materials at the atomic level. In 1933, they built a TEM that surpassed the resolution of light microscopes, opening a new era of high-resolution imaging. The shorter wavelengths of electrons than light allow them to generate images with a thousand times greater resolution. Ernst Ruska was awarded the Nobel Prize in Physics in 1986 for his work on electron microscopy.

According to the literature, Manfred von Ardenne was credited with using a scanning electron beam to image a sample in 1937. Still, Ernst Ruska and Bodo von Borries developed the first operational scanning electron microscope (SEM) in 1937. The first commercial TEM was produced by Siemensin in 1939. The SEM provides detailed imaging of the surface topology of the sample, morphology, and composition.

12.2.3 Compound Microscope and Its Parts

A compound microscope is an optical microscope used to magnify objects to reveal invisible details to the naked eye. It is called a "compound" microscope because it consists of multiple lenses, i.e., an objective and an eyepiece. Depending on the application, this lens system works together to provide a magnified image of the object onto the eye or the sensor. The different components of the compound microscope are demonstrated in Fig. 12.2. The main components of this instrument can be broadly categorized into two groups: (a) mechanical components and (b) optic components.

(a) *Mechanical Components*: Mechanical components include base, arm, head, stage, mechanical stage, revolving nosepiece, and coarse and fine focus knobs. These parts provide the physical support and handling of the microscope.
   - *Base*: The base is the bottom part of the microscope (Fig. 12.2) that offers stability and support for the whole instrument.
   - *Arm*: The arm is a curved structure that connects the base to the head (Fig. 12.2) and provides structural support, as well as is used to carry the microscope.
   - *Head*: This is the upper portion that bears the eyepieces and the objective lenses (Fig. 12.2). It may be tilted or adjusted for optimum viewing comfort.

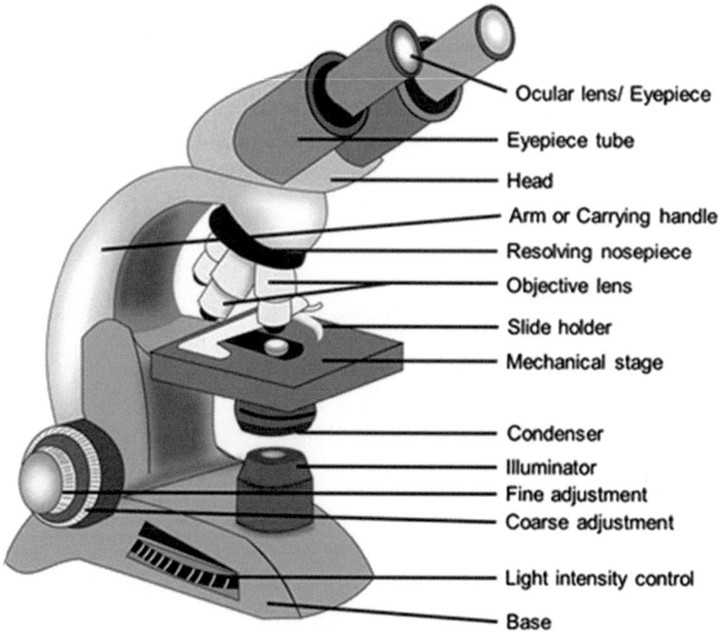

Fig. 12.2 Different parts of a compound microscope

- *Stage and Mechanical Stage*: The stage is a horizontal flat platform on which the prepared slide is placed for the observation of samples (Fig. 12.3). It is usually equipped with spring-loaded stage clips and an adjustable control system. It can be moved up and down to focus, and it has a hole in the center for light passage. Unlike a basic stage, which may require manual adjustment, a mechanical stage allows easy, precise movement of the slide by turning the knobs or levers. It has two perpendicular graduated scales that help users to record the position of objects on the slide. It is very useful for accurate positioning and quickly relocating an object.

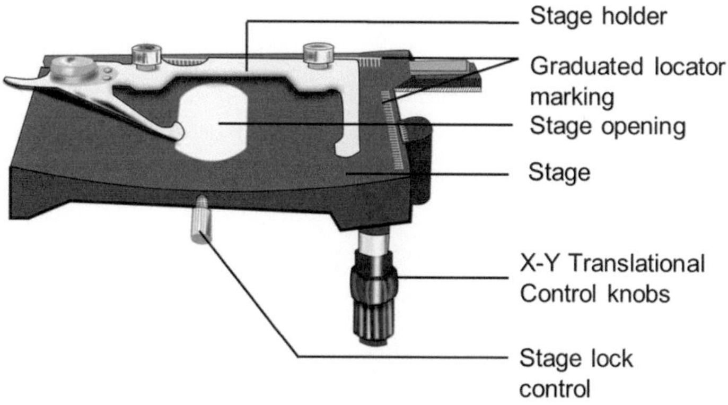

Fig. 12.3 Different components of stage and mechanical stage

- *Revolving Nosepiece or Turret*: The revolving nosepiece is a rotating mounting disc (Fig. 12.4) of modern microscopes. It is located at the lower part of the optical head of the microscope and holds multiple objective lenses with variable magnifications (Fig. 12.2). It enhances the functionality of the microscope by allowing the user to quickly and easily switching between different objective lenses by rotating them into alignment with the ocular lens.

Fig. 12.4 Resolving nosepiece

- *Coarse and Fine Focus Knobs*: The larger of these two knobs is called coarse knob and the smaller knob is are called fine focusing knob (Fig. 12.5). The coarse and fine focusing knobs are used to control the up-down movement of the stage and the objective lens to bring the sample into focus. A coarse adjustment is used for initial focus, while the fine-focusing knobs allow precise, and sharp focus. Special care should be taken when focusing an object under a 100X objective lens using coarse knobs, as this may damage the objective lens.

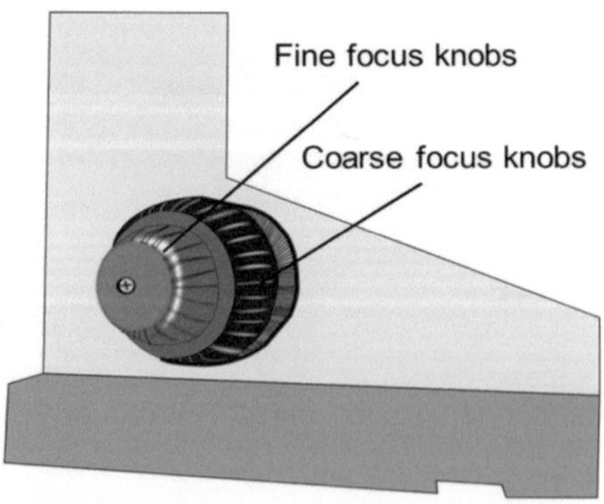

Fig. 12.5 Coarse and fine focus knobs

(b) *Optical Components*: The optical components manipulate the light to magnify the sample and improve the viewing experience.

- *Illuminator and Illumination System*: The illuminator and illumination system are fundamental components of a compound microscope. The effectiveness of objective and ocular lens systems is significantly influenced by the illumination conditions. The performance and quality of the magnified image produced by a microscope depend on the exposure of the sample to a suitable amount of light.

Compound microscopes often have a dedicated integral illuminator, which serves as the main light source for the microscope. Usually, the internal light source is located under the stage of the microscope. Advancements in technology have introduced various options for standard illuminators like halogen lamps, mercury-arc lamps, and light-emitting diodes (LEDs). Each type of illuminator has its advantages and disadvantages in terms of brightness, longevity, and heat production. This internal light source is essential for transmitting light through the sample, ensuring optimal clarity and detailed observation of the sample. The light intensity produced by these illuminators is crucial for optimizing image contrast and visibility. The light output from the illuminator can be regulated by an on- or off-switch. Furthermore, many microscopes have an adjustable knob (rheostat) on the side of the instrument. This knob allows the user to precisely control the intensity of light.

The illumination system encompasses the entire arrangement of components that work together to direct light onto the specimen. This system typically includes a combination of lenses, filters, and mirrors, all of which contribute to the quality of illumination. Selecting the appropriate illumination system for the microscope is crucial to ensure that the sample is exposed to sufficient light. The light source (bulb) must be centered on the sample to provide sufficient light to achieve this condition. Overexposed or insufficient light intensity both affect the image quality of the microscope. The appropriate amount of light should be decided by considering the application setup, the object being observed, and the desired results. For example, the magnification and resolving power of the lens also affect the amount of light required to view the sample.

The Koehler illumination method is a backlight illumination system that provides a bright and uniform illumination of the object and the image plane using two convex lenses: a collector and a condenser. Backlight illumination, also known as bright field illumination, is where the light source is situated behind the object, and the incident light from the light bulb floods the object being inspected. When observed under a microscope, samples with variability in opacity create a contrast between the different parts of the object and the light source. Dark-field illumination is more effective at distinguishing features in extremely transparent objects than backlight illumination due to their transparency. In contrast to bright-field illumination, dark-field illumination setups uses an optical system that prevents direct entry of the center of the light rays, creating a hollow inverted cone of light, and forces the oblique rays to illuminate the object in the object plane. It creates a high contrast between the transparent object and the light source, which makes it a better system for studying transparent objects. Epi-illumination is a valuable technique used for fluorescence imaging. In the epi-illumination technique, the light source is placed above the objective, and the light emitted by the source travels down the microscope's optical system and finally reaches the sample from above. The light emitted by the sample is collected through the objective lens located above the sample.

- *Mirror*: A mirror is typically located beneath the condenser and mounted on an adjustable arm (Figs. 12.2 and 12.6). It is used in a microscope to reflect a sufficient amount of natural or artificial external light from an external source upward through the condenser and finally onto the sample being observed by adjusting its angle. It plays an essential role, particularly in microscopes without built-in light sources.

Fig. 12.6 Mirror of compound microscope

- *Eyepiece*: The eyepiece is located on the top of the body tube of the microscope, near the eye or the sensor (Fig. 12.2). Each eyepiece contains an ocular lens that further magnifies the image produced by the objective lens and projects a virtual image of the sample. Generally, this lens has magnifications of 5×, 10×, or 15× (Fig. 12.7).

Fig. 12.7 Different types of eyepieces

- *Objective Lenses*: In a compound microscope, the set of objective lenses is usually located on the rotating nosepiece closest to the sample at the opposite end of the body tube (Fig. 12.2). These objective lenses are vital to initial magnification in a compound microscope. It also produces a real image of the object and transfers it to the eyepiece. Multiple objective lenses (Fig. 12.8) with different magnification powers (ranging from 4× to 100×) are used in compound microscopes. Ordinary objective lenses have magnification strengths of 4×, 10× (low power), 40× (high power), and 100× (oil-immersion lens). Most bright-field light microscopes have three objective lenses mounted on a rotating nosepiece base.

Fig. 12.8 Objective lens

They are the most complex component due to their multi-element design. Objective lenses come in two main types: refractive and reflective. They are further subdivided into finite and infinite conjugate types. Several specifications of the objective lens are shown on the body of the objective itself. It includes the objective design/standard, magnification, numerical aperture, working distance, lens-to-image distance, coverslip thickness correction, etc. Understanding each category and type is essential for selecting the right objective lens for a specific application.

Oil-immersion lenses are specialized objective lenses of compound microscopes with delicate glass surfaces and sensitive coatings. It is used in compound microscopes to produce highly magnified images with better resolution. However, high-magnification objective lenses have a higher numerical aperture than other objectives, which compromises image resolution by reducing light transmission through the objective lens. A significant difference in the refractive indices of air and glass causes the scattering of light rays in the air before it reaches the objective lens. This affects the image's resolution. To overcome this optical limitation and optimize the performance of the compound microscope, immersion oil is used that eliminates the air gap between the lens and cover glass by reducing the refractive index mismatch between the microscope slide, the sample, and the objective lens (Fig. 12.9). It minimizes the scattering and refraction of light (Fig. 12.9). Immersion oil is a transparent liquid with a refractive index very close to or similar to that of glass. It significantly increases the maximum angle at which light can strike the lens, dramatically increasing the amount of light collected and the camera's resolution. It substantially improves the quality, resolution, and clarity of high-magnification microscopic images by increasing light transmission, minimizing light refraction, reducing optical aberrations, enabling the use of an objective lens with a higher numerical aperture, enhancing contrast, and improving depth of field.

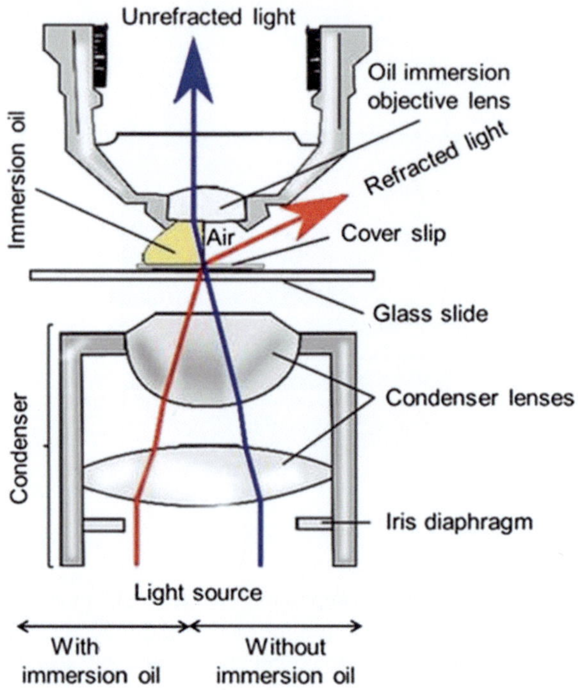

Fig. 12.9 Working principle of oil immersion lens

Careful attention to the compatibility of the objective lens with oil immersion and to the quality of immersion oil is a prerequisite for optimal performance and reliable observations under an oil-immersion objective. Use only high-quality immersion oil, as recommended. Any immersion oil with a refractive index significantly different from the one specified for the microscope system may distort the image and compromise its overall quality. Moreover, careful handling, proper and regular maintenance, and cleaning of the oil-immersion lens are essential for its longevity and optimal performance. Rough handling may damage the lens's sensitive coatings and alter its optical properties. Regular cleaning of the lens is essential to prevent contamination by residual oil, dust, debris, or microorganisms. These contaminants may interfere with the optical path, reduce light transmission, and significantly affect image quality. Precautions such as avoiding direct contact with the lens, using gentle, even pressure, and using lens paper, a clean, lint-free cloth, or the manufacturer's cleaning solutions may help prevent scratches or damage to the oil immersion lens during cleaning.

- *Condenser*: A condenser (Fig. 12.10) is a lens system located below the stage that enhances the illumination of the sample by concentrating parallel beams of light from the light source onto the sample. It ultimately improves the image quality. The condenser may move up and down to influence this focus.

Fig. 12.10 Condenser of compound microscope

- *Iris Diaphragm*: The iris diaphragm is within the condenser. It regulates the amount of light that reaches the condenser lens system and the sample by adjusting a small lever, which helps to optimize the contrast and resolution of the image.

Steps for Using Oil Immersion Lens in Compound Microscope

The item being observed is referred to as a specimen. The specimen is placed on a glass slide, which is then secured onto the stage, a platform in the microscope. After the slide is in place, the specimen is adjusted over the light source using the x-y mechanical stage knobs. Once the specimen is centered over the light, the entire stage can be adjusted up or down to bring the image into focus. The coarse-focusing knob is used for significant adjustments with 10X objective lenses, while the fine focusing knob is used for precise adjustments, particularly with 40X or 100X objective lenses.

(i) Start with lower magnification objectives (such as 40× or 10×) to locate and focus on the specimen.
(ii) Locate the area or region of the specimen of interest you want to observe and ensure it is well-focused at the lower magnification using the coarse and fine focus adjustments.
(iii) Rotate the nosepiece to switch to an oil immersion objective lens and refocus the specimen using the fine focus adjustment to achieve a clear image.
(iv) Apply a small drop of immersion oil without an air bubble on the coverslip.

(v) Carefully rotate and slowly lower the oil immersion lens into position so it is in direct contact with the oil and immersed in it. This allows the oil to spread evenly. Any objective lens not marked "oil" should not be used with oil.
(vi) Use the fine focus adjustment to bring the specimen into sharp focus.
(vii) Once the specimen is in focus, observe the specimen at high magnification.
(viii) Clean the microscope by rotating the nosepiece to switch back to a lower magnification objective and wiping excess immersion oil from the microscope stage, objective lenses, and other components using lens paper or a clean, lint-free cloth.

12.2.4 Numerical Aperture (NA)

The numerical aperture of a microscope objective represents its light-gathering ability. It is a function of the focal length and the diameter of the entrance pupil. NA directly affects the resolution, brightness (light capture), and contrast of microscopic observations and images. It is a dimensionless number determined by the design and characteristics of the objective lens. The numerical aperture is calculated by the formula: $NA = n \cdot \sin(\mu)$, where NA is the numerical aperture, n is the refractive index of the medium between the lens and the sample (usually air, so $n \approx 1$ or immersion oil, $n \approx 1.52$), and μ is the half-angle of the maximum cone of light that can enter or exit the lens (Fig. 12.11).

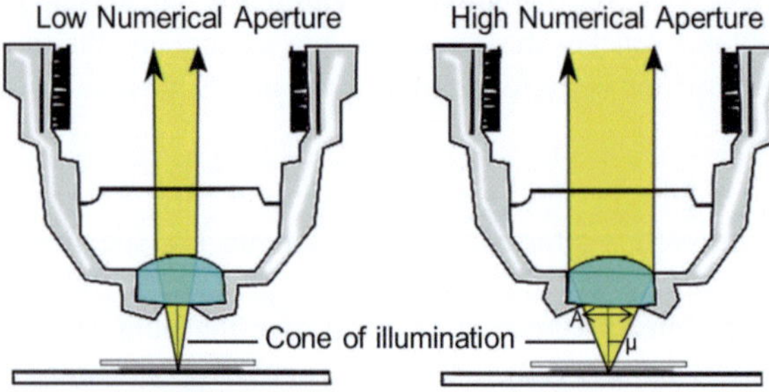

Fig. 12.11 Low and high numerical aperture of objective lens

12.2.5 Magnification of Microscope

Magnification and resolution are the two most important fundamental features of microscopes, which reveal the minute, invisible details of samples. The ability to enlarge an image beyond its original size is called magnification in a microscope. It has no unit, as it is expressed as a ratio indicating how many times larger the image appears than the actual size of the sample. Magnification is usually denoted by an "X" next to the numerical value.

The compound microscope usually has three or four objective lenses with variable magnification and resolving power. The magnification strengths of the objective and ocular lenses may vary from 4× to 100× and 10× to 15×, respectively. Most objectives contain a colored band around the entire circumference of the body that indicates their magnification (Fig. 12.8). For example, the yellow band denotes an objective with 10X magnification. In a compound microscope, the objective lens provides the initial magnification, and the eyepieces or ocular lenses add further magnification to the image. Thus, the microscope's total magnification is determined by multiplying the magnification of the eyepiece (ocular lens) by the magnification of the objective lens.

Total magnification = Magnification of objective lens × Magnification of ocular lens (eyepiece)

For example, if the eyepiece has a magnification of 10× and the objective lens has a magnification of 40×, the total magnification is 400×. The maximum magnification achieved by a compound binocular microscope is 1500× (15× eyepiece and 100× objective).

It should be remembered that just increasing the magnification alone does not necessarily improve the quality of the image. This may result in blurred or indistinct details in the magnified image. In addition, resolution, contrast, and numerical aperture are crucial factors in achieving optimal clarity and detail in magnified images. The area visible through a microscope during a given observation is termed the field of view. The relationship between magnification and field of view is negatively proportional. Higher magnification provides a narrow field of view, while lower magnification offers a broader field of view. The choice of magnification depends on the sample and the desired level of detail.

Digital Magnification In addition to optical magnification, some microscopes have digital magnification options, in which the camera captures the image and zooms it further through software. However, digital magnifications do not enhance optical resolution. While some microscopes (stereo microscopes) have zoom capability without changing the objective lens, zoom capability enables quick magnification adjustment without changing the objective lens.

12.2.6 Resolution of Microscope

The advantage of the microscope lies not only in its ability to magnify, as a highly magnified image may appear blurred without sufficient resolution if closely located points are not distinguishable from each other. The resolution of a microscope is its ability to differentiate two closely adjacent points as separate and distinct entities (Fig. 12.12). In other words, the closest or shortest distance between two points can be resolved as separate entities under the microscope. The human eye has a resolution of about 0.1 mm, or 100 microns, meaning that if two objects are closer than this distance, they will not be seen clearly with the naked eye. Higher-resolution microscopy enables clearer visualization and more precise differentiation of finer details in the samples. The resolving power of different types of microscopes varies significantly. A standard compound microscope is limited to a resolution of approximately 0.2 μm (200 nm). In contrast, both Scanning Electron Microscopes (SEM) and Transmission Electron Microscopes (TEM) offer substantially higher resolution. SEMs can resolve features to about 10 nm, while TEMs achieve even greater resolution, reaching approximately 0.2 nm.

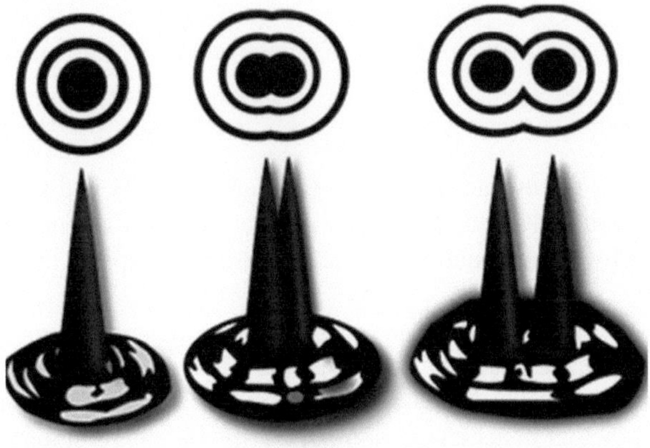

Fig. 12.12 Concept of resolution

The microscope's resolution is influenced by the physics of light and by practical limitations. Ernst Abbe recognized that the sample images contained overlapping, multiple-intensity, diffraction-limited points (or Airy discs). Ernst Abbe's formula $R = \lambda/(2 * \mathrm{NA})$ states that the resolution (R) of optical microscopy is directly proportional to the wavelength of light (λ) used for illumination and the numerical aperture (NA) of the optical system. According to this formula, improving by microscope resolution can be achieved by increasing the numerical aperture values. To achieve the maximum theoretical resolution, the microscope system should use all available optical components with the highest possible NA (a high refractive index imaging medium), shorter wavelength light, and correct system alignment. Additionally, high-quality lenses and optical components in microscopes, along with specific techniques like phase contrast, DIC, and fluorescence microscopy enhance contrast and detail visibility and reduce optical aberrations, significantly impacting resolution. Although resolution and magnification are not directly correlated, magnification increases the visibility of finer details and contributes to the perception of higher resolution. The lateral resolution of the optical microscope is limited to around 200 nm, whereas the axial resolution is around 500 nm.

For example, a compound microscope can achieve a maximum resolution of 203 nm using the shortest visible light wavelength of 400 nm, an oil-immersion objective with an NA of 1.45, and a condenser with an NA of 0.95. The maximum resolution of a standard bright-field compound light microscope is 0.2 μm or 200 nm, which means it can distinguish two points of ≥0.2 μm; any object closer than 0.2 μm appears as a single object. Therefore, the resolution limitation is 0.2 μm or 200 nm for the compound microscope.

There are two main aspects of microscope resolution:

(i) *Lateral resolution*: Lateral resolution refers to the smallest or minimum distance between two points in the same plane that can be distinguished. Diffraction of light primarily limits it. In general, light microscopes have a lateral resolution of approximately 200 nanometers. It can be calculated using the formula: Lateral resolution ≈ $\lambda/(2 * \text{NA})$, where λ represents the wavelength of the light (usually blue or violet light gives better resolution) and NA is the numerical aperture of the objective lens.

(ii) *Axial resolution*: Axial resolution refers to the smallest or minimum distance between two points within different planes that can be distinguished. This resolution is affected by the specimen's thickness and the properties of the objective lens and light source. Typical axial resolutions are around 500 nm.

12.2.7 Bright Field Microscopy

Bright field microscopes are one of the most fundamental and frequently used instruments in light microscopy that use two or more lenses to produce a dark image against a bright background. This is achieved by directly illuminating the samples with either natural or artificial light. Different structures within the sample differentially transmit, absorb, and reflect illuminating light to create contrast. Denser regions of the sample absorb more light and appear darker, while less dense areas allow light to travel more freely, giving a brighter appearance. Applying stains or dyes (artificial chromophores) during sample preparation often increases this inherent contrast. It allows for better visualization of complex sample structures that are differentially colored. Bright field microscopes are available in monocular (single eyepiece) and binocular (double eyepiece) configurations.

Bright-field microscopy requires intense illumination, usually provided by a high-intensity bulb beneath the microscope stage, to achieve high magnification. The light from this illuminator is transmitted through a condenser lens under the stage, focusing all light rays onto the sample to maximize illumination. The attached condenser focus knob may change the condenser's position. Once the optimum distance has been set, the condenser should not be moved to adjust brightness. If light levels are below optimal, the amount of light reaching the sample can be easily adjusted by opening or closing the diaphragm between the condenser and the sample. In some microscopes, a rheostat (dimmer switch) provides an alternative way to control the illuminator's intensity.

This bright field microscope's simplicity and user-friendly nature make it an ideal instrument in scientific research and education to observe a wide range of biological samples. It allows scientists to assess the fundamental structures of cells, tissues, and microorganisms. Additionally, it is used to diagnose and identify pathogens, such as bacteria and fungi, in patient samples. In terms of cost-effectiveness, a compound microscope is slightly more expensive than a simple microscope. However, due to its high magnification power, it is widely used in biological laboratories for studying cultures and individual cells.

12.2.8 Dark Field Microscopy

Dark-field microscopy is another technique of light microscopy that employs a specialized modified condenser to block direct illumination from reaching the objective lens. This technique involves placing a small opaque disk (about 1 cm in diameter) between the illuminator and the condenser lens to block most of the light from the illuminator, creating a hollow cone focused on the specimen. Only the specimen's scattered light (refracted or reflected) enters the objective lens. It makes an image that typically shows bright objects on a dark background.

Dark-field microscopy makes it helpful in visualizing transparent or unstained specimens or live organisms like bacteria or studying delicate structures within cells. This creates a dramatic effect by making unstained objects visible against the dark backdrop due to the scattered light (Fig. 12.13). With bright-field microscopy, the technique can highlight details within cells that might not be visible, allowing us to zoom in on specific features within unstained samples. Furthermore, dark-field microscopy is non-invasive, meaning it does not require staining or other sample preparation techniques that can alter the structure or behavior of the sample. In general, bright-field microscopy provides a general roadmap of the cellular landscape. In contrast, dark field microscopy allows us to create high-contrast, high-resolution images, zoom in on specific features within unstained specimens or samples, and offer a more nuanced view of the microscopic world.

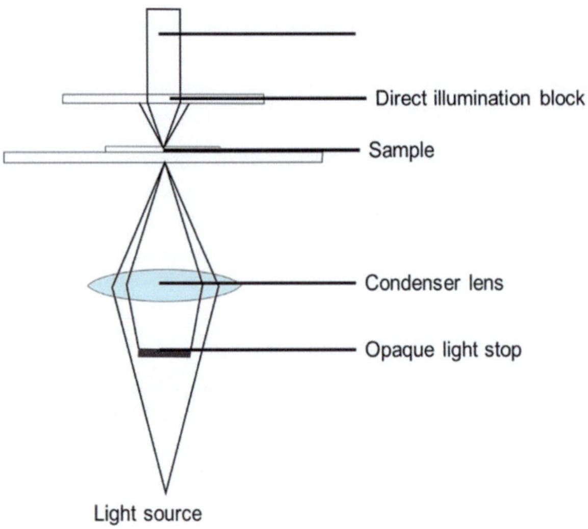

Fig. 12.13 This diagram of the light pathway of a dark field microscope

12.2.9 Phase-Contrast Microscopy

Phase-contrast microscopy allows scientists to examine cells and their processes in their natural state without killing and staining. Unlike bright field microscopy, this technique does not rely solely on differences in light absorption for contrast.

Variations in refraction and interference across a specimen's structures create subtle phase shifts in light waves, which are invisible to the naked eye. This technique uses an annular stop, a phase ring, and a condenser in the light path to deliberately convert small phase differences/shifts in light waves into visible intensity variations that create high-contrast, high-resolution images (Fig. 12.14).

The annular stop produces a hollow cone of light, focused on the specimen before it reaches the objective lens. The objective contains a phase plate with a phase ring, which causes waves travelling through the ring to be about one-half of a wavelength out of phase with those passing through the plate (Fig. 12.14). This results in structures that refract light, appearing dark against a bright background of only unrefracted light, producing high-contrast, high-resolution images.

This technique is perfect for studying cellular dynamics and processes, such as cell division or movement, in real time in living cells in their natural state, as it does not affect cell viability compared to bright field microscopy. It provides enhanced contrast for transparent specimens, which allows detailed visualization of internal organelles like mitochondria, vacuoles, and the nucleus.

However, like any tool, phase-contrast microscopy has its limitations. It works best with relatively thin specimens with variations in refractive index. Phase-contrast microscopy may produce blurred images due to different light interactions in thick samples with various depths, creating different phase shifts in areas above and below the focal plane. In addition, samples with very high or very low refractive indices compared to the surrounding medium can disrupt the phase shift calculations and lead to inaccurate image representation. Besides this, the phase-shift calculation process itself can introduce artifacts. Although phase-contrast microscopy offers better resolution than bright-field microscopy, it falls short compared to electron microscopy. It is slightly more expensive and complex to set up, handling specialized components require training and experience in data interpretation due to indirect visualization of cellular components.

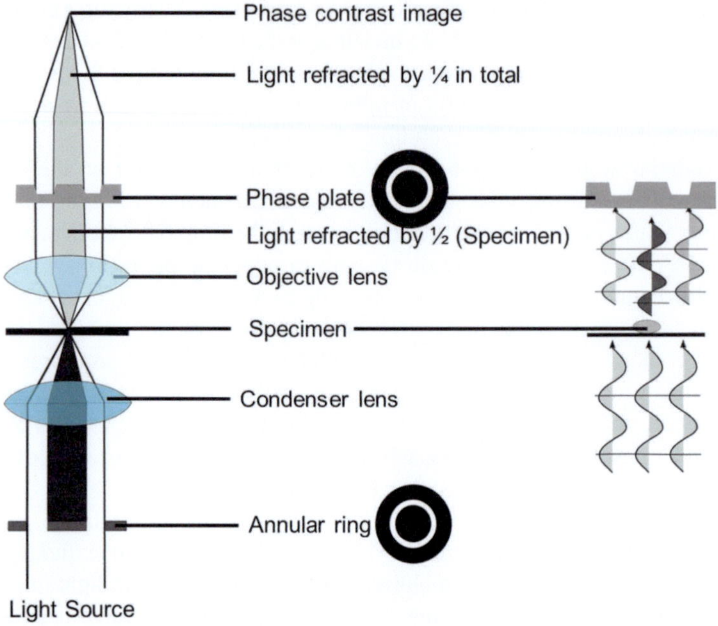

Fig. 12.14 Diagram of a phase-contrast microscope illustrates phase differences between light passing through the object and the background, which are enhanced by passing the light rays through different parts of a phase plate to create contrast

Despite these limitations, its advantages and diverse applications make phase-contrast microscopy a crucial tool for researchers in various biological fields, including cell biology (to study the morphology and dynamics of living cells, including cell division, organelle movement, and intracellular interactions), microbiology (for observing and identifying live bacteria, viruses, and other microorganisms without staining), parasitology (to examine the morphology and life cycle of parasites within host cells), and developmental biology (for tracking the development of embryos and observing cellular differentiation processes).

12.2.10 Fluorescence Microscopy

Fluorescence microscopy is a revolutionary imaging technique exploiting the specific properties of fluorescent chromophores, also known as fluorophores or fluorochromes, to visualize biological samples (Heit 2022). Unlike traditional microscopy, which relies on the natural contrast in the sample, fluorescence microscopy offers unmatched specificity, as only the fluorophore-tagged molecules will emit fluorescence, and remarkable sensitivity (it can detect a small amount of fluorescence) achieved through the use of specific fluorophore-conjugated antibodies or fluorescence dyes that selectively bind to target molecules in cells or tissues.

This imaging technique usually relies on high-intensity xenon arc or mercury-vapor lamps to excite the fluorophores. The excitation filter between the light source and the sample allows only specific wavelengths of light to pass through and excite the fluorophores. Fluorescent molecules or chromophores absorb light of specific shorter wavelengths (excitation wavelengths, generally ultraviolet or blue light) and then emit visible light of longer wavelengths (lower energy), a phenomenon called fluorescence. Fluorochromes include natural fluorescent substances (such as chlorophyll) and fluorescent stains (such as Texas red, 4′,6′-diamidino-2-phenylindole (DAPI), acridine orange, and FITC). The emitted fluorescence is transmitted to the detectors through the dichroic mirror and emission filter (Fig. 12.15). A dichroic mirror is a specialized mirror that reflects selected excitation wavelengths towards the sample while simultaneously transmitting the longer wavelengths emitted by the sample to the detector. The emission filter, located between the dichroic mirror and the detector, serves as the final checkpoint. It effectively blocks any residual excitation light or background noise, allowing only the desired fluorescence emission wavelengths to pass through the ocular lens. Modern fluorescence microscopes use high-sensitivity cameras (CCD or photomultiplier tubes) to capture the fluorescent light emitted (Heit 2022). These detectors convert light signals into an electronic signal that is then processed to produce digital images. The computer system processes and analyzes the captured image data using specialized software that allows brightness, contrast, and color adjustment to optimize the final image for visualization and analysis. This produces a picture of the specimen in bright colors against a dark background.

Fluorescence microscopy offers several advantages over traditional techniques, including simultaneous visualization of multiple targets using different colored fluorophores, detection of small quantities of fluorophores, and visualization of dynamic cellular processes in living cells. Unlike fixation techniques, targeted fluorescent tagging of specific molecules of interest in living cells allows researchers to visualize and study complex dynamic cellular behavior, and processes and to track specific molecules, organelles, and even pathogens.

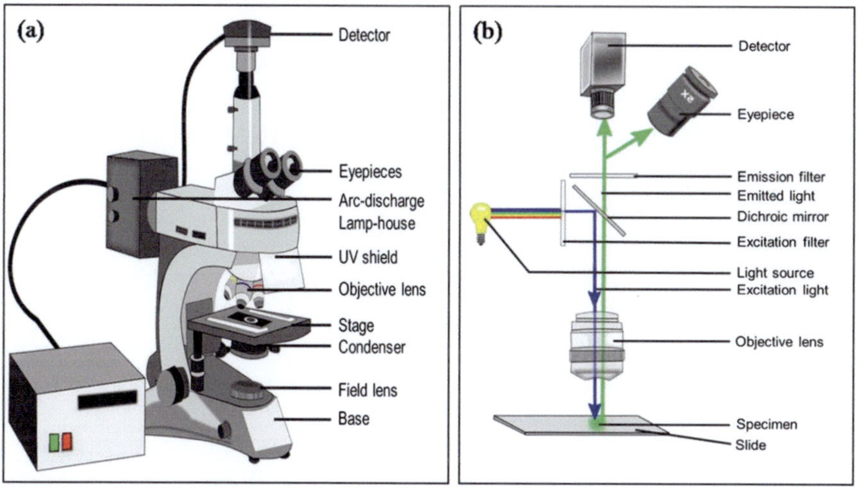

Fig. 12.15 Diagram of a fluorescence microscope demonstrating (**a**) basic components and (**b**) light path

It is an essential tool in several key applications, including cellular localization studies that help to understand the localization (identification of the distribution of specific proteins within a cell) and the interaction (interaction of particular proteins with other biomolecules) of a desired molecule by visualizing the co-localization of the molecules, gene expression visualization study of specific genes by tagging their mRNA or protein products with fluorophores, identification of specific cell types within the complex cellular environment by labeling desired cells with fluorophores, distinguishing living from dead cells, and cell-cell interaction study via labeling different cell types with unique fluorophores. Drug discovery monitors the effects of potential drugs on cellular processes by visualizing their interactions with specific molecules.

However, fluorescence microscopy has limitations, such as photo bleaching (continued exposure to light can damage fluorophores, leading to a loss of signal over time) and the need for careful planning and experimentation to select the appropriate fluorophore for specific targeting. Moreover, fluorescence microscopy is more expensive than traditional microscopy methods due to specialized equipment and techniques. Despite this, recent advancements in fluorophore development and imaging technologies promise to solve more complex biological mysteries in the future. Thus, it makesing fluorescence microscopy an invaluable tool for researchers in understanding the intricacies of cellular processes, especially in real time scenario of any dynamic processes (Stockert and Blázquez-Castro 2017).

12.2.11 Stereo Microscopy

Stereo microscopy (Fig. 12.16), also known as dissecting microscopy, provides a unique perspective in the field of microscopy. While other light microscopes focuse on the high magnification of thin samples, stereo microscopy provides a magnified three-dimensional (3D) view of larger specimens. This is achieved through a dual-objective lens system that replicates human binocular vision. In this technique, each objective lens captures a slightly different view of the specimen, and then they are merged by the eyepieces to create a perceptual depth in the final image that allows viewers to visualize the magnified 3D structure of the sample, including its depth, height, and spatial relationships between different parts. Different components of the stereo microscope are represented in Fig. 12.16.

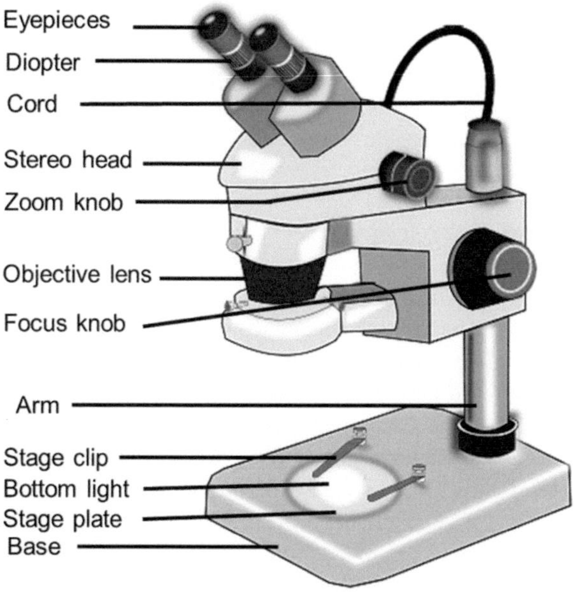

Fig. 12.16 Diagram of stereo microscope showing different components of it

Stereo microscopy offers several advantages over other microscopy techniques. These include a 3D view of larger specimens, a significant working distance between the objective lens and the sample, and nondestructive observation of specimens without extensive sample preparation. Despite its limitations in resolution, stereo microscopy remains the preferred choice for high-resolution, detailed, three-dimensional visualization of surface features and the topography of larger specimens. The extensive working distance facilitates dissections and micromanipulations, making it ideal for surgical and microdissection tasks. Notably, this technique enables nondestructive observation of specimens, minimizing the need for fixation or staining. This advantage opens doors to the study of the morphology and behavior of living organisms in three dimensions.

Stereo microscopy remains indispensable for researchers and professionals across diverse disciplines, including biology, forensics, and microelectronics. Its varied applications in various fields of biology highlight its importance for understanding the biological world in three dimensions. It is used for examining objects too large for high-power magnification, dissecting small animals or plant parts, studying plant morphology, and analyzing small fossils. Zoologists also use it to identify invertebrates and insects, botanists to analyze plant structures and in vitro cultures, geologists to examine fossils, rock formations, and mineral textures, and forensics to examine physical evidence. Additionally, in microelectronics, it is used to inspect circuit boards and components for defects. However, it has limitations such as lower resolution compared to high-power light microscopes and a shallow depth of field. It cannot resolve fine details of cellular structures. Due to its limited field depth compared to other techniques, this technique restricts focus to a specific portion of the specimen. Consequently, researchers must meticulously refocus at multiple depths within the thick sample for detailed, high-quality image. Moreover, it might not be suitable for advanced imaging techniques like specialized fluorescence or phase contrast. As we continue to explore the complexities of the world around us, stereo microscopy will remain a reliable companion, offering a clear and comprehensive view of the bigger picture.

12.2.12 Confocal Microscopy

Confocal microscopy enhances optical resolution and contrast in micrographs to a new level by using point illumination using a laser beam of light and a pinhole aperture (Fig. 12.17). Unlike traditional techniques that illuminate the entire sample simultaneously, a focused laser beam excites fluorophores at a specific point on a specimen labeled with a dye, causing them to emit fluorescent light. A pinhole aperture in front of the detector blocks out-of-focus light from other parts of the thicker specimens, allowing only fluorescence from the focal plane to reach the detector. Confocal microscopes are capable of collecting double and triple flurescent labels, which enables precise detection of co-localization of fluorescent markers. A complete, high-resolution, three-dimensional image is built up layer by layer based on the point-by-point scanning information.

The uniqueness of confocal microscopy includes its ability to acquire optical sections of a specimen by selectively illuminating a single plane at a time and blocking out-of-focus light (Fig. 12.17). This overcomes the blurry image-related limitations of conventional fluorescence microscopy and produces high-resolution images. By acquiring multiple optical sections, confocal microscopy allows for the reconstruction of a high-resolution 3D image of the specimen, providing a more comprehensive understanding of cellular architecture. The pinhole aperture significantly reduces background noise from out-of-focus areas, resulting in a crisp and clear image with improved contrast between the target fluorophores and the background. Confocal microscopy can image live cells by employing specific fluorophores that are non-toxic to the cell. This allows for the visualization of dynamic cellular processes in real time. Confocal microscopy uses a more focused beam than conventional fluorescence, minimizing photo bleaching (damage to fluorophores from light exposure). This allows for longer observation times of live cells.

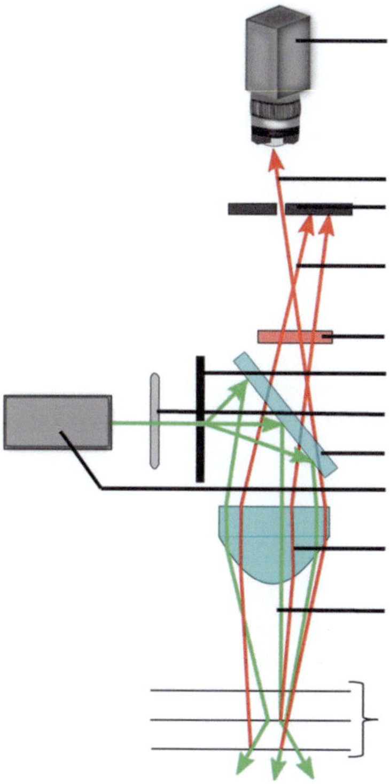

Fig. 12.17 Diagram of basic components of a confocal laser scanning microscope and light path through it

However, confocal microscopes are more expensive than traditional fluorescence microscopes due to complex optical components and can be time-consuming due to the point-by-point scanning process. Live cell imaging requires careful balancing of image quality with minimizing exposure times to maintain cell viability. While confocal microscopy offers improved imaging of thick samples compared to traditional techniques, it still has limitations for very thick specimens due to limitations in light penetration.

12.2.13 Expansion Microscopy

Since the 1990s, scientists in the field of microscopy have sought to overcome the diffraction limit—a physical constraint that restricts optical resolution to approximately 250 nm. This pursuit has led to the development of various types of super-resolution microscopy (SRM), including structured illumination microscopy (SIM), single-molecule localization microscopy (SMLM), and stimulated emission depletion microscopy (STED). Super-resolution microscopy enables the visualization of subcellular structures, such as nuclear pore complexes (NPCs) and centrioles, with unprecedented detail, achieving super-resolution ranging from 7 to 150 nm (Zhuang and Shi 2023).

In 2014, the Nobel Prize in Chemistry was awarded to three scientists for their contributions to super-resolution light microscopy. The following year, Edward Boyden and his team developed a chemical method to magnify the specimen itself, rather than amplifying its emitted signal (Chen et al. 2015). They named this chemical approach Expansion Microscopy (ExM). Over the years, expansion microscopy has established itself as an effective super-resolution technique for fixed specimens, enhancing the resolution of conventional microscopy systems by approximately fourfold. Users can achieve high-resolution images once a tissue is physically expanded, even with readily available conventional microscopy hardware (Chen et al. 2015). Thus, ExM represents a novel and powerful tool that facilitates high-resolution visualization and imaging of samples by physically expanding them in an isotropic manner, without compromising their ultrastructure.

Expansion Microscopy (ExM) utilizes a water-absorbing, swellable, polymer-based hydrogel to physically expand pre-fixed biological tissues. When the hydrogel permeates tissue samples, they become hydrated. Following this absorption of water, both the tissue and the hydrogel swell, allowing expansion. Notably, the hydrogel maintains the relative positions of molecules within the tissue by gently separating them. As a result, densely packed cellular components become distinct and separate when the expanded tissue is observed under a light microscope. This technique can be performed with standard laboratory equipment, making super-resolution imaging accessible to a wide range of research teams. The methods enable a linear expansion of up to 16-fold. Expansion microscopy significantly enhances imaging resolution to approximately 70 nm using standard microscopes, all while remaining cost-effective and preserving both spatial and molecular information (Kubalová et al. 2020).

The expansion microscopy protocol was initially developed using mouse brain tissue (Chen et al. 2015). Over the past decade, this technique has been adapted for various organisms, including fungi, humans, mice, fruit flies, zebra fish, and diverse soft tissues like the brain, skin, kidneys, and liver. However, in plant cells, the rigidity and structure provided by cell wall can hinder expansion during ExM. To overcome this challenge, researchers are using plant protoplasts as alternative samples, which allows for isotropic expansion without sacrificing important scientific details (Chakraborty 2025). Although the widespread application of this technique in plants may require additional time, its integration into plant karyotyping holds significant promise for enhancing both classical and molecular cytogenetics, ultimately benefiting human welfare.

Steps of Expansion Microscopy

The following steps outline the preparation and observation procedures for expansion microscopy (Fig. 12.18).

(a) *Fixation*: Sample fixation is a crucial initial step. A mixture of formaldehyde/acrylamide or glutaraldehyde/acrylamide is utilized to preserve the sample, ensuring that the fixatives do not compromise cellular structures. In some cases, detergents such as Triton X-100 are employed to enhance the penetration of reagents.

(b) *Labeling*: This step can occur before or after the expansion. In pre-expansion labeling, samples are tagged with fluorophores or other compatible markers. They are incubated with primary antibodies, followed by fluorescently labeled secondary antibodies or fluorescent probes, such as FISH probes. Additionally, counterstaining with DAPI may be performed for chromatin or chromosome visualization. In contrast, post-expansion labeling occurs at the end of the process, after expansion.

(c) *Anchoring or Gel Embedding*: The labeled sample is embedded in a polymer gel, specifically a polyelectrolyte hydrogel. Depending on the expansion microscopy protocol used, prior treatment may be necessary to facilitate anchoring the cross-linked samples to the gel.

(d) *Gelation*: To prevent cell damage, the specimen is immersed in a sodium acrylate monomer solution, followed by immersion in sodium polyacrylate, a highly penetrating hydrogel. Once inside the cells, these chemicals form a network of the desired polymer.

(e) *Homogenization*: Once the gel has formed, the sample undergoes homogenization to eliminate any intermolecular interactions that are not mediated by the gel, thereby facilitating isotropic expansion (Chen et al. 2015). This process may involve the use of broad-specificity proteases, such as proteinase K, for protein digestion. Simultaneously, nucleic acids are denatured through heat exposure but remain intact and are not cleaved by the proteases. Lipids are typically only partially fixed under most conditions (Humpfer et al. 2024) and are removed using a detergent. Achieving complete homogenization is essential for the samples, as any inadequacies in this step can lead to the tearing of the gel, particularly at higher expansion factors (Humpfer et al. 2024). It is important to emphasize that this step is vital for preventing sample distortion, and the choice of homogenization components should be guided by the nature of the specimen and the specific molecules to be visualized.

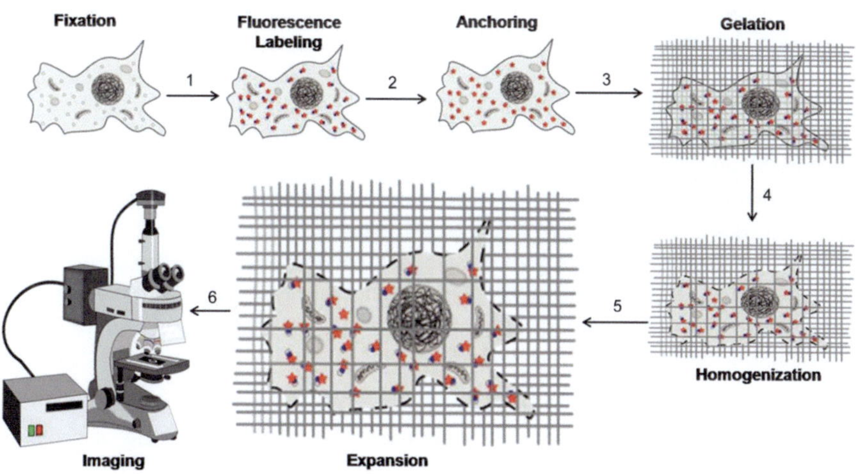

Fig. 12.18 Steps of expansion microscopy

(f) *Expansion*: Finally, the gel is immersed in double-distilled water, which diffuses into the polyelectrolyte hydrogel through osmotic force. This process encourages the expansion of the polymer, pulling apart anchored biomolecules in an isotropic manner and creating significant gaps between each biomolecule. The spatial organization of the expanded specimen is preserved.
(g) *Observation*: Sample imaged on any conventional microscope or a fluorescent microscope.

Advantages of Expansion Microscopy Over Traditional Microscopy

(i) *Super-resolution with standard equipment*: Expansion microscopy allows for nanoscale imaging of biological structures using standard fluorescence microscopes, effectively overcoming the diffraction limit of traditional light microscopy. This technique physically expands densely packed cellular or molecular features that may appear merged or indistinct under conventional microscopy.
(ii) *Cost-effectiveness*: The protocols for expansion microscopy utilize relatively inexpensive reagents and conventional fluorophores. Researchers can work with standard fluorescence microscopes rather than relying on specialized, costly, and complex super-resolution or electron microscopes.
(iii) *Multiplexing*: Expansion microscopy enables the simultaneous imaging of various biomolecules, including DNA, proteins, lipids, and RNA within expanded samples.
(iv) *Depth penetration*: This technique enhances clarity in dense chromatin regions by physically separating fibers, thereby improving access for probes and antibodies.
(v) *Preservation of molecular relationships*: The expansion process retains the relative spatial positions of molecules within the sample, allowing researchers to investigate the organization and interactions of proteins, nucleic acids, and other biomolecules at high resolution.
(vi) *Broad applicability*: Expansion microscopy is highly accessible and can be applied to various biological samples.

Applications of Expansion Microscopy

ExM is a novel and powerful tool that facilitates high-resolution visualization and imaging of samples by physically expanding them in an isotropic manner, without compromising their ultrastructure. The expansion microscopy protocol utilizes preserved tissues and was initially developed using cultured cells and brain tissue (Chen et al. 2015). Over the past decade, this technique has further evolved and been applied to various organisms, including fungi, humans, mice, fruit flies, zebrafish, and diverse soft tissues such as the brain, skin, kidneys, and liver. The following are the important applications of ExM.

- In neuroscience, ExM imaging has progressed remarkably.
- Newly developed ExM protocols can enable the practical imaging of proteins and RNAs (Wassie et al. 2019).

- Exfish, a technique that uses fluorescence in situ hybridization (FISH) to locate and identify RNA molecules, is now possible through ExM (Chen et al. 2016).
- Nanoscale visualization of chromatin structure is possible.
- Through the post-expansion method, ExM has been used to visualize centromeres and repetitive DNA sequences through FISH. After specific fluorescence labeling, it can detect histone variants (e.g., CENH3) and modifications at expanded centromeres.
- ExM's application in plants faces a hurdle. The cellulosic plant cell wall appears to be the first barrier. These cell walls inhibit the uniform expansion needed for practical imaging. Scientists are now considering using cell wall-free protoplasts to successfully apply expansion microscopy in plant systems (Chakraborty 2025).
- Applications of Exm in chromosome research, particularly in plant karyotyping, deserve future attention. It holds significant promise for enhancing both classical and molecular cytogenetics.
- ExM is a one-decade-old methodology still experiencing new developments. Scientists are developing a combination of multiple protocols to enable the simultaneous imaging of RNA, DNA, and proteins.

12.3 Refrigerators and Refrigeration

The refrigerator has become an indispensable part of modern life in homes, offices, shops, educational institutions, research laboratories, and many other settings. The advent of the contemporary refrigerator has transformed the way we store perishable items by maintaining lower temperatures that preserve freshness and extend shelf life. Refrigeration is the process by which a refrigerator extracts heat from an enclosed space or substance, thereby lowering its temperature. Any branded single- or double-doored refrigerator (within 250 L) is an indispensable item in a cytogenetics laboratory that helps to keep temperature-sensitive chemicals in solid or liquid form and helps to perform the pretreatment and fixation procedure for the chromosome preparation.

Exploring the early history of refrigeration reveals that substances such as ammonia, methyl chloride, and sulfur dioxide were employed initially as refrigerants. However, it became apparent that these chemicals posed significant health risks. Since the 1930s, there have been advancements in refrigeration technology that led to the adoption of and related compounds. Chlorofluorocarbons (CFCs) are colorless, odorless, non-flammable, and highly stable gases or liquids with low toxicity. They have also been useful as propellants in aerosols and in a variety of technical applications. Notably, their boiling points, low surface tension, and low viscosity enhance their effectiveness as refrigerants. Nevertheless, since the mid-1970s, CFCs have been identified as a major contributor to the degradation of Earth's ozone layer. While some developed countries have prohibited the use of CFCs in refrigeration, they remain prevalent in many applications. With ongoing technological advances and a focus on environmental health, alternatives with low global warming potential, such as HFO-1234yf (2,3,3,3-Tetrafluoroprop-1-ene), are now being utilized. Additionally, innovative refrigeration methods that employ solar, magnetic, and acoustic energy are being explored for the future.

12.3.1 Refrigerator (4 °C)

The 4 °C refrigerator is an indispensable cold storage device in any biological science laboratory (Fig. 12.19). It is designed to maintain an internal temperature of around 4 °C constantly. This is achieved through a sophisticated, complex interplay of components, including a thermostat, compressor, coils, insulation, and an air circulation fan. This specific temperature range is slightly above the freezing point of water. It is often considered optimal for storing plant samples and chemicals, and running temperature-controlled experiments. Despite its advantages, it has some limitations, such as higher energy consumption than standard refrigerators and unsuitability for specific biological samples. To prevent spoilage from freezer burn, most 4 °C refrigerators either have a limited frost buildup feature or utilize an automatic defrosting system.

Fig. 12.19 4 °C refrigerator

Components of 4 °C Refrigerator

The 4 °C refrigerator is distinguished by its advanced components, such as the compressor, condenser, expansion valve, and evaporator that combine to sustain a consistent temperature of 4 °C (Fig. 12.20). These components generally include:

(a) *Temperature control*: It has digital thermostats and sensors. A thermostat maintains the desired temperature within a permissible range inside the refrigerator. The sensor detects the temperature and switches the compressor to turn on or off when the desired temperature increases or decreases.
(b) *Compressor*: It is considered the "heart" of a refrigerator. It precisely regulates the circulation of refrigerant throughout the system. It is a motor-powered pump responsible for compressing low-pressure, low-temperature refrigerant gas into high-pressure, high-temperature gas by compression. It also transfers refrigerant through the coils, thereby facilitating the cooling process.
(c) *Condenser*: A condenser is a network of condenser coils fitted at the back or bottom of the refrigerator. It acts as a heat exchanger. It receives hot, vaporized refrigerant from the compressor after compression and releases the absorbed heat into the surrounding environment. It condenses the hot refrigerant vapor into a low-temperature liquid.
(d) *Evaporator coils*: These coils carry the refrigerant and absorb heat inside the refrigerator.
(e) *Expansion valve*: A narrow expansion valve rapidly reduces the pressure and temperature of high-pressure liquid refrigerant. This sudden pressure drop causes the refrigerant to evaporate partially, transforming into a low-pressure, cold gas mixture.

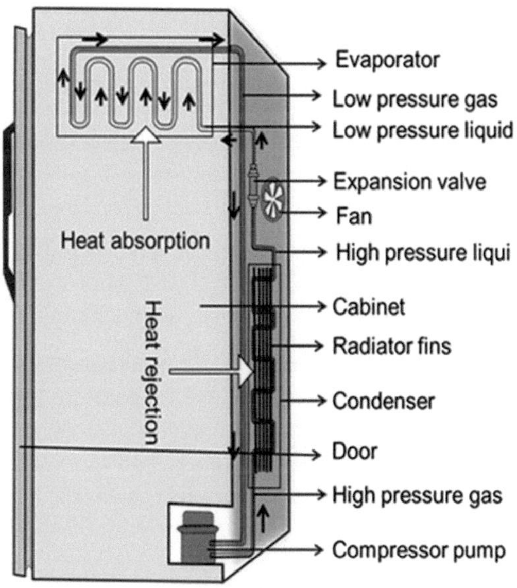

Fig. 12.20 Different components of 4 °C refrigerator

(f) *Insulation*: High-quality insulation materials are used in a 4 °C refrigerator to minimize heat transfer into the freezer compartment.
(g) *Evaporator*: It consists of evaporator coils and is located inside a refrigerator. These coils carry the cold refrigerant gas and absorb heat inside the refrigerator. This continuous heat absorption creates a cold environment within the freezer.
(h) *Air circulation fan*: Some models have an air circulation fan to ensure uniform cold air distribution throughout the evaporator's fins of refrigerator, maintaining a consistent temperature.
(i) *Humidity control*: Some 4 °C refrigerators have humidity control mechanisms to maintain optimal moisture levels for specific biological samples.
(j) *Specialized compartments*: Certain models may come with dedicated compartments designed for specific needs.

Uses The 4 °C refrigerator has a wide range of domestic, commercial, and scientific uses due to its ability to maintain a constant low temperature. In homes and restaurants, it is used to preservefood and beverages, slow spoilage, and ensure that products remain fresh and safe to consume. These refrigerators are critical for storing vaccines, medications, and biological samples in the medical field. The controlled environment helps preserve the efficacy of vaccines and medicines, while biological samples like blood, tissue, and DNA can be kept for analysis without degradation. Similarly, a 4 °C refrigerator is essential for storing various biological materials, including enzymes, bacteria, and cells, for experiments in research laboratories. The cool temperature slows down metabolic reactions, preserving the integrity of these materials.

12.3.2 Refrigerator (−20 °C)

A −20 °C refrigerator is designed to maintain an internal temperature of −20 °C. This type of refrigerator is a marvel of modern engineering and has uses in numerous fields, from food and medicine preservation to scientific research.

Components of –20 °C Refrigerator

The primary parts of a −20 °C refrigerator include the compressor, condenser, expansion valve, and evaporator (Fig. 12.21). These parts work together in a continuous compression, condensation, expansion, and evaporation cycle.

(a) *Temperature control*: It is equipped with digital thermostats acting as the temperature regulator. They monitor the desired temperature inside the freezer and maintain it precisely within a permissible range inside the refrigerator. The sensor detects the temperature inside the freezer and turns the compressor on to start cooling process when the temperature rises above 4°. It turns off when the desired temperature (−20 °C) is reached.

Fig. 12.21 −20 °C refrigerator

(b) *Compressor*: A motor-powered pump converts low-pressure, low-temperature refrigerant gas into high-pressure, high-temperature gas by compression. It is also responsible for pumping refrigerant through the coils, facilitating the cooling process. A higher-capacity compressor is necessary to generate the required pressure and cooling power.

(c) *Condenser*: A condenser is a network of coils located at the back or bottom of the refrigerator. It receives refrigerant flows after being compressed.

(d) *Evaporator and condenser coils*: These coils carry the refrigerant. The evaporator coil absorbs heat from inside the refrigerator, while the condenser coil releases this absorbed heat to the surrounding air, condensing it back into a high-pressure liquid.

(e) *Expansion valve*: An arrow expansion valve rapidly reduces the pressure and temperature of high-pressure liquid refrigerant. This sudden pressure drop partially causes the refrigerant to evaporate, forming a low-pressure, cold gas mixture.

(f) *Insulation*: A −20 °C refrigerator uses thicker, more specialized, high-quality insulation materials to prevent heat transfer and maintain the internal temperature.

(g) *Evaporator*: It consists of evaporator coils. These coils carry the cold refrigerant gas and absorb heat from inside the refrigerator. This continuous heat absorption creates a freezing cold environment within the freezer.

Enhanced evaporator design: The evaporator coils may be larger or have increased surface area for improved heat absorption.

(h) *Additional cooling fans*: Some models may employ internal fans to more effectively ensure uniform cold air distribution throughout the refrigerator and maintain a consistent temperature.
(i) *Humidity control*: Some 4-degree refrigerators have humidity control mechanisms to maintain optimal moisture levels for specific biological samples.
(j) *Specialized compartments*: Certain models may have dedicated compartments designed for specific needs.

Working Principle
A −20 °C refrigerator, also known as a deep freezer, operates on the same basic principles as a regular refrigerator, but with some modifications to achieve and maintain colder temperatures. Different parts work together in a continuous cycle of compression, condensation, expansion, and evaporation. These fundamental processes in a constant cycle include-

(i) The refrigerator's compressor pump receives the low-pressure refrigerant gas and compresses it, increasing the pressure and temperature of the refrigerant gas.
(ii) This high-pressure, high-temperature gas flows into the condenser, where it loses heat and turns into a high-pressure liquid.
(iii) A high-pressure liquid moves to the expansion valve, where the pressure and temperature of the liquid refrigerant rapidly reduce. This liquid refrigerant quickly turns into a cold gas in the evaporator.
(iv) The cold refrigerant gas absorbs heat from the freezer compartment, evaporating completely.
(v) The evaporator absorbs heat from the surroundings, cooling the refrigerator's interior to maintain the desired temperature.
(vi) The thermostat controls the compressor, maintaining the desired −20 °C temperature.

Uses In medicine, these refrigerators play a critical role in preserving certain pharmaceuticals, biological samples, and vaccines that require very low temperatures to maintain their efficacy. For instance, some COVID-19 vaccines required storage at temperatures as low as −20 °C, making these refrigerators indispensable in the global battle against the pandemic. In scientific research, −20 °C refrigerators are often used to store biological samples, chemicals, and other materials that require a stable, low-temperature environment. Meanwhile, in the food industry, these refrigerators help prolong the shelf life of certain food items, thus reducing waste and increasing profitability.

12.4 Temperature-Controlled Incubator

An incubator is standard laboratory equipment for any laboratory handling biological samples (Fig. 12.22). In a cytogenetics laboratory, it is an alternative to a refrigerator where anyone can perform pretreatment more accurately at a particular temperature. Germination of seeds is also possible inside a temperature-controlled incubator. It is essential to carry out enzymatic digestion of root tips at 37 °C, a critical step of chromosome preparation for any molecular cytogenetic experiment. This type of incubator is built with appropriate insulation to maintain a particular range of temperature, humidity, and other environmental conditions to perform various experiments. Generally, it can perform well between 4 and 40 °C. Besides cytogenetics, it is used in fields like microbiology, molecular biology, and environmental science, as it helps cultivate microorganisms, cells, tissues, and even plants and insects like *Drosophila*. The operation of a lab incubator is based on the thermoelectric principle, in which a thermostat maintains a stable temperature by generating a thermal gradient.

Fig. 12.22 Incubator

12.5 Weighing Balance

Weighing balances (Fig. 12.23) are vital instruments in any laboratory for determining the mass of objects. They come in various types, including precision and analytical balances, which are designed to measure significantly smaller masses than standard scales. Among these, the analytical balance stands out as the most

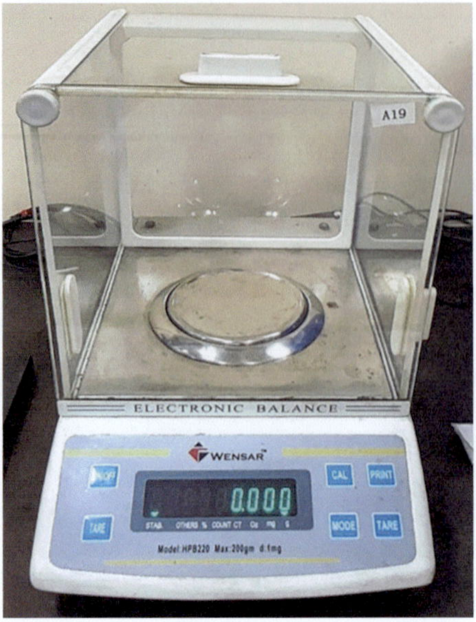

Fig. 12.23 Weighing balance

accurate, offering readability and measurement precision up to 0.0001 g. It is specifically engineered to measure small masses with exceptional accuracy. In contrast, a precision balance—also called a top-loading balance—can weigh up to 200 g with a readability ranging from 1 mg to 1 g. Generally, precision balances are more affordable than analytical balances.

Maintenance of the Balance

Regular cleaning and maintenance of the balance are required to obtain error-free results. Don't forget to practice the following things:

- Keep the balance on a stable surface and check its alignment.
- After weighing any chemical or samples, remove sample residue particles using a soft tissue paper or brush.
- The balance needs to have a cover to protect it from dust.
- Placing the balance in a temperature-controlled room is beneficial for its long-term service.

12.6 pH Meter

A pH meter is a relatively inexpensive and easy-to-use scientific instrument or device that measures the solution's pH (hydrogen ion concentration) by measuring the electrical potential between two electrodes.

pH pH stands for "potential of hydrogen" and is a quantitative measure of the molar concentration of hydrogen ions in the solution (H^+). If we express the hydrogen ion concentration of an aqueous solution in its molecular value, we derive a scale of 1 (100) via 10^{-7} to 10^{-14} mole L^{-1}. When written as a function of its negative logarithm, it creates a real and simple scale ranging from 0 to 14.

Thus, pH = −log of concentration of $[H^+]$ = log (1/concentration of $[H^+]$)
*Concencentration of H^+ expressed in mole L^{-1}

H^+ ion concentration and pH have a logarithmic relationship. It means that a change of one pH unit corresponds to a tenfold change in H^+ ion concentration. For example, a solution with a pH of 6 has ten times the H^+ ion concentration of a solution with a pH of 7.

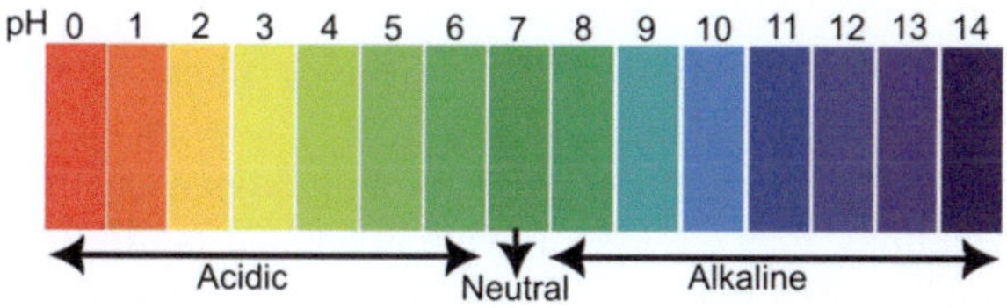

Fig. 12.24 pH Scale with indicator color

pH Scale The pH scale ranges from 0 to 14 (Fig. 12.24) and expresses the acidity or alkalinity of a solution.

- *Neutral*: A pH of 7 indicates a neutral solution. It means the concentration of H^+ ions is equal to that of hydroxide ions (OH^-), and the solution is neither acidic nor basic.
- *Acidic*: Solutions with a pH below 7 are acidic, and it means they have higher H^+ ions than OH^- ions. The lower the pH, the more acidic the solution and the higher the H^+ ion concentration.
- *Basic*: Solutions with a pH above 7 are basic, and it means they have a higher concentration of OH^- ions than H^+ ions. The higher the pH, the more basic the solution and the lower the H^+ ion concentration.

Components of a pH Meter with Combination Electrode

A pH measurement system consists of three basic components: a reference electrode, a measurement or sensing electrode, and a highly sensitive voltmeter (Fig. 12.25a, b). Since 1947, pH sensor manufacturers have combined these two electrodes into a single unit called a combination electrode. These types of pH sensors are currently used exclusively in laboratories and industries.

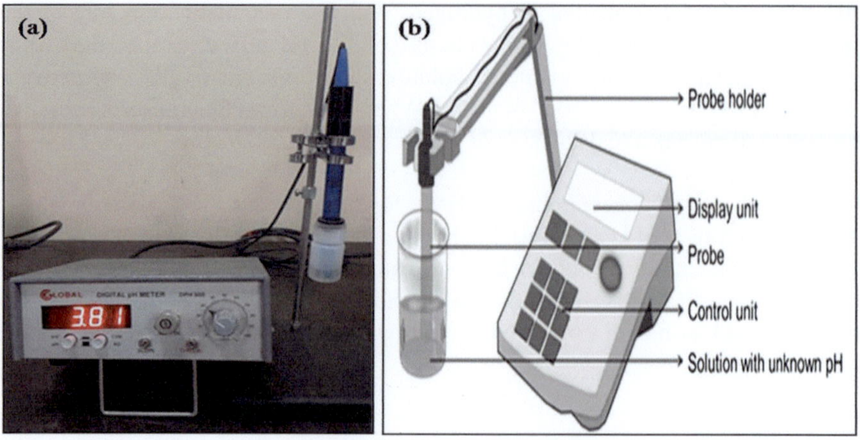

Fig. 12.25 (**a**) pH meter and (**b**) diagram of pH meter showing different components of pH meter

Despite the incredible accuracy of platinum or hydrogen electrodes for measuring the pH of solutions, they are used a minimal amount in pH meters due to their complicated handling. The glass membrane electrode revolutionized pH measurement and is commonly used in pH meters. It provides more reliability and accuracy, combined with stable electronic amplification.

(a) *Measurement electrode or sensing electrode*: The measurement electrode or sensing electrode is a specialized sensor that determines the pH value of an unknown solution. It may be a platinum/hydrogen electrode or a glass electrode type. The glass electrode consists of a silver wire coated with thin silver chloride (Ag-AgCl), an internal buffer solution, a glass shaft, and a hemispherical, special pH-sensitive glass membrane (Fig. 12.26). Ag-AgCl is a conducting electrode, and the Ag-AgCl wire is connected to a pH meter via a coaxial pH cable. It is partially filled with an internal buffer solution at pH value of 7 and a specific amount of potassium chloride. The glass shaft protects the conducting electrode. When immersed, the membrane glass forms a thin gel layer between the glass surface and the solution.

(b) *Reference electrode*: The reference electrode follows the Nernst equation and produces a stable voltage potential. Mercury-mercurous chloride (calomel) and silver-silver chloride reference (Ag-AgCl) electrodes are pH meters' most reliable reference systems. However, silver-silver chloride (Ag/AgCl) electrodes are applied most frequently as reference systems due to their practical acceptance, ease of manufacture, high potential stability, and accuracy under wide temperature fluctuations and at high temperatures. The internal reference electrode is immersed in an internal buffer solution (reference electrolyte—potassium chloride) with a known pH (Fig. 12.26). A porous liquid junction (diaphragm) at the bottom of the electrolyte chamber serves as the pathway between the reference electrolyte solution and the measured liquid medium. It allows the potassium chloride electrolyte to create an electrical path between the electrode and the measurement electrode. It provides an electrolytic interface between the Ag-AgCl conducting system and the measured liquid solution. The diaphragm usually consists of a porous ceramic disc fused into the glass wall at the lower end of a reference electrode. In a combination pH electrode, the concentric space surrounding the measurement electrode is filled with KCl electrolyte and contains the internal reference system. The reference electrode often determines the overall accuracy of the pH measurement.

(c) *Meter*: The meter measures the difference in potential between the measurement electrode and the reference electrode and converts this voltage difference into a pH reading.

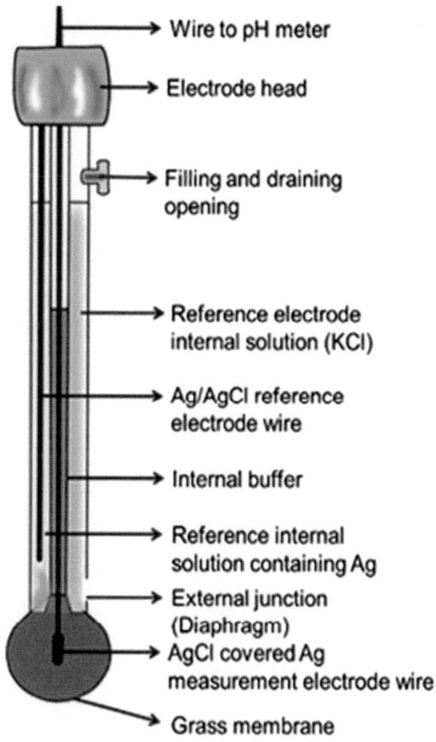

Fig. 12.26 Detailed structure of probe of pH meter

Working Principle

A pH meter operates by measuring the electrical potential between two electrodes—the measurement electrode and the reference electrode. The reference electrode is usually a silver-silver chloride with a steady or constant potential. On the other hand, the measurement electrode is a glass electrode sensitive to hydrogen ions. Its potential fluctuates depending on the pH of the solution.

When the measurement electrode is immersed in a solution, a thin gel layer is formed at the inner side of the glass membrane between the glass surface and the solution. The formation of this layer is influenced by the quality and composition of the membrane glass, the temperature, and the pH value of the measured solution. A continuous exchange of H^+ ions in the gel layers and H^+ ions of the solutions occurs on both sides of the membrane. This creates a potential difference between the two electrodes, which is detected by the pH meter and converted into a pH reading value.

The system reaches equilibrium when the concentration of H^+ ions in the solution and the gel layers is equal, resulting in a 0 mV potential difference. If there is a difference in H^+ ion concentration difference between the inner buffer and the outer solution, a potential difference develops that is proportional to the pH difference. To measure membrane potential, the membrane must be conductive, which is achieved by the mobility of alkaline ions in the membrane glass (Li^+ ions in most glasses today or Na^+ ions in older membrane glasses). The gel layer of a glass electrode significantly controls a pH meter's response time and slope. It takes 1–2 days for a gel layer to develop fully, so electrodes must be hydrated for at least 24 h before use. Most manufacturers deliver hydrated electrodes, which are kept wet with KCl solution in a plastic cap, making them ready for immediate use.

Calibration of pH Meters

Calibrating the pH meter using buffer solutions (pH 4.01, pH 7.00, and pH 10.01) is a crucial aspect of testing equipment. It ensures that the electrode within the pH probe remains accurate and free from damage or drift. Inaccurate pH readings often result from inadequate calibration, improper storage, and insufficient cleaning of the pH meter. Generally, the calibration procedure is conducted with two buffer solutions of pH 7 and 4.

Calibration with pH 7

- Rinse the pH probe with distilled or reverse osmosis water.
- Take the buffer solution 7 in a small beaker and place the pH probe inside the beaker, and wait for its stabilization. If necessary, use the meter's calibration settings to fix the pH values of the buffer solutions.
- Once the pH readings have stabilized, issue the midpoint calibration command: *cal, 7.*

Calibration with pH 4

- Rinse the pH probe with distilled or reverse osmosis water.
- Take the buffer solution 4 in a small beaker and place the pH probe inside the beaker, and wait for its stabilization. If necessary, use the meter's calibration settings to fix the pH values of the buffer solutions.
- Once the pH readings have stabilized, issue the midpoint calibration command: *cal, 4.*

Measurement of pH of a Solution

Requirements: (i) pH meter, (ii) unknown solution, (iii) distilled water, and (iv) beakers.

Procedure

- Rinse the pH meter electrodes with distilled water and dry them gently.
- Take the unknown solution into a clean beaker.
- The pH meter electrode is dipped into the sample solution, ensuring the fully submerged tip of the electrode.
- Gently stir the sample solution to ensure uniform mixing and temperature equilibration.
- Allow the pH meter to stabilize and display the pH reading.
- Record the pH value of the unknown solution.
- Rinse the pH meter electrodes with distilled water and dry them gently after each measurement.

Maintenance of pH Meters

It is crucial to care for and maintain the pH meter properly to ensure a long lifespan and accurate readings. The electrodes should be cleaned regularly with distilled water both before and after each use, and they should be stored in an appropriate storage solution to preserve their functionality. Regular calibration checks are also important to ensuring the accuracy of the pH meter. Follow the manufacturer's instructions for using and maintaining the meter.

Uses of pH Meters

pH meters are used in a wide variety of applications in chemistry, biology, environmental science, food science, and industries for measuring the pH of any prepared solutions, plant extracts, water, soil and other environmental samples, beverages, food processing, etc.

12.7 Micropipette

A micropipette is a common yet essential laboratory instrument designed to accurately and precisely measure and transfer small volumes of liquid, typically in the microliter (μL) range. Different variants of micropipettes, including various designs and sizes, are used in molecular biology, immunology, cell culture, analytical chemistry, biochemistry, and microbiology laboratories. These instruments are essential for dispensing precise liquid volumes and are available in volumes ranging from 0.2 μL to 1,000 μL (Fig. 12.27).

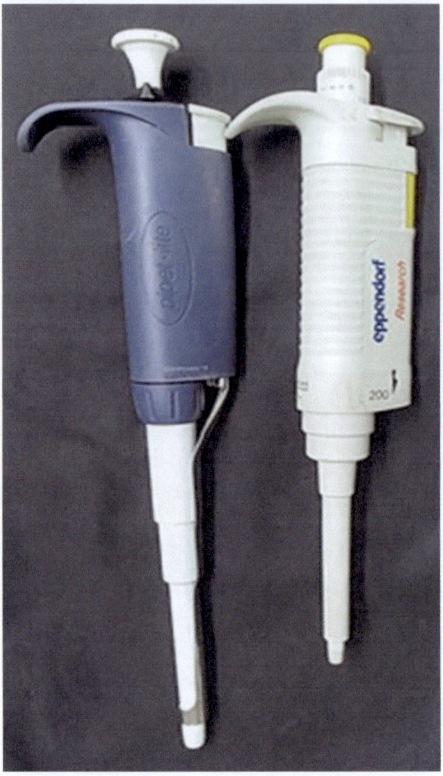

Fig. 12.27 Micropipette

Types of Micropipette

Based on volume/capacity, micropipettes are of two types—fixed volume and variable volume.

(a) *Variable volume micropipette*: This type of micropipette has a specific minimum and maximum volume range within which the volume of the liquid to be aspirated or dispensed can be adjusted depending on the user's requirements (Fig. 12.28a).

(b) *Fixed volume micropipette*: This type of micropipette keeps the volume of liquid to be aspirated or dispensed fixed. It is used when the same volume of liquid is to be dispensed multiple times (Fig. 12.28b).

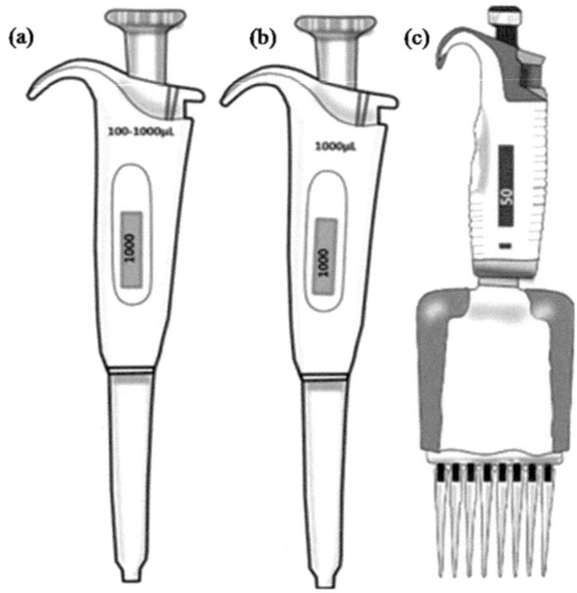

Fig. 12.28 (**a**) variable volume micropipette, (**b**) fixed volume micropipette, and (**c**) multichannel micropipettes

Based on number of channels, micropipettes are of two types—single-channel micropipettes and multi-channel micropipettes.

(a) *Single-channel micropipettes*: These pipettes have a single tip-holding channel to aspirate or dispense one volume of fluid at a time (Fig. 12.28a, b).
(b) *Multichannel micropipettes*: Available multi-channel micropipette variants (8-channel, 12-channel and 16-channel) have multiple tip-holding channels to measure and transfer various volumes of liquid at a time. It reduces the workload of a single-channel micropipette when working with large volumes of samples (Fig. 12.28c).

Based on the working principle micropipettes are two types—air displacement micropipettes and positive displacement micropipettes.

(a) *Air displacement micropipette*: This type of micropipette works on the air displacement principle. It consists of a piston that aspirates and dispenses liquid samples as the air pocket moves up and down, respectively. The internal mechanism of the pipette does not come in direct contact with the liquid sample. Instead, a disposable micropipette tip is used to draw the liquid into and dispense it from the pipette.
(b) *Positive displacement micropipette*: The piston comes in direct contact with the sample in these pipettes. The disposable tip in a positive displacement micropipette is a microsyringe composed of a capillary and a piston (movable inner part), which directly displaces the liquid.

Based on the operating mechanism, micropipettes are of two types—mechanical and electronic.

(a) *Mechanical micropipette*: These pipettes are operated manually using a piston-shaft spring mechanism.
(b) *Electronic micropipette*: An electronic micropipette is mainly automated. The one-touch buttons perform the aspirating and dispensing of liquid instead of pressing and depressing the manual plunger. Electronic pipettes also often enable the user to create custom programs on the device, allowing the pipettes to suit diverse application needs.

Components of a Micropipette

Although different variants of micropipettes are available, basic components are standard for all micropipettes, which are briefly discussed below.

- *Plunger button*: The plunger is a piston that aspirates and dispenses liquid for a specific volume (Fig. 12.29).
- *Volume adjustment thumbwheel or knob*: The volume adjustment thumb-wheel or knob is rotated clockwise or anticlockwise to decrease or increase the volume setting and set the desired volume of liquid to be aspirated or dispensed (Fig. 12.29). A distinct click sound at every volume change ensures a precise volume setting and prevents accidental changes.
- *Body*: The main body of the pipette houses the plunger (Fig. 12.29).
- *Finger support part*: It helps to hold the micropipette easily.
- *Volumetric digital display*: This shows the volume of the liquid to be aspirated or dispensed.
- *Universal tip-holding cone*: The universal tip-holding cone is the part of the pipette (Fig. 12.29) that enhances the compatibility of the instrument to hold most of the standard micropipette tips.

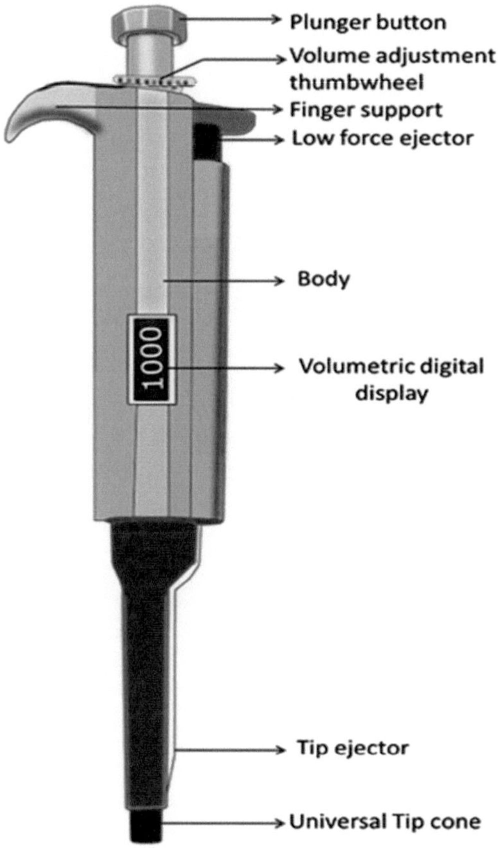

Fig. 12.29 Components of a micropipette

- *Tip ejection system*: The internal mechanism of the micropipette does not come in direct contact with the liquid sample. Instead, a disposable micropipette tip is used to draw the liquid into and dispense it from the micropipette. Micropipettes have a tip ejection system, consisting of an ejector button and tip ejector, that allows the safe, effortless, and quick ejection of micro tips without touching them. The tips can be easily removed from the micropipette by pressing the tip ejector button.

Micropipette Tips Micropipette tips (Fig. 12.30) are disposable plastic tubes attached to the tip-holding cone of the pipette. They are available in various sizes to accommodate different volumes of liquid.

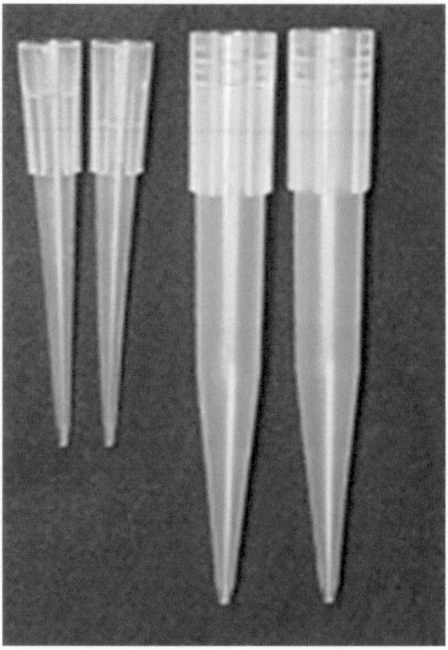

Fig. 12.30 Micropipette tips

Additionally, certain micropipettes have a calibration tool and a micropipette stand as an accessory.

Calibration of a Micropipette

Micropipettes should be calibrated regularly (every month or in a few-month gap) to ensure they measure and transfer liquid accurately and precisely. Usually, the interval at which a micropipette needs to be calibrated depends on the frequency of pipette use, type of liquid dispensed by the pipettes, handling and care of the pipette, and accuracy of the applications. A qualified laboratory technician typically calibrates according to ISO 8655 standards and requires special ambient conditions, as air pressure, humidity, temperature, and even altitude affect the pipetting results.

Maintenance of Micropipette

Micropipettes should be adequately cared for and maintained, including following the manufacturer's instructions to extend their lifespan. Additional care includes storing them in a clean and dry place, regularly cleaning them with a mild soap and water solution, and avoiding dropping or bumping them. Care and maintenance of micropipettes are recommended.

12.8 Eppendorf Centrifuge Tubes

Eppendorf centrifuge tubes are small, conical-shaped, autoclavable tubes made up of polypropylene or polyethylene (Fig. 12.31). They are commonly used in laboratories for various functions, including centrifugation, mixing, and storage. These tubes come in multiple sizes, ranging from 0.5 to 50 mL. Their high-quality materials that are resistant to breakage and leakage, durability to withstand high speeds and temperatures, and safety features such as a secure lid and a g-Safe® ring that prevents the tube from breaking during centrifugation make them popular among scientists and researchers worldwide. These tubes are available in sterile and non-sterile forms and are easy to open and close, even with one hand. In EMA-based chromosome preparation, enzymatic digestion of plant roots is carried out within it. Moreover, Eppendorf centrifuge tubes are commonly used for centrifugation, mixing solutions, storing samples, PCR, DNA purification, RNA extraction, ELISA, and Western blot.

Fig. 12.31 Eppendorf centrifuge tubes

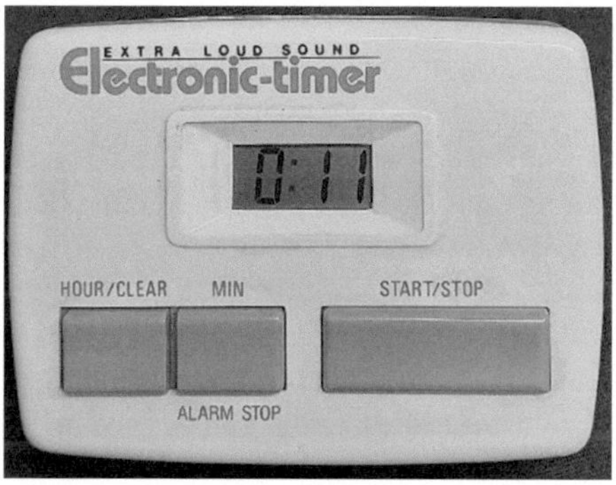

Fig. 12.32 Digital timer

12.9 Digital Timer

A digital timer is an electronic device that allows users to accurately monitor and measure time intervals (Fig. 12.32). These timers typically feature a user-friendly interface with buttons to start, stop, and reset the timer. While they are not essential, digital timers can be very useful in any laboratory setting. Alternatively, similar timing functions can be performed using a mobile device.

12.10 Tweezers

In the molecular cytogenetics method of chromosome preparation, tweezers with a precise, fine-pointed tip (Fig. 12.33) are essential for macerating the enzymatically digested root tip on a glass slide, facilitating the retrieval of cytoplasm-free chromosomal plates. These tweezers are composed of high-quality stainless surgical steel, making them easy to disinfect, sterilize, or autoclave. It is advisable to use the protective cap after each use consistently.

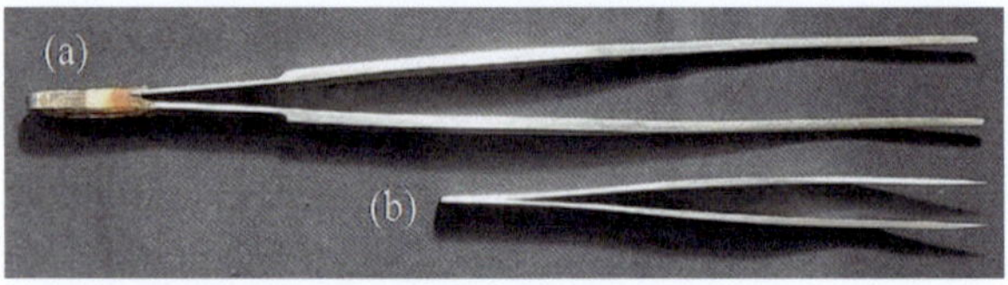

Fig. 12.33 (**a**) Forceps and (**b**) tweezers

What distinguishes tweezers from forceps?

The term "forceps" is often used interchangeably with "tweezers," as both tools are utilized in various laboratory settings. However, there are key differences between them: tweezers are typically used to grasp and pull small objects, while forceps are designed to hold and manipulate larger objects or tissue.

12.11 Hand Blower

A hand blower is a small, portable, handheld device that generates airflow (Fig. 12.34). In EMA methods, Giemsa dye is used for chromosome staining. Following Giemsa staining, the slides are rinsed with distilled water to remove any excess stain. Then, the Giemsa-stained slides are placed on a clean, flat surface, and a hand blower is used to gently blow air onto the slides until they are scorched. This air-drying process is vital for eliminating any residual stains and water particles on the slide and maintaining the integrity of chromosome morphology, which is crucial for accurate analysis. Once the slide is dry, it can be examined under a microscope.

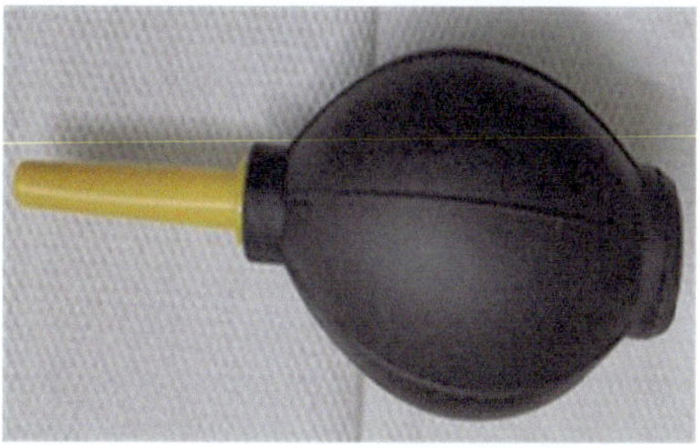

Fig. 12.34 Hand blower

12.12 Frosted Slides

Frosted slides are microscope slides designed with a frosted area at one end (Fig. 12.35). This frosted area is usually produced by chemically etching the glass, which results in a smooth and matte surface. This space is used to label important information, such as the sample name and date, using slide-marking pencils or pens. This feature of easy labelling is utilized in histology, cytology, and microbiology for various microscopy-based applications. Different-colored frosted slides are also used for organization and simple sample-type differentiation.

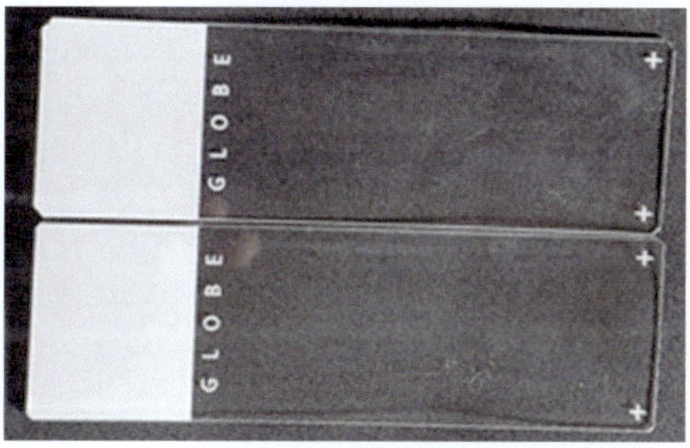

Fig. 12.35 Frosted slides

12.13 Cover Glasses

Chromosomal study greatly depends on microscopic visualization techniques using sophisticated microscopes and staining protocols. Cover glasses are transparent, thin, flat, round (not preferred for cytogenetics studies), square, or rectangular in shape used for microscopic visualization of samples. They are placed over the prepared sample to flatten it, hold it in place, protect it from environmental damage, and facilitate high-resolution microscopic visualization and imaging. Despite their simple nature, cover glasses play a crucial role in achieving optimal microscopic visualization of samples.

Although cover glass with various shapes, sizes, and optical properties is available in the market, selecting appropriate high-quality cover glass with thickness, flatness, chemical inertness, and UV transparency is essential for maximizing the reliability and optimal visualization of cytogenetic preparation. High transparency with a refractive index near 1.5 and devoid of any inclusions is necessary for clear visualization under the microscope. Consistent, and standardized thickness of cover glass is essential for proper focusing and image quality, especially at high magnifications, and should be selected according to the specifications of the microscope objective. Quality cover glass is chemically inert or resistant to laboratory solvents and staining solutions used in cytogenetic procedures. Although various types of cover glasses are available in the market, we will focus on the two most commonly used cover glasses in cytogenetics.

Cover Glasses for Conventional Cytogenetics During conventional preparation of somatic and gametic chromosomes, students are advised to use a standard conventional thin, flat, and high-quality cover glass without any distortions and imperfections. This type of cover glass is available in round and square shapes with various dimensions, like 18 mm × 18 mm and 22 mm × 22 mm (Fig. 12.36). They may vary in thickness. The most commonly used thickness is 1 (85–115 μm). For chromosomal studies, users are advised to use square cover glasses measuring 22 mm × 22 mm.

Fig. 12.36 Cover glasses for conventional cytogenetics

Cover Glasses for Molecular Cytogenetics Unlike conventional cytogenetics, the enzymatic maceration and air drying (EMA)-based chromosome preparation and Giemsa staining in molecular cytogenetics do not require the use of cover glass. However, cover glasses are necessary when observing and documenting the chromosome plates under the oil immersion objective lens. As chromosome maceration takes place, wider areas of the slide require larger and wider versions of cover glasses for microscopy. These cover glasses are wider than conventional cover glasses and are available in a variety of sizes (24 × 32 mm, 24 × 50 mm, 24 × 40 mm, and 32 × 50 mm) and thicknesses (No. 0: 85–115 μm to No. 5: 0.5–0.6 mm) to meet the specific requirements (Fig. 12.37). Its wider size provides more working space and accommodates more samples compared to conventional cover glass, thereby improving image resolution for microscopy techniques. They are made up of borosilicate glass, quartz, and plastic. Borosilicate cover glass is the most commonly used due to its strength, durability, and resistance to scratching. In contrast, quartz cover glass is more expensive and has a higher refractive index. Plastic cover glass is the most affordable but not so strong. Wide-field cover glasses are compatible with a range of microscopy techniques, including fluorescence, phase contrast, and confocal microscopy.

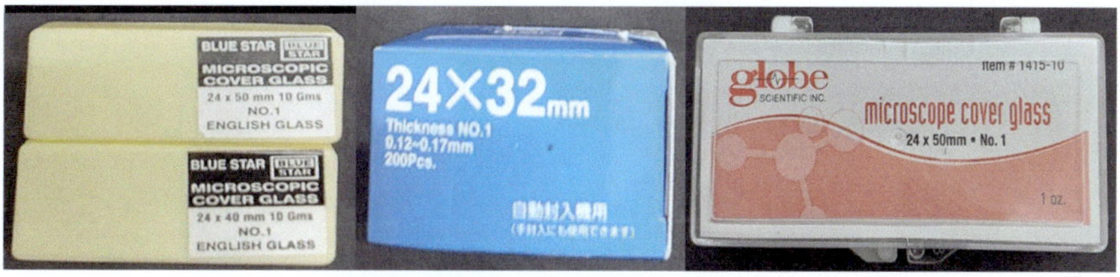

Fig. 12.37 Cover glasses for molecular cytogenetics

References

Chakraborty S (2025) Super-califragilisticexpialidocious-resolution microscopy: how expansion microscopy can be applied to plants. Plant Cell 37(2). https://doi.org/10.1093/plcell/koaf002

Chen F, Tillberg PW, Boyden ES (2015) Expansion microscopy. Science 347:543–548

Chen F, Wassie AT, Cote AJ, Sinha A, Alon S, Asano S, Daugharthy ER, Chang JB, Marblestone A, Church GM, Raj A (2016) Nanoscale imaging of RNA with expansion microscopy. Nat Methods 13(8):679–384

Heit B (2022) Fluorescent microscopy. Humana Press

Humpfer N, Thielhorn R, Ewers H (2024) Expanding boundaries – a cell biologist's guide to expansion microscopy. J Cell Sci 137(7):jcs260765

Kubalová I, Schmidt Černohorská M, Huranová M, Weisshart K, Houben A, Schubert V (2020) Prospects and limitations of expansion microscopy in chromatin ultrastructure determination. Chromosom Res 28(3):355–368

Nechyporuk-Zloy V (2022) Principles of light microscopy: from basic to advanced. Springer

Stockert JC, Blázquez-Castro A (2017) Fluorescence microscopy in life sciences. Bentham Science Publishers

Wassie AT, Zhao Y, Boyden ES (2019) Expansion microscopy: principles and uses in biological research. Nat Methods 16:33–41

Zhuang Y, Shi X (2023) Expansion microscopy: a chemical approach for super-resolution microscopy. Curr Opin Struct Biol 81:102614

Index

T. B. Jha, M. Halder, *Classical and Molecular Plant Cytogenetics*,
https://doi.org/10.1007/978-981-95-8265-5

MIX
Papier aus verantwortungsvollen Quellen
Paper from responsible sources
FSC® C105338

If you have any concerns about our products,
you can contact us on
ProductSafety@springernature.com

In case Publisher is established outside the EU,
the EU authorized representative is:
Springer Nature Customer Service Center GmbH
Europaplatz 3, 69115 Heidelberg, Germany

Printed by Libri Plureos GmbH
in Hamburg, Germany